T0182156

NEUROMETHODS

Series Editor
Wolfgang Walz
University of Saskatchewan
Saskatoon, SK, Canada

For further volumes:
http://www.springer.com/series/7657

Prion Diseases

Edited by

Pawel P. Liberski

Department of Molecular Pathology,
Medical University of Lodz, Lodz, Poland

Editor
Pawel P. Liberski
Department of Molecular Pathology
Medical University of Lodz
Lodz, Poland

ISSN 0893-2336 ISSN 1940-6045 (electronic)
Neuromethods
ISBN 978-1-4939-8417-6 ISBN 978-1-4939-7211-1 (eBook)
DOI 10.1007/978-1-4939-7211-1

Softcover reprint of the hardcover 1st edition 2017

Printed on acid-free paper

This Humana Press imprint is published by Springer Nature
The registered company is Springer Science+Business Media LLC
The registered company address is: 233 Spring Street, New York, NY 10013, U.S.A.

Dedication

This book is dedicated to the memory of D. Carleton Gajdusek (1923–2008), one of the greatest scientists of the twentieth and twenty-first centuries who first discovered kuru for the Western medicine and paid attention to prion diseases in men, at his time known as slow virus disorders. I was privileged to work for him as his postdoc (1986–1989) and many years after this time. He was not only a great scientist but also a fascinating person and a compulsive talker of utmost interesting stories. If he were alive, I would have asked him to write on kuru not myself, as he did in the past in my other book (Gajdusek D.C. Transmissible brain amyloidoses of the brain. In: Light and Electron Microscopic Neuropathology of Slow Virus Disorders. Ed. P.P. Liberski, CRC Press, Boca Raton, 1993: 1–31). The last time I met him was at the kuru meeting ('The end of kuru: 50 years of research into an extraordinary disease' organized by John Collinge and Michael P. Alpers).

D.C. Gajdusek and I in Paris, 2008.

Preface to the Series

Experimental life sciences have two basic foundations: concepts and tools. The *Neuromethods* series focuses on the tools and techniques unique to the investigation of the nervous system and excitable cells. It will not, however, shortchange the concept side of things as care has been taken to integrate these tools within the context of the concepts and questions under investigation. In this way, the series is unique in that it not only collects protocols but also includes theoretical background information and critiques which led to the methods and their development. Thus it gives the reader a better understanding of the origin of the techniques and their potential future development. The *Neuromethods* publishing program strikes a balance between recent and exciting developments like those concerning new animal models of disease, imaging, in vivo methods, and more established techniques, including, for example, immunocytochemistry and electrophysiological technologies. New trainees in neurosciences still need a sound footing in these older methods in order to apply a critical approach to their results.

Under the guidance of its founders, Alan Boulton and Glen Baker, the *Neuromethods* series has been a success since its first volume published through Humana Press in 1985. The series continues to flourish through many changes over the years. It is now published under the umbrella of Springer Protocols. While methods involving brain research have changed a lot since the series started, the publishing environment and technology have changed even more radically. Neuromethods has the distinct layout and style of the Springer Protocols program, designed specifically for readability and ease of reference in a laboratory setting.

The careful application of methods is potentially the most important step in the process of scientific inquiry. In the past, new methodologies led the way in developing new disciplines in the biological and medical sciences. For example, Physiology emerged out of Anatomy in the nineteenth century by harnessing new methods based on the newly discovered phenomenon of electricity. Nowadays, the relationships between disciplines and methods are more complex. Methods are now widely shared between disciplines and research areas. New developments in electronic publishing make it possible for scientists that encounter new methods to quickly find sources of information electronically. The design of individual volumes and chapters in this series takes this new access technology into account. Springer Protocols makes it possible to download single protocols separately. In addition, Springer makes its print-on-demand technology available globally. A print copy can therefore be acquired quickly and for a competitive price anywhere in the world.

Saskatoon, Canada ***Wolfgang Walz***

Preface

Prion diseases represent one of the most exciting and intriguing fields in biomedical sciences, with three Nobel Prizes awarded so far—D. Carleton Gajdusek in 1976, Stanley B. Prusiner in 1997, who coined the term "prion" in 1982 [1], and Kurt Wuthrich in 2002. I became interested in the prion field in 1977, when I read an article, entitled "Epidemiology of Creutzfeldt-Jakob disease in England and Wales," by Walter B. Matthews, in the *Journal of Neurology Neurosurgery and Psychiatry* [2], where for the first time I was confronted by "scrapie," a term previously unknown to me. On immediately searching this key word, I found the Nobel Lecture by Gajdusek [3]. At that time, I was a student with an interest in electron microscopy, which I began to learn from Prof. Michal Karasek (1937–2009), Dr. Iwona Giryn (deceased 2011), and Dr. Barbara Mirecka (deceased 1992). My skills were expanded through a British Council Fellowship at the MRC and AFRC Neuropathogenesis Unit, under Dr. Peter Gibson. These research training experiences paved the way to a Fogarty Fellowship in the Laboratory of Central Nervous System Studies, under the direction of D. Carleton Gajdusek, a Nobel laureate, and Dr. C. J. Gibbs Jr., at the National Institute of Neurological Disorders and Stroke, of the National Institutes of Health. I developed and printed there 17,000 electron micrographs. Professionally, this period was the most exhilarating time of my life.

The proposal of this book came as a little surprise from Wolfgang Walz, the Editor of this series at Springer. I then approached several of my expert colleagues and we started to write. Based on my professional experience, this book is somewhat biased toward morphological approaches. Chapters by Diane Ritchie and James Ironside, by Martin Jeffrey and Gilian McGovern, and by Frank Bastian and me provide in-depth coverage of different aspects of neuropathology, immunohistochemistry, electron microscopy, and immune-gold electron microscopy of prion diseases. These colleagues are the best in the world in this research area. It is noteworthy that a substantial portion of the book is devoted to electron microscopy, which has become merely a shadow of the importance it played in the biomedical sciences when I started my career, now being replaced by molecular biology.

The next section provides the reader with detailed information about the clinical description of prion diseases and the detection of prion protein and biomarkers, written by Richard Knight; Byron Caughey, Christina D. Orru, Bradley R. Groveman, Matilde Bongianni, Andrew G. Hughson, Lynne D. Raymond, Matteo Manca, Allison Kraus, Gregory J. Raymond, Michele Fiorini, Maurizio Pocchiari, and Gianluigi Zanusso; Elizaveta Katorcha and Ilya Baskakov; and Joanna Gawinecka, Matthias Schmitz, and Inga Zerr.

Separate chapters cover kuru. Among the authors, two were privileged to see real kuru—Shirley Lindenbaum and David Asher. The chapter by Pedro Piccardo and Luisa Gregori provides a broad overview of prion research in nonhuman primates. This chapter is unique as higher primates such as chimpanzees are no longer used in experimental research. There are not many in the world, who studied kuru in its natural environment. For me, I could study only one of the last kuru brains.

Finally, we return to molecular biology, which is now the ultimate answer to any scientific question, in contrast to what Gajdusek used to say: "Molecular biology became molecular technology and because of this is no longer a science." To this group belong the exquisite chapters by Giuseppe Legname on synthetic prions, Abigail Diack and Jean Manson, and Glenn C. Telling and Julie Moreno on transgenic mouse models of prion disease. I am more than happy to include their chapters.

Last but not least, I would like to thank Michael Alpers for his invaluable discussions of kuru, Ewa Skarżyńska for her enormous help in handling the multiple tasks associated with this book, and Anna Rakovsky, from Springer Verlag, for her unending patience.

This work was supported in part by National Science Centre Poland, grant no UMO-2012/04/M/NZ4/00232.

Lodz, Poland *Paweł P. Liberski*

References

1. Prusiner SB (1982) Novel proteinaceous infectious particles cause scrapie. Science 216(4542), 136–144
2. Matthews WB (1975) Epidemiology of Creutzfeldt-Jakob disease in England and Wales. J Neurol Neurosurg Psychiatry 38(3), 210–213
3. Gajdusek DC (1977) Unconventional viruses and the origin and disappearance of kuru. Science 197(4307), 943–960

Contents

Dedication ..... *v*
Preface to the Series ..... *vii*
Preface ..... *ix*
Contributors ..... *xiii*

1 Kuru: Introduction to Prion Diseases 1
Paweł P. Liberski and Agata Gajos

2 Anthropological Methods Used in Kuru Research 33
Shirley Lindenbaum

3 Nonhuman Primates in Research on Transmissible Spongiform Encephalopathies 49
David M. Asher, Pedro Piccardo, and Luisa Gregori

4 Clinical Features and Diagnosis of Human Prion Diseases 65
Richard Knight

5 Neuropathology, Immunohistochemistry, and Biochemistry in Human Prion Diseases 79
Diane L. Ritchie and James W. Ironside

6 Subcellular and Molecular Changes Associated with Abnormal PrP Accumulation in Brain and Viscera of Classical and Atypical Prion Diseases 99
Martin Jeffrey and Gillian McGovern

7 Electron Microscopy of Prion Diseases 123
Paweł P. Liberski

8 Cell Death and Autophagy in Prion Diseases 145
Pawel P. Liberski

9 Methods for Isolation of *Spiroplasma* sp. from Prion-Positive Eye Tissues of Sheep Affected with Terminal Scrapie 159
Frank O. Bastian

10 Detection and Diagnosis of Prion Diseases Using RT-QuIC: An Update 173
Byron Caughey, Christina D. Orru, Bradley R. Groveman, Matilde Bongianni, Andrew G. Hughson, Lynne D. Raymond, Matteo Manca, Allison Kraus, Gregory J. Raymond, Michele Fiorini, Maurizio Pocchiari, and Gianluigi Zanusso

11 Analysis of Charge Isoforms of the Scrapie Prion Protein Using Two-Dimensional Electrophoresis 183
Elizaveta Katorcha and Ilia V. Baskakov

12 Exosomes in Prion Diseases 197
Alexander Hartmann, Hermann Altmeppen, Susanne Krasemann, and Markus Glatzel

13 Synthetic Mammalian Prions 209
Fabio Moda, Edoardo Bistaffa, Joanna Narkiewicz, Giulia Salzano, and Giuseppe Legname

14 Biomarkers in Cerebrospinal Fluid 229
Joanna Gawinecka, Matthias Schmitz, and Inga Zerr

15 The Use of Transgenic and Knockout Mice in Prion Research 253
Abigail B. Diack and Jean C. Manson

16 Transgenic Mouse Models of Prion Diseases 269
Julie Moreno and Glenn C. Telling

Index *303*

Contributors

HERMANN ALTMEPPEN • *Institute of Neuropathology, University Medical Center Hamburg-Eppendorf, Hamburg, Germany*
DAVID M. ASHER • *Laboratory of Bacterial and Transmissible Spongiform Encephalopathy Agents, Division of Emerging and Transfusion-Transmitted Diseases, Office of Blood Research and Review, Center for Biologics Evaluation and Research, United States Food and Drug Administration, Silver Spring, MD, USA*
ILIA V. BASKAKOV • *Department of Anatomy and Neurobiology, Center for Biomedical Engineering and Technology, University of Maryland School of Medicine, Baltimore, MD, USA*
FRANK O. BASTIAN • *Department of Animal Science, Louisiana State University Agricultural Center, Baton Rouge, LA, USA*
EDOARDO BISTAFFA • *IRCCS Foundation Carlo Besta Neurological Institute, Milan, Italy; Scuola Internazionale Superiore di Studi Avanzati (SISSA), Trieste, Italy*
MATILDE BONGIANNI • *Department of Neurological and Movement Sciences, University of Verona, Verona, Italy*
BYRON CAUGHEY • *Laboratory of Persistent Viral Diseases, Rocky Mountain Laboratories, National Institute for Allergy and Infectious Diseases, National Institutes of Health, Hamilton, MT, USA*
ABIGAIL B. DIACK • *The Roslin Institute & R(D)SVS, University of Edinburgh, Easter Bush, UK*
MICHELE FIORINI • *Department of Neurological and Movement Sciences, University of Verona, Verona, Italy*
AGATA GAJOS • *Department of Extrapyramidal Diseases, Medical University of Lodz, Lodz, Poland*
JOANNA GAWINECKA • *Department of Neurology, Clinical Dementia Center, National Reference Center for TSE, Georg-August University, Gö ttingen, Germany*
MARKUS GLATZEL • *Institute of Neuropathology, University Medical Center Hamburg--Eppendorf, Hamburg, Germany*
LUISA GREGORI • *Laboratory of Bacterial and Transmissible Spongiform Encephalopathy Agents, Division of Emerging and Transfusion-Transmitted Diseases, Office of Blood Research and Review, Center for Biologics Evaluation and Research, United States Food and Drug Administration, Silver Spring, MD, USA*
BRADLEY R. GROVEMAN • *Laboratory of Persistent Viral Diseases, Rocky Moutain Laboratories, National Institute for Allergy and Infectious Diseases, National Institutes of Health, Hamilton, MT, USA*
ALEXANDER HARTMANN • *Institute of Neuropathology, University Medical Center Hamburg-Eppendorf, Hamburg, Germany*
ANDREW G. HUGHSON • *Laboratory of Persistent Viral Diseases, Rocky Mountain Laboratories, National Institute for Allergy and Infectious Diseases, National Institutes of Health, Hamilton, MT, USA*

JAMES W. IRONSIDE • *National CJD Research & Surveillance Unit, Centre for Clinical Brain Sciences, School of Clinical Sciences, University of Edinburgh, Western General Hospital, Edinburgh, UK*

MARTIN JEFFREY • *Animal and Plant Health Agency, Lasswade Veterinary Laboratory, Pentlands Science Park, Penicuik, Scotland, UK*

ELIZAVETA KATORCHA • *Department of Anatomy and Neurobiology, Center for Biomedical Engineering and Technology, University of Maryland School of Medicine, Baltimore, MD, USA*

RICHARD KNIGHT • *National CJD Research & Surveillance Unit, Centre for Clinical Brain Sciences, School of Clinical Sciences, University of Edinburgh, Western General Hospital, Edinburgh, UK*

SUSANNE KRASEMANN • *Institute of Neuropathology, University Medical Center Hamburg-Eppendorf, Hamburg, Germany*

ALLISON KRAUS • *Laboratory of Persistent Viral Diseases, Rocky Mountain Laboratories, National Institute for Allergy and Infectious Diseases, National Institutes of Health, Hamilton, MT, USA*

GIUSEPPE LEGNAME • *Scuola Internazionale Superiore di Studi Avanzati (SISSA), Trieste, Italy*

PAWEŁ P. LIBERSKI • *Laboratory of Electron Microscopy and Neuropathology, Department of Molecular Pathology and Neuropathology, Medical University Lodz, Lodz, Poland*

SHIRLEY LINDENBAUM • *City University of New York, New York, NY, USA*

MATTEO MANCA • *Laboratory of Persistent Viral Diseases, Rocky Mountain Laboratories, National Institute for Allergy and Infectious Diseases, National Institutes of Health, Hamilton, MT, USA*

JEAN C. MANSON • *The Roslin Institute & R(D)SVS, University of Edinburgh, Easter Bush, UK*

GILLIAN MCGOVERN • *Animal and Plant Health Agency, Lasswade Veterinary Laboratory, Pentlands Science Park, Penicuik, Scotland, UK*

FABIO MODA • *IRCCS Foundation Carlo Besta Neurological Institute, Milan, Italy*

JULIE MORENO • *Department of Microbiology, Immunology, and Pathology, Cell and Molecular Biology Graduate Program, Prion Research Center, Colorado State University, Fort Collins, CO, USA*

JOANNA NARKIEWICZ • *Scuola Internazionale Superiore di Studi Avanzati (SISSA), Trieste, Italy*

CHRISTINA D. ORRU • *Laboratory of Persistent Viral Diseases, Rocky Mountain Laboratories, National Institute for Allergy and Infectious Diseases, National Institute of Health, Hamilton, MT, USA*

PEDRO PICCARDO • *Laboratory of Bacterial and Transmissible Spongiform Encephalopathy Agents, Division of Emerging and Transfusion-Transmitted Diseases, Office of Blood Research and Review, Center for Biologics Evaluation and Research, United States Food and Drug Administration, Silver Spring, MD, USA*

MAURIZIO POCCHIARI • *Department of Cell Biology and Neurosciences, Instituto Superiore di Sanità, Rome, Italy*

GREGORY J. RAYMOND • *Laboratory of Persistent Viral Diseases, Rocky Mountain Laboratories, National Institute for Allergy and Infectious Diseases, National Institutes of Health, Hamilton, MT, USA*

LYNNE D. RAYMOND • *Laboratory of Persistent Viral Diseases, Rocky Mountain Laboratories, National Institute for Allergy and Infectious Diseases, National Institutes of Health, Hamilton, MT, USA*

DIANE L. RITCHIE • *National CJD Research & Surveillance Unit, Centre for Clinical Brain Sciences, School of Clinical Sciences, University of EdinburghWestern General Hospital, Edinburgh, UK*

GIULIA SALZANO • *Scuola Internazionale Superiore di Studi Avanzati (SISSA), Trieste, Italy*

MATTHIAS SCHMITZ • *Department of Neurology, Clinical Dementia Center, National Reference Center for TSE, Georg-August University, Göttingen, Germany*

BEATA SIKOSKA • *Laboratory of Electron Microscopy and Neuropathology, Department of Molecular Pathology and Nueropathology, Medical University LodzLodz, Poland*

GLENN C. TELLING • *Department of Microbiology, Immunology, and Pathology, Cell and Molecular Biology Graduate Program, Prion Research Center, Colorado State University, Fort Collins, CO, USA*

GIANLUIGI ZANUSSO • *Department of Neurological and Movement Sciences, University of Verona, Verona, Italy*

INGA ZERR • *Department of Neurology, Clinical Dementia Center, National Reference Center for TSE, Georg-August University, Göttingen, Germany*

Chapter 1

Kuru: Introduction to Prion Diseases

Paweł P. Liberski and Agata Gajos

Abstract

Kuru, the first human transmissible spongiform encephalopathy, was transmitted to chimpanzees in the D. Carleton Gajdusek (1923–2008) laboratory. In this review, we briefly summarize the history of this seminal discovery along with its epidemiology, clinical picture, neuropathology, and molecular genetics. The discovery of kuru opened new windows into the realms of human medicine, and was instrumental in the later transmission of Creutzfeldt-Jakob disease and Gerstmann-Sträussler-Scheinker disease, as well as the relevance that bovine spongiform encephalopathy had for transmission to humans. The transmission of kuru was one of the greatest contributions to biomedical sciences of the twentieth century.

Key words Kuru, Prion diseases, Neuropathology, D. Carleton Gajdusek

1 Introduction

Kuru is a disease that is linked forever with the name of D. Carleton Gajdusek, who initiated the study of "prion diseases" in humans [1–11]. The disease was first reported in Western medicine by D. Carleton Gajdusek and Vincent Zigas in 1957 [12, 13].

Kuru was the first human prion disease transmitted to chimpanzees, classified as a transmissible spongiform encephalopathy (TSE), said to be caused by a slow virus. The recognition of kuru as transmissible (i.e., infectious) [14–18], a form of Creutzfeldt-Jakob disease (CJD) [19] which was transmitted by cannibalism (see Lindenbaum, this volume), provides the example of a disease caused by a novel class of pathogens. Kuru earned a Nobel Prize for D. Carleton Gajdusek in 1976 for *"for [...] discoveries concerning new mechanisms for the origin and dissemination of infectious diseases,"* followed by the award of a second Nobel Prize to Stanley B. Prusiner in 1997, for *"his discovery of Prions—a new biological principle of infection."* Kuru was linked indirectly to a third Nobel Prize when Kurt Wüthrich who determined the structure of the prion protein [20]. As Gajdusek stressed for the

Pawel P. Liberski (ed.), *Prion Diseases*, Neuromethods, vol. 129,
DOI 10.1007/978-1-4939-7211-1_1, © Springer Science+Business Media LLC 2017

last time in his life at the Royal Society meeting on kuru [21], the solving of the kuru riddle contributed to developing ideas of molecular casting and to further understanding of such diverse areas as dermatoglypes and osmium shadowing in electron microscopy. Recently, kuru research had an impact on the concepts of nucleation-polymerization, and has led to a concept of "conformational disorders" [22–24] or prionoids [25].

2 Background and Ethnographic Setting

"Kuru" in the Fore language of Papua New Guinea means to tremble from fear or cold [13, 26–39]: *"The natives of almost all of the Fore hamlets have stated that it has been present for a 'long time'; but they soon modify to mean that in recent years it has become an increasingly severe problem and that in the early youth of our oldest informants there was no kuru at all"* [13].

Kuru was restricted to people belonging to the Fore linguistic group in Papua New Guinea's Eastern Highlands and neighboring linguistic groups (Auiana, Awa, Usurufa, Kanite, Keiagana, Iate, Kamano, Gimi; Figs. 1 and 2). Those groups with which kuru-affected peoples did not intermarry, such as the Anga (Kukukuku), separated from the Fore by Lamari River, and the remote Iagaria, Kamano, and Auiana people, were not affected. Zigas and Gajdusek [40, 41] noticed that when Fore of Kasarai in the South Fore moved temporarily to live with the Yar people and settled there for about a decade, they still had cases of kuru. It seems that kuru first was said to appear at, or shortly after, the turn of the twentieth century [42–44] in Uwami village (a fact that was denied by Uwami informants—[45]; the other places were also mentioned: Kasokana in 1922, Kamila and Wanikanto in 1923) and spread from there to Awande in the North Fore where the Uwami had social contacts. Within 20 years it had spread further into Kasokana (in 1922, according to Lindenbaum [44]) and to Miarasa villages of North Fore, and a decade later it had reached the South Fore at Wanikanto and Kamira villages. The slow march of kuru was inconsistent with a contemporary genetic hypothesis of Bennett et al. [46, 47], but it was consistent with a slow infectious disease.

Kuru became endemic in all villages which it entered, and became hyperendemic in the South Fore region. All native informants stressed the relatively recent origin of kuru. Interestingly, when kuru first appeared, its symptoms were considered poetically by Fore to be similar to "the swaying of *casuarinas tree*" and kuru was labeled *cassowary disease* to stress the similarity between *cassowary* quills and "waving *casuarinas* fronds."

Fig. 1 Members of Fore tribe. Courtesy of late D. Carleton Gajdusek; code dcg-57-ng-335

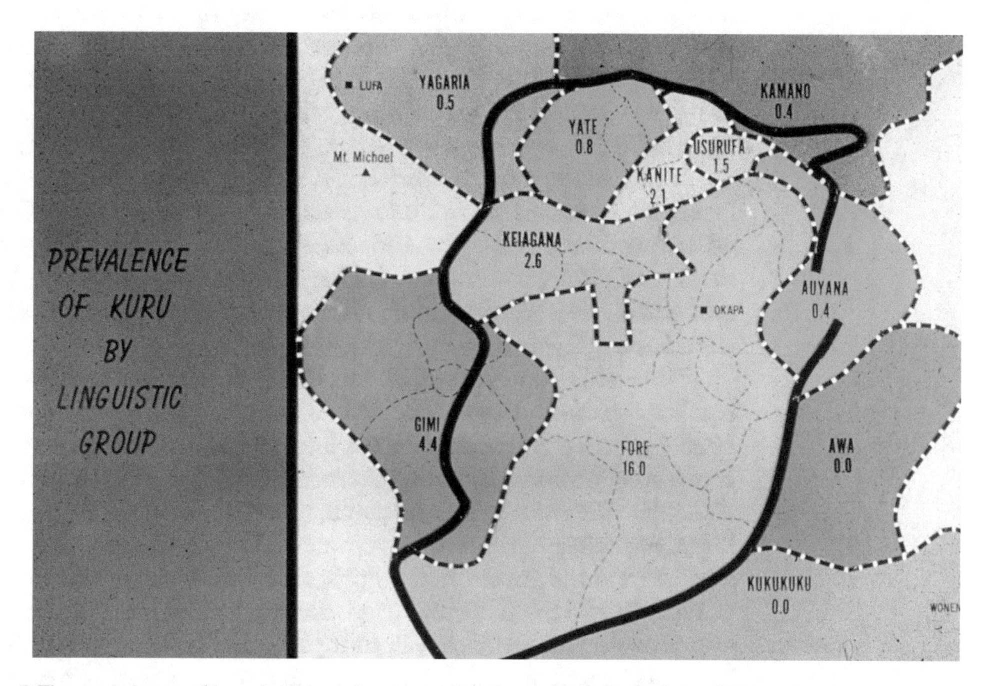

Fig. 2 The prevalence of kuru by linguistic group. Courtesy of late D. Carleton Gajdusek

3 Cannibalism

"We are often asked when and how we first came upon the idea that cannibalism was involved in the spread of disease. It is useless to speculate about the origin of this idea; I know of few Europeans who did not arrive at such a conjecture." "With a the disappearance of kuru from the youngest age-group in 1964, and then progressively from older age-group, cannibalism of those who died of kuru was established as the unique means of dissemination of disease—involving the contamination particularly of woman and children, who preferentially participated in mourning rites and the associated cannibalism. Transmission being interrupted by the cessation of cannibalism during the 1950s, no one born since then has developed kuru" [29]. Alpers presented this idea in 1967 at the International Academy of Pathology, Washington [28, 48].

Ritualistic endocannibalism (eating relatives as part of a mourning ritual in contrast to eating enemies, i.e., exocannibalism) was practiced not only in the kuru area but in many surrounding Eastern Highland groups in which kuru never developed [49–54]. *"When a body was considered for human consumption, none of it was discarded except the bitter gall bladder. In the deceased's old sugarcane garden, maternal kin dismembered the corpse with a bamboo knife and stone axe. They first removed hands and feet, then cut open the arms and legs to strip the muscles. Opening the chest and belly, they avoided rupturing the gall bladder, whose bitter content would ruin the meat. After severing the head, they fractured the skull to remove the brain. Meat, viscera, and brain were all eaten. Marrow was sucked from cracked bones, and sometimes the pulverized bones themselves were cooked and eaten with green vegetables. In North Fore but not in the South, the corpse was buried for several days, then exhumed and eaten when the flesh had 'ripened' and the maggots could be cooked as a separate delicacy"* [49].

The first Europeans who entered the kuru area were the Ashton brothers, around 1934, and Ted Ubank, a gold prospector, in 1936 [45, 49]. In the late 1930s and 1940s, many gold miners, Protestant missionaries, and government officials became familiar with the presence of endocannibalism of Eastern Highland. However, missionaries and government officials tried to stop cannibalism and the government imprisoned some who participated in it. Early in 1951 and 1953, kuru was observed by the anthropologists Ronald and Catherine Berndt [49], and the first mention of kuru (*skin-guria* in Pidgin) was included in reports of patrol officers in 1953. Zigas was told about kuru in 1955, and he was joined by D. C. Gajdusek 2 years later. I asked Gajdusek in the 2000s when the hypothesis of cannibalism as a vehicle to spread kuru was first envisaged. His response was that *"even completely drunk would come to the conclusion that a disease endemic among cannibals must be spread through eating corpses."* For instance, in a letter dated 15 March 1957 to Smadel, Gajdusek reported a case of kuru *"in the center of tribal groups of cannibals"* [...] who *feeding the children the body of kuru case"* [29]. Gajdusek said this some 50 years after the discovery of kuru. In a letter Smadel wrote to Morris in 1957 [29], he said, *"Gajdusek is now among cannibals studying a most important new disease with neurological manifestations."* The first to investigate and formally publish the hypothesis that kuru spreads through cannibalism were Glasse and Lindenbaum [54–56].

In 1962 and 1963 [57, 58] Glasse sent reports of their fieldwork reports to John Gunther, the Director of New Guinea Department of Public Health, which provided evidence that kuru was of recent origin, and an account of the data they were gathering about consumption of the dead, with its apparent association with cannibalism. They reported their findings at a kuru workshop in Adelaide in 1962, and in 1963 they spoke about the cannibalism hypothesis to a group of scientists who visited them in the South Fore. Their hypothesis was published in 1967 [56] and 1968 [54]. They also wrote about it in 1993 [55]. Lindenbaum discussed the hypothesis again in later publications [49, 52, 53]. However, although Gajdusek said that the hypothesis of cannibalism was taken for granted, it is also true that in his Nobel Prize lecture he said that kuru spread by *"conjunctival, nasal and skin contamination with highly infectious brain tissue"*; thus at that time Gajdusek was still skeptical about the cannibalism hypothesis which he regarded as exotic. Robert Glasse quoted this *"as the only mention of cannibalism he found in our publishing writing."* Some authors even denied the very existence of cannibalism [59] but the denial clearly belongs to another mythology. Cannibalism among the Fore had ended long ago, although a court case elsewhere in Papua New Guinea was reported in 1978 [44].

During the last 50 years, the number of kuru cases has steadily declined, with the youngest patients becoming progressively older, and the disease is now extinct. However, we cannot be sure that a limited number of cases with incubation period in excess of 50 years may appear in the future.

Among the Fore, kuru was believed to result from sorcery [49]. The victim was said to have been chosen because of some real or imaginary fault [41, 60]. To cause kuru, a would-be sorcerer would need to obtain a part of the victim's body (nail clippings, hair) or excreta, particularly feces- or urine-soaked vegetation, saliva, blood, or partially consumed food, such as peelings from sweet potato eaten by the victim, or clothing. These were packed with leaves and made into a *"kuru bundle"* and placed partially submerged in swampy land. Subsequently, the sorcerer shook the package daily until the tremor characteristic of kuru was induced in his victim. As a result, the kin of a kuru victim attempted to identify and subsequently kill ("tukabu") a suspected sorcerer if they could not bribe or intimidate him to release a victim from the kuru spell [49].

Divination rituals helped to identify a sorcerer. One method was to collect water for the kuru victim from different sources; if one "induced" vomiting, it was considered to identity the sorcerer's residence. Another method was to place hair clippings from a kuru victim in a bamboo cylinder, and a freshly killed possum in another cylinder. Calling the name of a suspected sorcerer while shaking the cylinders, a member of the victim's family placed the possum-containing cylinder into a fire. The sorcerer was identified if the liver of the possum, believed to be the residence of his soul, remained uncooked. Still another involved the roasting of small rats, in separate bamboo cylinders, each one having been given the name of a hamlet or village in which the suspected sorcerer lived. Careful inspection of the rat's viscera helped to identify the sorcerer.

Killing a sorcerer—*tukabu* (or *tokabu*)—was a ritualistic form of vendetta called *sangguma* in Tok Pisin; it included crushing with stones the bones of the neck, arm, and thigh, as well as the loins, biting the trachea, and grinding the genitalia with stones and clubs. Most cases ascribed to Tukabu were sudden deaths associated with disease such as a stroke. Tukabu allegations were frequent, but physical assassinations were rare.

4 Kuru Etiology: The Insight into a Novel Class of Pathogens

Although on epidemiological grounds the etiology of kuru was thought to be infectious, patients had no meningoencephalitic signs or symptoms (fever, seizures, or coma), no cerebrospinal fluid pleocytosis or elevated protein level, and, on autopsy, no perivascular cuffings or other signs of inflammatory brain pathology. However, even before seeing the disease, Gajdusek knew that kuru was an "encephalitis-like disease" based on reports sent by Dr. Zigas to Dr. John Gunther, then Assistant Adminstrator of the Trust Territory [29]. Zigas wrote to Gunther that kuru was *"a probably new form of encephalitis" [...] "started with fever, somnolence, muscular pain and weakness, headache [...] vertigo [...] mask-like faces, flexed arms and wrists, unsteady walk, ocular disorders such as diplopia, strabismus, nystagmus, tremor of fingers and hands."* Neither environmental [61–63] nor then available genetic studies [27, 64–72] provided any clues. Moreover, all attempts to transmit kuru to small laboratory animals, or to isolate any microorganism including a virus, using tissue cultures or embryonated hen's eggs, were successful. In other wide-ranging investigations, such as exhaustive genetic analyses, the search for nutritional deficiencies or environmental toxins did not result in a tenable hypothesis [62, 63].

On July 21, 1959, while in New Guinea, Gajdusek received a letter from the American veterinarian, William Hadlow, at the Rocky Mountain Laboratory in Hamilton, Montana [35, 73–77], which pointed out the analogies between kuru and scrapie, a slow neurodegenerative disease of sheep and goats known to be endemic in the United Kingdom since the eighteenth century [78] and experimentally transmitted to goats in 1936 [79, 80]. Having seen photographs of kuru plaques at a Wellcome Medical Museum exhibition in London, he enclosed a copy of a letter pointing to this similarity to the editor of Lancet [73, 76]. Hadlow based his observations not merely on the presence of amyloid plaques, but mainly because of the presence of vacuolated neurons:

"I've been impressed with the overall resemblance of kuru, and an obscure degenerative disorder of sheep called scrapie [...] The lesions in the goat seem to be remarkably like those described for Kuru. [...] All this suggests to me that an experimental approach similar to that adopted for scrapie might prove to be extremely fruitful in the case of kuru. [...] because I've been greatly impressed by the intriguing implication, I've submitted a letter to The Lancet."

Gajdusek commented on this in 1981: *"Much more compelling for us than popular speculations about cannibalism was Dr. William Hadlow suggestion, in 1959 that scrapie, and infectious chronic disease of sheep and goats, was clinically and pathologically similar to kuru.* [29] *When we become aware in this way that viruses could produce a chronic neurological disease, we obviously pursued this lead experimentally with animal models. If kuru were infectious, it was obvious to us that cannibalism would be implicated as a mode of transmission; the real question, however was whether it was infectious in such a way that standard indices of infectivity were inadequate."*

A similar observation was made by veterinary neuropathologist Innes [81] during his visit to the Gajdusek laboratory [82, Gajdusek—telephone conversation, 2008]. Hadlow, in his recollection of that seminal observation, had pointed out intracellular vacuoles as those neuropathological changes that attracted his attention some 40 years ago [83, 84]. Such intracellular vacuoles in scrapie were first described by Besnoit and his colleagues in 1898 [79]. Dr. Gajdusek replied that *"[As you may have been able to gather from our articles on kuru, we are pursuing the matter of possible infectious etiology extensively—I am, in fact, a virologist by training. However, we have thus far had poor luck with inoculation experiments and the possibility of doing more extensive inoculation works has, until now, been small. We are, however, proceeding accordingly at the present time and frozen and fresh material are being injected into a number of animal hosts during this years work on kuru. In your note to LANCET, which I am deeply grateful to you for bringing to my attention, I note that you have probably not seen our extensive pathological description of kuru which includes some features which were little stressed in the report you have quoted],"* and took up Hadlow's recommendation to hold small laboratory rodents and (especially) apes and monkeys for longer periods of observation than had thus far been carried out. He also renewed attempts to obtain optimal tissue for inoculation from rapidly autopsied kuru patients (letter from D. C. Gajdusek dated August 6th, 1959).

In 1961, Gajdusek presented a lecture at the Xth Pacific Science Congress in Honolulu entitled *"Kuru: an appraisal of five years of investigation. With a discussion of the still undiscardable possibility of infectious agent"* in which he said: *"In spite of all the genetic evidence, both the pathological picture and the epidemiological peculiarities of the disease persistently suggest that some yet-overlooked, chronic, slowly progressive, microbial infection may be involved in kuru pathogenesis. Similar suspicion prevails in our current etiological thinking about a number of less exotic and less rare chronic, progressive degenerative diseases of the central nervous system in man. Thus, [...], amyotrophic lateral sclerosis, Schilder disease, leukoencephalitis, Koshevnikoff's epilepsy syndrome in the Soviet Union, the Jakob-Creutzfeldt syndromes, acute and chronic cerebellitis, and even many forms of Parkinsonism, especially the Parkinsonism dementia encountered among the Chamorro population in Guam, continue to suggest the possibility that in man there may be infections analogous to the slow infections of the nervous system of animals which were intensively studied by Bjorn Sigurdsson, the Icelandic investigator who formulated the concept of 'slow virus infections.'"* This contention preceded the discovery of kuru transmissibility by more than 4 years [85]. Parenthetically, many of the diseases mentioned by Gajdusek are now grouped together under the umbrella of "protein conformational disorders" [22, 24, 86, 87]. Finally, in 1965, in a monograph "Slow, Latent and Temperate Virus Infections" which resulted from a meeting convened in 1964, Gibbs and Gajdusek [88] wrote in an addendum, *"although several of the inoculated primates died of acute infection during the period of observation, [...] none has developed signs suggestive of chronic neurological disease until the recent onset in two chimpanzees. The first of these, inoculated 20 months previously with a suspension of frozen brain material from a kuru patient, has developed progressive incapacitating cerebellar signs with ataxia and tremor; the second, similarly inoculated with a suspension of brain material from another kuru patient, has developed, 21 months after inoculation, slight wasting lassitude, and some tremor which appeared to be progressive. Whether these syndromes are spontaneous or related to the inoculation remains to be determined."*

5 Epidemiology of Kuru: A Strong Support of the Cannibalism Theory

Kuru incidence increased in the 1940s and 1950s [12, 18, 89–92] and approached a mortality rate in some villages of 35/1000 among a population of some 12,000 Fore people [82, 93]. This mortality rate distorted certain populational parameters: in the South Fore, the female:male ratio was 1:1.67 in contrast to 1:1 ratio in unaffected Kamano people. This ratio increased to 1:2 to even 1:3 among South Fore. Gajdusek even calculated the female deficit in the population to be 1676 persons [36]. The almost total absence of kuru cases in South Fore among children born after 1954, and the rising of age of kuru cases year by year, suggested that transmission of kuru to children had stopped in the late 1950s [94–96] when cannibalism ceased to be practiced among the Fore people. Also, brothers with kuru tend to die at the same age, which suggested that they were infected at similar age but not at the same time. The assumption that affected brothers were infected with kuru at the same age led to a calculation of minimal age of exposure for males to be in a range of 1–6 years, with a mean incubation period of 3–6 years, and the maximum incubation period of 10–14 years [42].

Alpers and Gajdusek wrote a year before the transmission of kuru was published [89], *"The still baffling, unresolved problem of the etiology of kuru in the New Guinea Highlands has caused as to wonder whether or not any or many of the unusual features of its epidemiological pattern and its clinical course may not be changing with time, or even altering drastically under the impact of extensive rapid cultural change, the result of ever increasing inroads of civilization upon the culture of the Fore people."* This was indeed the case. The comparison of total number of deaths from kuru in the period 1961–1963 and 1957–1959 showed a 23% reduction, and among children, 57% reduction, and the kuru mortality rates dropped from 7.64 to 5.58 deaths per thousand. These alterations were not uniform, the North Fore reduction exceeded the South Fore reduction, and it is worth recalling that South Fore kuru deaths accounted for 60% of the total. This trend continued until the disappearance of kuru epidemic in 2009 [97].

The almost "formal" proof that kuru was indeed transmitted by cannibalism was provided by Klitzman et al. [98] who studied clusters of kuru patients who participated in a limited number of kuru feasts in 1940s and 1950s. Three clusters were identified, one of which will be recalled here. Two brothers, Ob and Kasis from the North Fore village of Awande, developed signs and symptoms of kuru in 1975, 21 or 27 years after the latest or the earlier exposure, respectively. They participated in 2 feasts for kuru victims. In those feasts, the close relatives were the major mourners who actively participated in the consumption of the dead.

Of interest, Klitzman et al. [98] noticed that taking into consideration the fact that Fore women participated in numerous kuru feasts, it is strange that any of them survived into the 1970s. Modern molecular genetics explained this fact in terms of the codon 129 polymorphism of the *PRNP* gene. In the younger patients, homozygotes $129^{\text{Met Met}}$ predominate; the latter finding is reminiscent of variant CJD (vCJD) [99], and suggests the increased susceptibility of $129^{Met\ Met}$ individuals, with a shorter incubation period than other *PRNP* codon 129 genotypes.

6 Transmission Experiments

The transmission of kuru to chimpanzees won a Nobel Prize for Gajdusek in 1976 [14, 16, 37, 38, 85, 100–105]. The list of nonprimate host range for kuru transmission is given in Table 1. Table 2 contains the list of nonhuman primates to which kuru was transmitted over the years. They include Rhesus monkeys [106], marmosets [107], and gibbon and sooty mangabey monkeys [108]. Detailed description of experimental kuru in 41 chimpanzees was published in 1973 [2]. The incubation period varied from 11 to 39 months (the average 23 months for the first passage; 12 months for the second passage; 13 months for the third passage and the same for the fourth passage) and the clinical course was divided into three stages:

6.1 Early Stage (I)

(a) Prodromal period characterized by earliest alterations in behavior: animals became inactive, sometimes "extremely dirty" and submissive. *"Vicious and aggressive animals became passive and withdrew from competition with their normal cagemates, allowing smaller chimpanzees to tease and take food from them [...] periods of sullen apathy were often interrupted by outbursts of furious screaming."*

Table 1
Nonprimate host range for kuru transmission; incubation period in months

Species	Incubation period (months)
Goat (*Capra hircus*)	(104)+
Guinea pig (*Cavia porcellus*)	(27)
Opossum (*Didelphis marsupialis*)	(22+)
Domestic cat (*Felis domesticus*)	(59)
Gerbil (*Meriones unguiculatus*)	(24)+
Hamster (*Mesocricetus auratus*)	(28)
Mous (*Mus musculus*)	22.5
Ferret (*Mustela putorius*)	18–70.5
Mink (*Mustela vison*)	45
Sheep (*Ovis aries*)	(63)+

Number in parenthesis—number of months elapsed since the inoculation, during which the animals remained asymptomatic

Table 2
A host range of the primates susceptible for kuru

Species	Incubation period (months)
Apes	
Chimpanzee (*Pan troglodytes*)	10–82
Gibbon (*Hylobates lar*)	+ (10)
New world monkeys	
Capuchin (*Cebus albifrons*)	10–92
Capuchin (*Cebus apella*)	11–71
Spider (*Ateles geoffroyi*)	10–85.5
Marmoset (*Saguinus sp*)	1176
Wolly (*Lagothrix lagotricha*)	33
Old world monkeys	
African Green (*Cercopithecus aethiops*)	18
Baboon (*Papio anubis*)	(130)
Bonnet (*Macaca radiate*)	19–27
Bushbaby (*Galago senegalensis*)	(120)
Cynomolgus macaque (*Macaca fascicularis*)	16
Patas (*Erythrocebus patas patas*)	(136)
Pigtailed macaque (*Macaca nemestrina*)	70
Rhesus (*Macaca mulatto*)	15–102
Sooty mangabey (*Cercocebus atys*)	+(2)
Talapoin (*Cecopithecus talapin*)	(1+)

(b) Period of minimal disabilities characterized by minor motor dysfunction: animals did not want to go outside cages, "to run or to climb." They were slow and fell with forced movements; the movements were *"like [...] in slow-motion cinema."*

6.2 Intermediate Stage (II)

The onset of this stage was characterized by difficulties exhibited when a chimp tried to rise from a supine position; gait became ataxic but animals still could sit. The gait of chimpanzees is *quadrupedal*, "knuckle walking", where animals placed hands on the ground not with palms but with knuckles and this pattern is preserved, but the gait itself is grossly ataxic and dysmetric. Truncal titubation, so characteristic for human kuru, is present since stage II. Passive muscle tone is increased and flexion contractures may develop if an animal lives long enough. Severe coarse tremor is seen, choreiform movements are observed, and the negligence develops. Difficulties in seeing, lateral nystagmus, and intermittent left strabismus were seen. "Babinsky" sign was occasionally observed.

6.3 Late Stage (III)

Characterized by severe neurological deficits: the animals could not rise by themselves from a supine position, they could not sit but placed themselves in one position, and decubitus ulcers were common. They ate inedible objects. A severe startled response comprising flexion of all extremities accompanied by violent coarse trembling of all limbs was a characteristic finding. *"Silly smiles, with grimacing, were prominent. Fixed and pained faces and slow, clumsy, voluntary motion [...] were prominent."*

Kuru neuropathology in chimpanzees was described by late Beck and Daniel [109–115]. The neuropathological picture was practically identical to that of natural kuru except for the absence of amyloid plaques. In the cerebral cortex, the spongiform change and intraneuronal vacuoles were the most prominent lesions, accompanied by a severe astrocytic gliosis. Binucleated neurons were prominent; the same type of neuronal lesions were also seen in the spider monkey [116].

7 Clinical Manifestations

"I was still very young when I saw [kuru] and even after we treated it there was no help. Everyone was falling apart. [Kuru victims] were aware there was no cure and that they would die. It wasn't just one person that this sickness came to—there were about three in a house line and then after they died there would be another three. It was ... ongoing... there were many deaths. Once a [person] ... was affected by kuru [their] family would think that the clan had poisoned [them] and they would start ... shooting at each other and that made it worse. It was chaos! (Taurubi)" [117].

Kuru is an invariably fatal cerebellar ataxia accompanied by tremor, choreiform, and athetoid movements [18, 28, 32, 35, 48, 78, 117–126]. In contrast to the neuropathological picture, neurological signs and symptoms are highly uniform. Dementia, typical for most subtypes of sporadic CJD (see Knight—this volume), was reported to be barely noticeable, and if it was present, then it appeared only late, during the evolution of the illness. However, in the most recent study, dementia was definitely observed [78]. In contrast, kuru patients often displayed emotional alterations, including inappropriate euphoria and compulsive laughter (the journalistic notorious "laughing death" or "laughing disease"), or apprehension and depression. Kuru is divided into three clinical stages: ambulant, sedentary, and terminal (the Pidgin expressions, *wokabaut yet*, i.e., is still walking; *sindaun pinis*, i.e., is able only to sit; and *slip pinis*, i.e., is unable to sit up) [118, 127]. The duration

Fig. 3 Early stage of kuru in a victim who is supported by sticks. Courtesy of late D. Carleton Gajdusek; code: dcg-57-ng-359

of kuru, as measured from the onset of prodromal signs and symptoms until death was about 12 months (3–23 months) [118, 127] and 10–25 months in a recent study [78].

There is an ill-defined prodromal period (*kuru laik i-kamp nau*, i.e., "kuru is about to begin") characterized by the presence of headache and limb pains, frequently in the joints; knees and ankles came first, followed by elbows and wrists; sometimes, interphalangeal joints were first affected, and abdominal pains and loss of weight. This period lasted approximately a few months. Fever and other signs of infectious disease are never seen, but the patient's general feeling was reported as reminiscent of that accompanying acute respiratory infection. Some patients even said that they expected a cough to come and when it did not they began to fear incoming kuru.

The prodromal period is followed by the "ambulant stage," the end of which is defined when the patient is unable to walk without a stick (Fig. 3). The patients were psychologically supported by a community of kin; one of the very important signs of this was the search for a sorcerer who, as already mentioned, they believed caused kuru. This period is characterized by the onset of subtle signs of gait unsteadiness that are usually only self-diagnosed, but which over a period of a month or so progress to severe astasia and ataxia. Incoordination of the trunk muscles and lower limbs followed. As patients were well aware that kuru heralded death in about a year, they became withdrawn and quiet. A fine "shivering" tremor, starting in the trunk, amplified by cold and associated with "goose flesh," is often followed by titubation and other types of abnormal movements. Attempts to maintain balance resulted in clawing of the toes and curling of the feet; this "clawing response" is regarded by Collinge et al. [78] as pathognomonic of kuru. Plantar reflex is always flexor, while clonus, in particular ankle clonus but also patellar clonus, is a hallmark of the clinical picture. However, clonus may be present for only a limited period of time. The ankle clonus was in most cases the most enhanced, but patellar clonus and clonus of fingers and toes were also readily elicited.

In the early stages ataxia could be demonstrated only when the patient stood on one leg; the Romberg sign was almost always negative (2 of 34 kuru cases in Alpers series [127]), but with the progression of disease, ataxia became marked and the Romberg sign became positive; indeed, the patient cannot stand with his or her feet close together. Ataxia in the upper limbs followed that in the lower limbs; dysmetria was usually the first sign of the upper limb ataxia. Intention tremor was found in 19 of 34 cases in Alpers series in the first stage of kuru but was constantly present in the second stage. Dysarthria appeared early. Resting tremor is a cardinal sign of kuru. According to Alpers *"[it is] difficult to describe and analyze. It appears to include the following components: a shivering component, an ataxic component, and, in the latter stages and certain cases only, the extrapyramidal component. A fine shivering-like tremor may be present from the onset of disease [...] it is potentiated by cold and thus may not be found in the heat of the day; a sudden drop in temperature not sufficient to make others shiver will induce it in kuru patients. As ataxia increases a more obvious ataxic component is added and the shaking movements become wilder and more grotesque."* The major component in kuru kinetic tremor is the ataxic one: it is enhanced by the muscular activity, and when the patient becomes motionless, it subsides. *"It often seemed to be triggered by minor movement, an adjustment of posture, stretching out the arm in greeting, or even a sudden turning of the eyes."* Patients learn how to control the tremor. A child was trembling violently. Zigas and Gajdusek [40, 41] found that tremor could be almost completely alleviated when the child lay curled into a flexed, fetal position in the mother's lap.

Photophobia was frequently observed [78]. A horizontal convergent strabismus is a typical sign, especially in younger patients; nystagmus was common; in the terminal stage, typically it was a pendular nystagmus, but the papillary responses were preserved. Facial hemispasm and supranuclear facial palsies were also common.

The second "sedentary" stage begins when the patient is unable to walk without constant support and ends when he or she is unable to sit without it. *"The gait was, by definition non-existent. However, if a patient was 'walked' between two assistants a caricature of walking was produced, with marked truncal instability, weakness at hips and knees and heavy leaning on one or other assistant for support; but steps could be taken, and were characterized by jerky flinging, at times decomposed movements, which led to a high-steppage, stamping gait.*" Postural instability, severe ataxia, tremor, and dysarthria progress endlessly through this stage. Deep reflexes may be increased, but the plantar reflex is still flexor. "Jerky" ocular movements were characteristic. Opsoclonus, a chaotic saccades in any direction following voluntary conjugate movements of the eyes, was occasionally noticed. Zigas and Gajdusek [40, 41] reported a peculiar, jerking, clonic movements of the eyelids and eyebrows in patients confined to the dark indoors of huts and then transported outdoors to the light. Two cases of 34 showed signs of dystonia.

In the third stage, the patient is bedridden and incontinent (Fig. 4), with dysphasia and primitive reflexes, and eventually succumbs in a state of advanced starvation. *"The patient at the beginning of the third stage usually spent the day supported in the arms of*

Fig. 4 The female kuru patient dying in the terminal stage. Courtesy of late D. Carleton Gajdusek

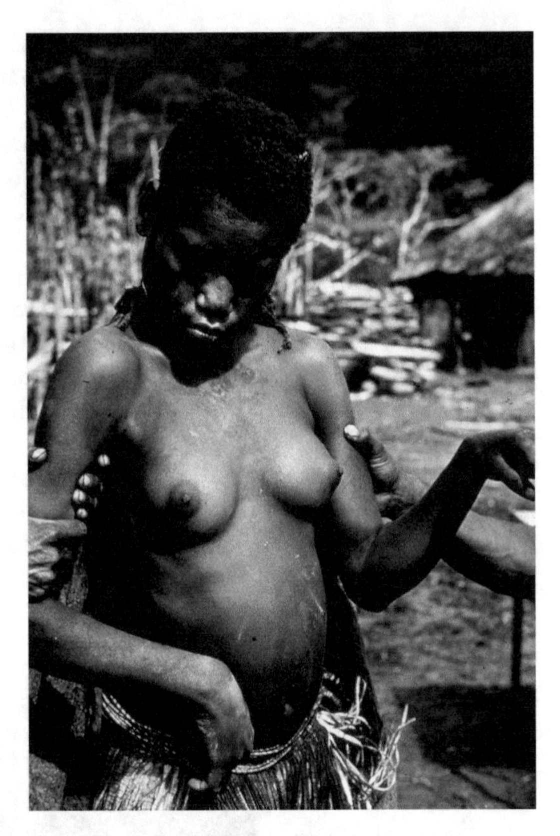

Fig. 5 Dystonic postures in a female affected with kuru. Courtesy of late D. Carleton Gajdusek; code: dcg-57-ng-346A

a close relative." Extraocular movements were either jerky or, to the contrary, slow and rigid. Deep reflexes were exaggerated, but Babinsky sign was never noticed. Generalized muscle wasting became evident and fasciculation, spontaneous, or evoked by tapping, was seen. Some symptoms of dementia were also observed. Even in terminal stages, they tried to accommodate the request of the examiner, even when they could move only the head and the eye. A strong grasp reflex occurred as well as fixed dystonic postures (Fig. 5), athetosis, and chorea. In one case "*almost constant small involuntary movements, involving mouth, face, neck, and hands*" were seen.

Terminally, "*the patient lies moribund inside her hut surrounded by a constant group of attending relatives. [...] She barely moves and is weak and wasted. Her pressure sores may have spread widely to become huge rotting ulcers which attract a swarm of flies. She is unable to speak. The jaws are clenched and have to be forced open in order to put food or fluid in. [...] Despite her mute and immobile state she can make clear signs of recognition with her eyes and may even attempt to smile.*"

It is worth mentioning the incredibly strong support given by Fore to dying kinsmen. *"The family members live with the dying patient, siblings sleep closely huddled to their brother or sister in decubitus, parents sleep with their Kuru-incapacitated child cuddled to them and a husband will patiently lie beside her terminal, uncommunicative, incontinent, foul smelling wife"* [41].

8 Neuropathology

The first systematic examination of kuru neuropathology (12 cases) was published by Klatzo et al. in 1959 [128, 129]. Macroscopically the brain is normal (Fig. 6). Neuronal alterations he described were totally nonspecific in nature, but nonetheless sufficient to draw a parallel between kuru and Creutzfeldt-Jakob disease.

Neurons were shrunken and hyperchromatic or pale, with dispersion of Nissl substance or containing intracytoplasmic vacuoles similar to those already described in scrapie. In the striatum, some neurons were vacuolated to such a degree that they looked "motheaten." Neuronophagia was observed. A few binucleated neurons were visible and torpedo formation was noticed in the Purkinje cell layer, along with empty baskets that marked the presence of degenerated Purkinje cells (Fig. 7). In the medulla, neurons of the vestibular nuclei and the lateral cuneatus were frequently affected; the spinal nucleus of the trigeminal nerve and nuclei of VIth and VIIth, and motor nucleus of the VIth cranial nerves were affected less frequently, while nuclei of the XIIth cranial nerve, the dorsal nucleus of Xth cranial nerve, and nucleus ambiguous were relatively spared. In the cerebral cortex, the deeper layers were affected more than the superficial layers, and neurons in the hippocampal

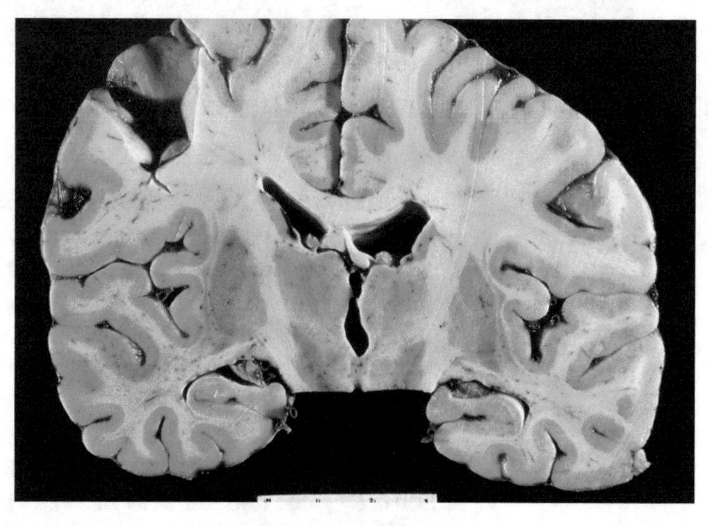

Fig. 6 Macroscopic section of the kuru brain [141]

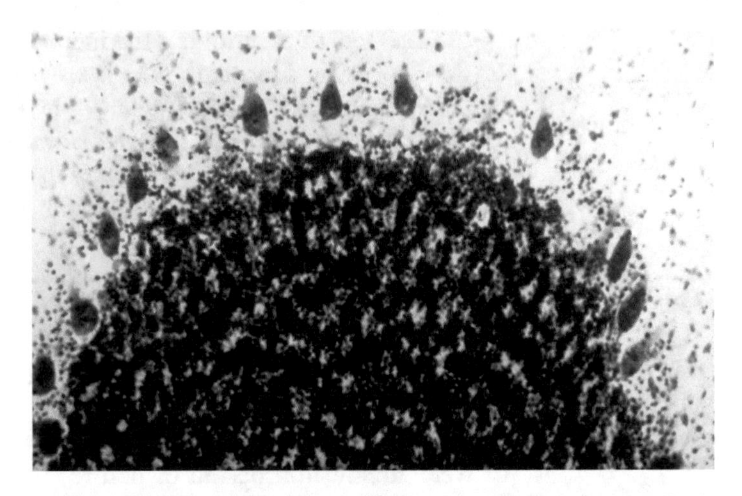

Fig. 7 Depopulated layer of Purkinje cells. Courtesy of late D. Carleton Gajdusek

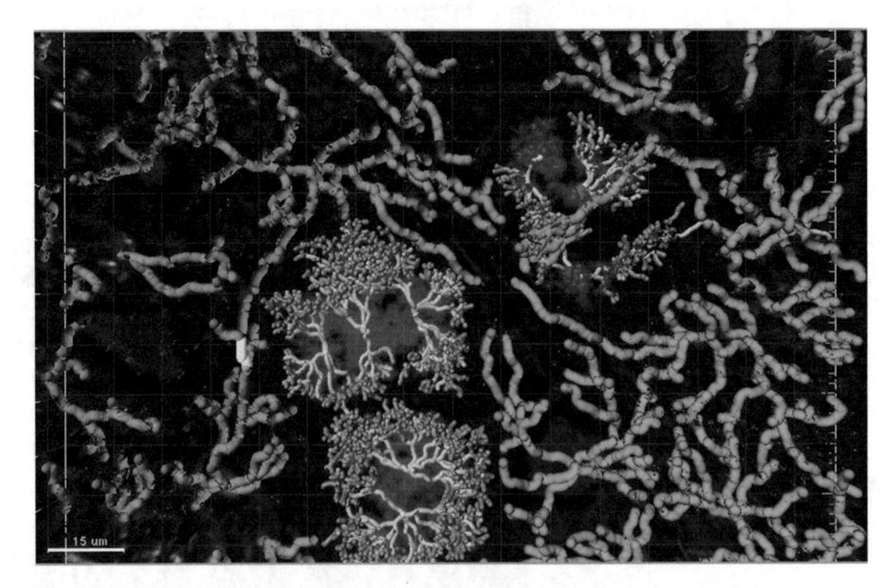

Fig. 8 Confocal laser microscopy image of kuru plaques surrounded by numerous neuritis

formation were normal. In the cerebellum, the paleocerebellar structure (vermis and flocculonodular lobe) was most severely affected, and spinal cord pathology was most severe in the corticospinal and spinocerebellar tracts. Astro- and microglial proliferation was widespread; the latter formed rosettes and appeared as rod or amoeboid types or as macrophages (gitter cells). Myelin degradation was observed in 10 of 12 cases. Interestingly, the significance of vacuolar changes was not appreciated by Klatzo et al. [128, 129], but *"small spongy spaces"* were noted in 7 of 13 cases studied by Beck and Daniel [111–115].

The most striking neuropathologic feature of kuru was the presence of numerous amyloid plaques (Figs. 8, 9, 10 and 11). In a first

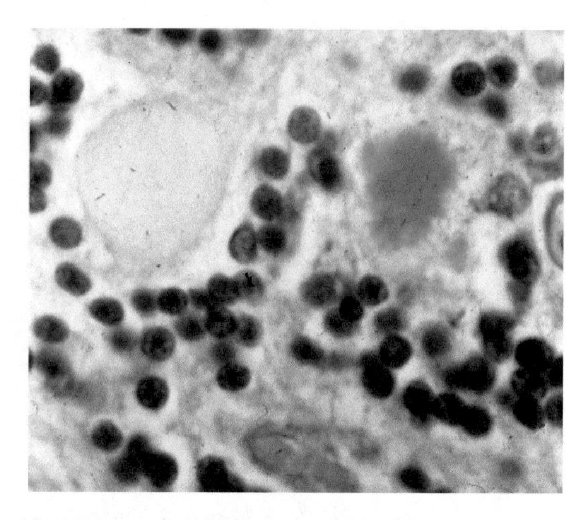

Fig. 9 Typical kuru plaque in H&E. Courtesy of late D. Carleton Gajdusek

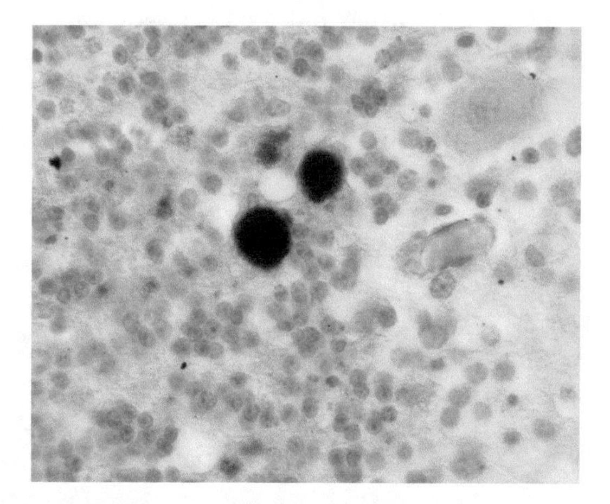

Fig. 10 Kuru plaques stained immunohistochemically with antibodies against PrP

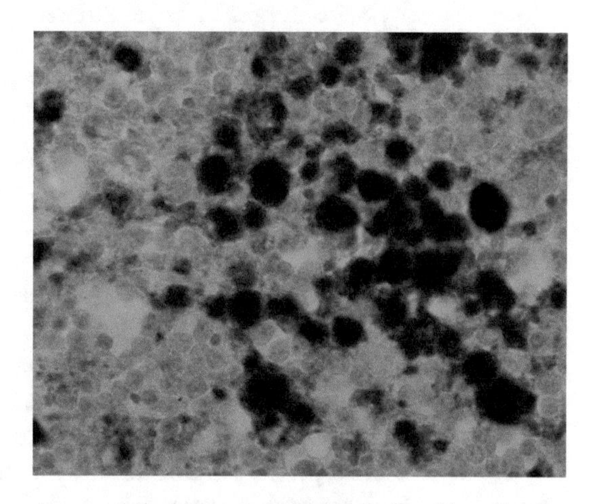

Fig. 11 Kuru plaques stained immunohistochemically with antibodies against PrP

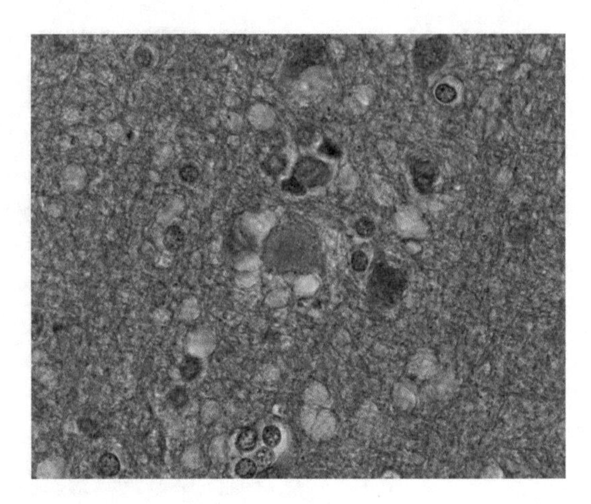

Fig. 12 A kuru plaque stained with Alcian blue

description of a 50-year-old female patient Yabaiotu, Klatzo stated, *"Numerous round or asteroid bodies are present, which are brightly anisotropic in polarized light. These probably represent neurofibrillary degeneration of Purkinje cells"* [29]. They were found in 6 of 12 cases studied by Klatzo et al. [128, 129], and in "about three-quarters" of the 13 cases studied by Beck and Daniel [111, 115]; they became known as "kuru plaques" [130–136].

These measured 20–60 μm in diameter, were round or oval, and consisted of a dark-stained core with delicate radiating periphery surrounded by a pale "halo." Kuru plaques were most numerous in the granular cell layer of the cerebellum, basal ganglia, thalamus, and cerebral cortex in that order of frequency. Kuru plaques are metachromatic and stain with PAS, Alcian blue (Fig. 12), and Congo red, and a proportion of them are weakly argentophilic when impregnated according to Belschowsky or von Braunmühl techniques. Of historical interest, another unique disease reported by Seitelberger [137] as *"A peculiar hereditary disease of the central nervous system in a family from lower Austria"* (germ. Eigenartige familiar-hereditare Krankenheit des Zentralnervensystems in einer niederoosterreichen Sippe) was mentioned by Neuman et al. [138] who was thus the first person to suggest a similarity between kuru and GSS. Indeed, the latter was transmitted to nonhuman primates in 1981 [139]. In a letter of Klatzo to Gajdusek in 1957, Klatzo stated, *"It seems to be definitely a new conditions without anything similar described in the literature. The closest condition I can think of is that described by Jakob and Creutzfeldt"* [29].

Renewed interest in kuru pathology has been provoked by the appearance of a novel form of CJD, variant CJD, characterized by numerous amyloid plaques, including "florid" or "daisy" plaques—a kuru plaque surrounded by a rim of spongiform vacuoles [136]. To this end, a few papers reevaluating historic material have been published [140]. We [141] studied by PrP-immunohistochemistry the case of a young male kuru victim of the name Kupenota from the South Fore region whose brain tissue had transmitted disease to chimpanzees, and McLean et al. [142, 143] examined a series of 11 archived cases of kuru. In contrast to the classical studies described above, all papers stressed the presence of typical spongiform change present in deep layers (III–V) of the cingulate, occipital, enthorrinal, and insular cortices, and in the subiculum. Spongiform change was also observed in the putamen and caudate, and some putaminal neurons contained intraneuronal vacuoles. Spongiform change was prominent in the molecular layer of the cerebellum, in periaqueductal gray matter, basal pontis, central tegmental area, and inferior olivary nucleus. The spinal cord showed only minimal spongiform change.

There are no ultrastructural observations on kuru in humans, except those reported in a paper by Peat and Field [144] who described "intracytoplasmic dense barred structures," otherwise normal structures [145], and Field et al. [146], who described the typical ultrastructure of the kuru plaques and "herring-bone" structures, again as the normal structure of the neuron or Hirano bodies [147]. In kuru in chimpanzees, Lampert et al. [148] and Beck et al. [116] found severe confluent spongiform change corresponded to typical membrane-bound vacuoles. Neurites showed dystrophic changes. Our studies on formalin-fixed paraffin-embedded kuru specimens reversed to electron microscopy revealed typical plaques composed of amyloid fibrils (Fig. 13).

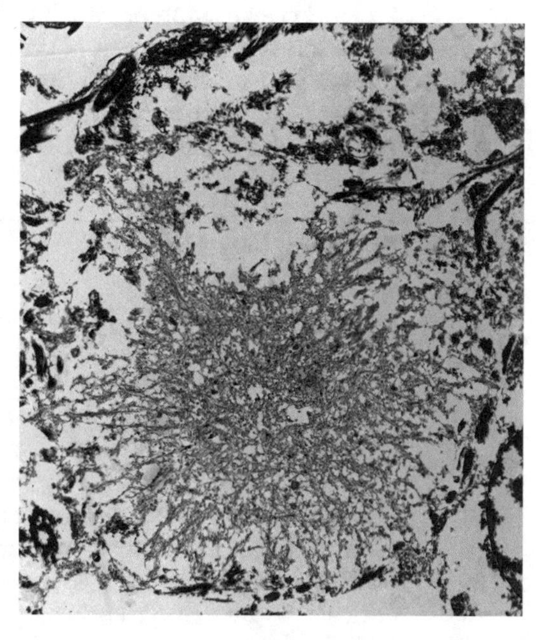

Fig. 13 A kuru plaque retrieved from paraffin-embedded block. Original magnification, ca 3000×

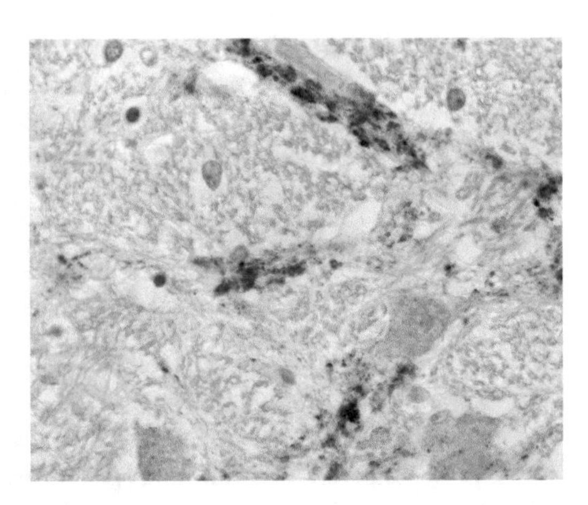

Fig. 14 Numerous PrP deposits on neuronal processes. Immunohistochemistry with usage of 3F4 antibody

Immunohistochemical studies revealed that misfolded PrP^{Sc} was present not only as kuru plaques but also in synaptic and perineuronal sites (Fig. 14) [133, 141], and in the spinal cord the *substantia gelatinosa* was particularly affected, as in iatrogenic CJD cases following peripheral inoculation [149]. Brandner et al. [150] studied one very recent case of kuru and basically confirmed the findings of Hainfellner et al. [141]. The latter case has been neuropathologically compared with known subtypes of CJD and it seems the most similar to type 3 129 MV of CJD of the Collinge et al. [151] classification or type 2 CJD of the Parchi et al. [152] classification. Of note, immunocytochemistry with 12F10 antibodies revealed a stronger signal than that using 3F4 anti-PrP antibodies [142]. No positive results were obtained by staining a tonsil biopsy with anti-PrP antibodies [153].

9 Genetics and Molecular Biology of Kuru

Even after 40 years, the summary of the genetics of kuru written by Michael P. Alpers [89] is still valid: *"it was recognized that a strong familial association of disease does not necessarily prove that the cause is genetic. Furthermore, it was hard to see how a disease so prevalent and at the same time so lethal could have become established in the population by purely genetic means, unless there was some immense associated heterozygote advantage."* At the beginning, it was demonstrated that two kuru cases were $129^{Met\ Met}$ [154]. Further studies found that individuals of $129^{Val\ Val}$ and $129^{Met\ Val}$ genotype were susceptible to kuru, but those of $129^{Met\ Met}$ genotype were overrepresented in the younger age group while those of $129^{Val\ Val}$ $129^{Met\ Val}$ were overrepresented in much older age group [54, 99, 155–158]. In contrast, those people who survived the epidemic were characterized by almost the total absence of $129^{Met\ Met}$ homozygotes. The more recent case studies by Lantos et al. [159], McLean et al. [143], and us [141] were all $129^{Met\ Met}$ homozygotes. Recent genome-wide studies confirmed a strong association of kuru with a SNP localized within the codon 129 but also with two other SNPs localized within genes *RARB* (the gene encoding retinoic acid receptor beta) and STMN2 (the gene encoding SCG10) [155].

The practice of endocannibalism underlying the kuru epidemic created a selective pressure on the prion protein genotype [160, 161]. As in CJD, homozygosity at codon 129 ($129^{Met\ Met}$ or $129^{Val\ Val}$) is overrepresented in kuru [54, 99, 155–158]. Furthermore, Mead et al. [156, 157] found that among Fore women over 50 years of age, there is a remarkable overrepresentation of heterozygosity ($129^{Met\ Val}$) at codon 129, which is consistent with the interpretation that $129^{Met\ Val}$ makes an individual resistant to TSE agents and that such a resistance was selected by cannibalistic rites. Another protective polymorphism G127 V located in a highly conserved region of PrP was discovered by the Collinge's group [162, 163]. This 127^{Val} was not found in any of kuru patients. Because of this $129^{Met\ Val}$ heterozygote advantage, it has been suggested that the heterozygous genotype at codon 129 has been sustained by a widespread ancient practice of human cannibalism [164]. Furthermore, there is a hypothesis that extinction of Neanthertals who coexisted with *Homo sapiens* some 45,000–30,000 years ago is connected to the appearance of "kuru-like" epidemics spread by cannibalism [165, 166]. Collinge et al. [123] suggested that the survival advantage of the *PRNP* $129^{Met\ Met}$ heterozygotes provides a basis for a selection pressure not only in Fore but also in those human populations that practiced cannibalism. Of note this was preconceived by Alpers and Gajdusek in 1965: *"In order to explain the combination of high incidence and high lethality, which at first glance might seem to entirely rule out a genetic cause unless there was an immense heterozygote advantage, we postulated that environmental change, of relatively recent origin, has given a lethal expression to a previously benign gene mutation established in the Fore population as a genetic polymorphism"* [89].

The molecular strain typing of kuru cases was performed by the Collinge's group [167, 168]. This typing is based on the electrophoretic mobility of de-, mono-, and diglycosylated bands of PrP^{Sc} following digestion with proteinase K [151]. The four major types of PrP^{Sc} were found. The human PrP^{Sc} type 1 and 4 occur only in individuals of the codon $129^{Met\ Met}$ of the *PRNP* gene; type 3 is seen in individuals with at least one Val at this codon and type 2 occurs in all codon 129 variants. There is another classification based on only 2 PrP^{Sc} types [152], and the agreement between supporters of either classification has not yet been achieved. The kuru specimens revealed type 2 ($PrP^{Met\ Met}$) or 3 ($129^{Val\ Val}$) PrP^{Sc} patterns and the glycoform ratio was similar to that of sporadic CJD but not typical for vCJD [169–171]. In primates inoculated with kuru and sCJD VV2 and sCJD MV 2 K, the "b" pattern of pathology, i.e., coarser vacuoles situated in the subcortical structures and in deeper layers of the cerebral cortex, was observed. PrP^{Sc} consisted of doublet of 20 and 21 kDa. The latter notion is supported by the fact of a similar transmission rate of kuru to transgenic mice lacking mouse *PrP* gene but expressing human *PrP 129^Val Val^* gene [167, 168]. In contrast, kuru was reported as not transmissible to normal wild-type mice but it was later shown that it transmits to CD-1 mice with unique clinical and neuropathological patterns in infected animals [172]. Of interest, the robust presence of PrP^{Sc} in follicular dendritic cells in the spleen suggests a possibility of the spreading of the kuru agent via the bloodstream. Collectively, those data suggest that kuru is unique and different from either sporadic CJD or variant CJD.

10 Conclusions and Speculations

Kuru, an extinct exotic disease transmitted by cannibalism in Papua New Guinea, still impacts on many aspects of neurodegeneration research. First, it shows that a human neurodegenerative disease can result from an infection with an infectious agent, once called a "slow virus" [104]. This discovery opened a window into the new class of human diseases that included Creutzfeldt-Jakob disease, Gerstmann-Sträussler-Scheinker disease, and, recently, fatal familial insomnia. Parenthetically, CJD was said to be a possible analogue on kuru based on nonspecific neuropathological findings, but Gerstmann-Sträussler-Scheinker disease was similarly identified because of the presence of numerous amyloid plaques not unlike kuru plaques. The kuru plaque became a link to Alzheimer disease and, as Gajdusek suggested [23], all amyloidoses share a common pathogenetic mechanism—processing of a normal protein into an amyloid deposit [173–175]. This event underlies all "conformational disorders," including pathogenetically novel classes of neurodegenerations like α-synucleinopathies, tauopathies, and expanded triplet disorders. In his last paper, Gajdusek [21] wrote that kuru opened new vistas like molecular casting, osmium shading of electron microscopy, formation of prion-like infectious nucleants, and stalagmite and stalactite.

Acknowledgements

I am immensely indebted to late Dr. D. Carleton Gajdusek and Dr. Clarence J. Gibbs Jr. for generously providing me unique illustrations I have used through the text. I thank Prof. Shirley

Lindenbaum and Prof. Michael Alpers for exciting discussion and helpful criticism, and Prof. James W. Ironside, the National CJD Surveillance Unit, for reading and correcting the MS in totally impossibly express tempo.

References

1. Asher DM, Gibbs CJ Jr, Sulima MP et al (1993) Transmission of human spongiform encephalopathies to experimental animals: comparison of chimpanzee and Squirrel monkey. In: Brown P (ed) Transmissible spongiform encephalopathies—impact on animal and human health, developments of biological standardization, vol 80. Karger, Basel, pp 9–13
2. Asher DM (2008) Kuru: memories of the NIH years. Philos Trans R Soc Lond Ser B Biol Sci 363(1510):3618–3625
3. Gajdusek DC (1962) Kuru: an appraisal of five years of investigation. Eugen Q 9:69–74
4. Gibbs Jr CJ, Gajdusek DC (1969) Infection as the etiology of spongiform encephalopathy (Creutzfeldt-Jakob disease). Science 165(3897):1023–1025
5. Hornabrook RW (1975) Kuru. P N G Med J 4:203–206
6. Hornabrook RW (1979) Kuru and clinical neurology. In: Prusiner SB, Hadlow WJ (eds) Slow transmissible diseases of the nervous system. Academic Press, New York
7. Liberski PP, Brown P (2007) Kuru—fifty years later. Neurol Neurochir Pol 41(6): 548–556
8. Liberski PP, Brown P (2004) Kuru: a half-opened window onto the landscape of neurodegenerative diseases. Folia Neuropathol 42(Suppl A):3–14
9. Liberski PP, Brown P (2009) Kuru: its ramifications after fifty years. Exp Gerontol 44(1–2):63–69
10. Liberski PP, Brown P (2006) Prion disease: from ritualistic endocannibalism to cellular endocannibalism—from Kuru to autophagy. In: Karasek M (ed) Aging and age-related disease: the basics. Nova Science Publisher, Inc., New York
11. Liberski PP, Gajdusek DC (1997) Kuru: fourty years later, a historical note. Brain Pathol 7(1):555–560
12. Gajdusek DC, Zigas V (1959) Kuru. Clinical, pathological and epidemiological study of an acute progressive degenerative disease of the central nervous system among natives of the Eastern Highlands of New Papua. Am J Med 26(3):442–469
13. Gajdusek DC, Zigas V (1961) Studies of kuru. I. The ethnologic setting of kuru. Am J Trop Med Hyg 10:80–91
14. Gajdusek DC, Gibbs Jr CJ, Alpers PM (1967) Transmission and passage of experimental kuru to chimpanzees. Science 155(3759):212–214
15. Gajdusek DC, Gibbs Jr CJ (1971) Transmission of two subacute spongiform encephalopathies of man (kuru and Creutzfeldt-Jakob disease) to New World monkeys. Nature 230(5296):588–591
16. Gajdusek DC, Gibbs CJ, Alpers PM (1966) Experimental transmission of a kuru-like syndrome to chimpanzees. Nature 209(5025):794–796
17. Gajdusek DC, Rogers NG, Basnight M et al (1969) Transmission experiments with kuru in chimpanzees and the isolation of latent viruses from the explanted tissues of affected animals. Ann N Y Acad Sci 162(1):529–550
18. Gajdusek DC, Zigas V (1957) Degenerative disease of the central nervous system in new Guinea. The endemic occurrence of "kuru" in the native population. N Engl J Med 257(20):974–978
19. Gibbs Jr CJ, Gajdusek DC, Asher DM et al (1968) Creutzfeldt-Jakob disease (spongiform encephalopathy): transmission to chimpanzee. Science 161(3839):388–389
20. Jaskolski M, Liberski PP (2002) Kurt Wuthrich—co-winner of the Nobel Prize in Chemistry, 2002. Acta Neurobiol Exp 62:288–289
21. Gajdusek DC (2008) Kuru and its contribution to medicine. Philos Trans R Soc Lond Ser B Biol Sci 363(1510):3697–3700
22. Beyreuther K, Masters CL (1997) βA4-amyloid domain is essential for axonal sorting of APP: implications for Alzheimer's disease. In: Abstracts of the satellite meeting "Brain Tumors and Alzheimer's Disease, From Neuropathology to Molecular Biology", Bali, Indonesia, 3–5 Sept 1997
23. Gajdusek DC (1996) Infectious amyloids: subacute spongiform encephalopathies as transmissible cerebral amyloidosis. In: Fields BN, Knippe DM, Howley PM (eds) Fields virology, 3rd edn. Lippincott-Raven Publishers, Philadelphia

24. Gajdusek DC (2001) Molecular casting of infectious amyloids, inorganic and organic replication: nucleation, conformational change and self-assembly. In: Aggeli A, Boden B, Shuguang Z (eds) Self-assembling peptide systems in biology, medicine and engineering. Springer, New York
25. Ashe KH, Aguzzi A (2013) Prions, prionoids and pathogenic proteins in Alzheimer disease. Prion 7(1):55–59
26. Gajdusek DC, Alpers MP, Gibbs CJ Jr (1976) Kuru: epidemiological and virological studies of unique New Guinean disease with wide significance to general medicine. In: Hornabrook RW (ed) Essays on Kuru. E.W. Classey Ltd., Faringdon, Berks
27. Gajdusek DC, Alpers MP (1972) Genetic studies in relation to kuru. I. Cultural, historical, and demographic background. Am J Hum Genet 24(Suppl):1–38
28. Gajdusek DC, Alpers MP (1966) Kuru in childhood: disappearance of the disease in the younger age group. 7th Annual Meeting of American Pediatric Society. J Pediatr 69:886–887
29. Gajdusek DC, Farquhar J (eds) (1981) Kuru. Early letters and field-notes from the collection of D. Carleton Gajdusek. Raven Press, New York
30. Gajdusek DC, Gibbs CJ (1978) Unconventional viruses causing the spongiform virus encephalopathies. A fruitless search for the coat and core. In: Kurstak E, Maramosch K (eds) Viruses and environment. Academic Press, New York
31. Gajdusek DC, Reid HL (1961) Studies of kuru. IV: the kuru pattern in Moke, a representative Fore village. Am J Trop Med Hyg 10:628–638
32. Gajdusek DC (2008) Early images of kuru and the people of Okapa. Philos Trans R Soc Lond Ser B Biol Sci 363(1510):3636–3643
33. Gajdusek DC (1973) Kuru in the new Guinea highlands. In: Spillane J (ed) Tropical neurology. Oxford University Press, London
34. Gajdusek DC (1979) Le kuru. Colloque sur les Viruses lents. Organize le 17 Septembre 1978, a Talloires, France. Collection Foundation Merieux. pp 25–57
35. Gajdusek DC (1979) Observations on the early history of kuru investigation. In: Prusiner SB, Hadlow WJ (eds) Slow transmissible diseases of the nervous system, vol 1. Academic Press, New York, pp 70–36
36. Gajdusek DC (1967) Slow virus infections of the nervous system. N Engl J Med 276(7): 392–400
37. Gajdusek DC (1977) Unconventional viruses and the origin and disappearance of kuru. In: Les Prix Nobel en 1976. Nobel Foundation, PA Norstedt & Soner, Stockholm, pp 167–216
38. Gajdusek DC (1977) Unconventional viruses and the origin and disappearance of kuru. Science 197(4307):943–960
39. Gajdusek DC (1977) Urgent opportunistic observations: the study of changing, transient and disappearing phenomena of medical interest in disrupted primitive human communities. In: Health and disease in tribal societies. Elsevier, Amsterdam, pp 69–102
40. Zigas V, Gajdusek DC (1957) Kuru. Clinical study of a new syndrome resembling paralysis agitans in natives of the eastern Highlands of Australian New Guinea. Med J Aust 44(21): 745–754
41. Zigas V, Gajdusek DC (1959) Kuru. Clinical, pathological and epidemiological study of a recently discovered acute progressive degenerative disease of the central nervous system reaching "epidemic" proportions among natives of the Eastern Highlands of New Guinea. P N G Med J 3:1–31
42. Mathews JD (2008) The changing face of kuru: a personal perspective. Philos Trans R Soc Lond Ser B Biol Sci 363(1510): 3679–3684
43. Mathews JD (1967) The epidemiology of kuru. P N G Med J 10:76–82
44. Lindenbaum S (2009) Cannibalism, kuru and anthropology. Folia Neuropathol 47(2): 138–144
45. Mathews JD (1971) Kuru. A puzzle in cultural and environmental medicine. Thesis submitted to the University of Melbourne towards the Degree of Doctor of Medicine. Melbourne
46. Bennett JH, Rhodes FA, Robson HN (1959) A possible genetic basis for kuru. Am J Hum Genet 11(2, part 1):169–187
47. Bennett JH, Rhodes FA, Robson HN (1958) Observations on kuru. I. A possible genetic basis. Australas Ann Med 7(4):269–275
48. Alpers MP (1968) Kuru: implications of its transmissibility for the interpretation of its changing epidemiologic pattern. In: Bailey OT, Smith DE (eds) The central nervous system, International Academy of Pathology Monograph No. 9. Williams & Wilkins Company, Baltimore, pp 234–250
49. Lindenbaum S (2013) Kuru Sorcery. Disease and danger in the New Guinea Highlands, 2nd edn. Paradigm Publishers, Boulder, p 224
50. Lindenbaum S (2008) Understanding kuru: the contribution of anthropology and medicine. Philos Trans R Soc Lond Ser B Biol Sci 363(1510):3715–3720
51. Lindenbaum S (1998) Images of catastrophe: the making of an epidemic. In: Singer M (ed)

The political economy of AIDS. Baywood Publishing Co Inc, Amityville
52. Lindenbaum S (2001) Kuru, prions, and human affairs. Annu Rev Anthropol 30: 363–385
53. Lindenbaum S (2004) Thinking about cannibalism. Annu Rev Anthropol 33:475–498
54. Matthews JD, Glasse RM, Lindenbaum S (1968) Kuru and cannibalism. Lancet 2(7565):449–452
55. Glasse R, Lindenbaum S (1993) Fieldwork in the South Fore: the process of ethnographic inquiry. In: Prusiner SB, Collinge J, Powell J, Anderton B (eds) Prion diseases of humans and animals. Ellis Horwood, New York
56. Glasse R (1967) Cannibalism in the Kuru region of New Guinea. Trans N Y Acad Sci 29(6):748–754
57. Glasse, R. (1962) Fieldwork report to the Department of Public Health. South Fore cannibalism and kuru. Reprinted by the National Institutes of Health
58. Glasse, R. (1963) Fieldwork Report to the Department of Public Health. The spread of kuru in the kuru region. Reprinted by the National Institutes of Health
59. Arens W (1979) The man-eating myth: anthropology and anthropophagy. Oxford University Press, Oxford
60. Zigas V (1979) Origin of investigations on slow virus infections in man. In: Prusiner SB, Hadlow WJ (eds) Slow transmissible diseases of the nervous system, vol 1. Academic Press, New York, pp 3–6
61. Hamilton L, Gajdusek DC (1969) Nutrition in the kuru region. II. A nutritional evaluation of tradition Fore diet in Moke village in 1957. Acta Trop 26(4):331–345
62. Sorenson ER, Gajdusek DC (1969) Nutrition in the kuru region. I. Gardening, food handling, and diet of the fore people. Acta Trop 26:281–330
63. Sorenson ER, Gajdusek DC (1966) The study of child growth and development in primitive cultures. A research archive for ethnopediatric film investigations of styles in the patterning of the nervous system. Pediatrics 37(suppl):149–243
64. Curtain CC, Gajdusek DC, Zigas V (1961) Studies on kuru. II. Serum proteins in natives from the kuru region of New Guinea. Am J Trop Med Hyg 10:92–109
65. Kitchin D, Bearn AG, Alpers MP, Gajdusek DC (1972) Genetic studies in relation to kuru. III. Distribution of the inherited serum group-specific protein (Gc) phenotypes in New Guineans: an association of kuru and the Gc Ab phenotype. Am J Hum Genet 24(6 Pt 2):S72–S85
66. Mbaginta'o IG (1976) Medical practices and funeral ceremony of the Dunkwi Anga. J Soc Océan 32(53):299–305
67. Plato CC, Gajdusek DC (1972) Genetic studies in relation to kuru. IV. Dermatoglyphics of the Fore and Anga populations of the Eastern Highlands of New Guinea. Am J Hum Genet 24:S86–S93
68. Simmons RT, Gajdusek DC (1960) Blood group genetical studies on kuru-afflicted natives of the eastern Highlands of New Guimea, and comparison with unaffected neighboring tribes in Papua New Guinea. In: Proceedings of the 8th Congress of International Society of Blood Transfusions, Tokyo, pp 255–259
69. Simmons RT, Graydon JJ, Gajdusek DC et al (1972) Genetic studies in relation to kuru. II. Blood-group genetic patterns and populations of the eastern Highlands of New Guinea. Am J Hum Genet 24:S39–S71
70. Simmons RT, Graydon JJ, Zigas V et al (1961) Studies on kuru. V. A blood group genetical survey of the kuru region and other parts of Papua New Guinea. Am J Trop Med Hyg 10:639–664
71. Simmons RT, Graydon JJ, Zigas V et al (1961) Studies on kuru. VI. Blood groups in kuru. Am J Trop Med Hyg 10:665–668
72. Wiesenfeld SL, Gajdusek DC (1975) Genetic studies in relation to kuru. VI. Evaluation of increased liability to kuru in Gc, Ab-Ab indyviduals. Am J Hum Genet 27:498–504
73. Gajdusek DC (1993) Kuru and scrapie. In: Prusiner SB, Collinge J, Powell J, Anderton B (eds) Prion diseases of humans and animals. Ellis Horwood, New York
74. Hadlow WJ (2008) Kuru likened to scrapie: the story remembered. Philos Trans R Soc Lond Ser B Biol Sci 363(1510):3644
75. Hadlow WJ (1995) Neuropathology and the scrapie-kuru connection. Brain Pathol 5(1):27–31
76. Hadlow WJ (1959) Scrapie and kuru. Lancet 2:289–290
77. Hadlow WJ (1993) The Scrapie-kuru connection: recollections of how it came about. In: Prusiner SB, Collinge J, Powell J, Anderton B (eds) Prion diseases of humans and animals. Ellis Horwood, New York
78. Collinge CJ, Whitfield J, McKintosch E et al (2008) A clinical study of kuru patients with long incubation periods at the end of the epidemic in Papua New Guinea. Philos Trans R Soc Lond Ser B Biol Sci 363(1510): 3725–3739
79. Besnoit C (1899) La tremblante ou nevrite peripherique enzootique du mouton. Rev Vet (Tolouse) 24:265–277

80. M'Gowan JP (1914) Investigation into the disease of sheep called "scrapie" (Traberkrankheit, La tremblante). With especial reference to its association with Sarcosporidiosis. William Blackwood and Sons, Edinburgh
81. Innes JMR, Saunders LZ (1962) Comparative neuropathology. Academic Press, New York
82. Goldfarb LG, Cervenakova L, Gajdusek DC (2004) Genetic studies in relation to kuru: an overview. Curr Mol Med 4(4):375–384
83. Zlotnik I (1957) Significance of vacuolated neurones in the medulla of sheep infected with scrapie. Nature 180(4582):393–394
84. Zlotnik I (1957) Vacuolated neurons in sheep affected with scrapie. Nature 179(4562):737
85. Gajdusek DC, Gibbs CJ (1964) Attempts to demonstrate a transmissible agent in kuru, amyotrophic lateral sclerosis, and other subacute and chronic system degenerations of man. Nature 204:257–259
86. Carrell RW, Lomas DA (1997) Conformational disease. Lancet 350(9071): 134–138
87. Gajdusek DC (1988) Transmissible and nontransmissible dementias: distinction between primary cause and pathogenetic mechanisms in Alzheimer's disease and aging. Mt Sinai J Med 55(1):3–5
88. Gibbs CJ Jr, Gajdusek DC (1965) Attempts to demonstrate a transmissible agent in kuru, amyotrophic lateral sclerosis, and other subacute and chronic progressive nervous system degenerations of man. In: Gajdusek DC, Gibbs CJ Jr, Alpers M (eds) Slow, latent, and temperate virus infections, NINDB Monograph No. 2. US Department of Health, Education, and Welfare, Washington, DC
89. Alpers MP, Gajdusek DC (1965) Changing pattern of kuru: epidemiological changes in the period of increasing contact of the Fore people with western civilization. Am J Trop Med Hyg 14(5):852–879
90. Alpers MP (1979) Epidemiology and ecology of kuru. In: Prusiner SB, Hadlow WJ (eds) Slow transmissible diseases of the nervous system, vol 1. Academic Press, New York, pp 67–90
91. Alpers MP (2008) The epidemiology of kuru: monitoring the epidemic from its peak to the end. Philos Trans R Soc Lond Ser B Biol Sci 363(1510):3707–3713
92. Gajdusek DC (1996) Kuru in childhood: implications for the problem of whether bovine spongiform encephalopathy affects humans. In: Court L, Dodet B (eds) Transmissible subacute spongiform encephalopathies: prion diseases. Elsevier, Amsterdam
93. Goldfarb LG (2002) Kuru: the old epidemic in a new mirror. Microbes Infect 4(8): 875–882
94. Gajdusek DC, Zigas V, Baker J (1961) Studies on kuru. III. Patterns of kuru incidence: demographic and geographic epidemiological analysis. Am J Trop Med Hyg 10:599–627
95. Mathews JD (1967) A transmission model for kuru. Lancet 1(7494):821–825
96. Matthews JD (1976) Kuru as an epidemic disease. In: Hornabrook RW (ed) Essays on kuru, Monographs no 3. Papua New Guinea Institute of Medical Research, E. W. Glassey, Faringdon
97. Alpers MP (2008) Some tributes to research colleagues and other contributors to our knowledge about kuru. Philos Trans R Soc Lond Ser B Biol Sci 363(1510):3614–3617
98. Klitzman RL, Alpers MP, Gajdusek DC (1984) The natural incubation period of kuru and the episodes of transmission in three clusters of patients. Neuroepidemiology 3:3–20
99. Cervenakova L, Goldfarb LG, Garruto R et al (1998) Phenotype-genotype studies in kuru: implications for new variant Creutzfeldt-Jakob disease. Proc Natl Acad Sci U S A 95(22):13239–13241
100. Brown P, Gibbs CJ, Rodgers-Johnson P et al (1994) Human spongiform encephalopathy: the National Institutes of Health series of 300 cases of experimentally transmitted disease. Ann Neurol 35(5):513–529
101. Gibbs Jr CJ, Amyx HL, Bacote A et al (1980) Oral transmission of kuru, Creutzfeldt-Jakob disease, and scrapie to nonhuman primates. J Infect Dis 142(2):205–208
102. Gibbs Jr CJ, Gajdusek DC, Amyx H (1979) Strain variation in the viruses of Creutzfeldt-Jakob disease and kuru. In: Prusiner SB, Hadlow WJ (eds) Slow transmissible diseases of the nervous system, vol 1. Academic Press, New York, pp 87–110
103. Gibbs CJ Jr, Gajdusek DC (1979) Transmission and characterization of the agents of spongiform virus encephalopathies: kuru, Creutzfeldt-Jakob disease, scrapie and mink encephalopathy. In: Immunological disorders of the nervous system, Res Publ ARNMD, vol 49, pp 383–410
104. Gibbs CJ (1993) Spongiform encephalopathies—slow, latent, and temperate virus infections—in retrospect. In: Prusiner SB, Collinge J, Poqwell J, Anderon B (eds) Prion diseases of humans and animals. Ellis Horwood, New York
105. Liberski PP (1981) Wirusy powolne układu nerwowego człowieka i zwierząt. Część I. Kuru. (Slow viruses of the nervous system of man and animals. part I. Kuru). Postepy Hig Med Dosw 35:471–493

106. Gajdusek DC, Gibbs Jr CJ (1972) Transmission of kuru from man to Rhesus monkeys (Macaca mulatto) 8.1/2 years after inoculation. Nature 240(5380):351
107. Peterson DA, Wolfe LG, Deinhardt F et al (1974) Transmission of kuru and Creutzfeldt-Jakob disease to Marmoset monkeys. Intervirology 2(1):14–19
108. Masters CL, Alpers MP, Gajdusek DC et al (1976) Experimental kuru in the gibbon and Sooty mangabey and Creutzfeldt-Jakob disease in the pigtailed macaque. J Med Primatol 5(4):205–209
109. Beck E, Daniel MP (1979) Kuru and Creutzfeldt-Jakob disease: neuropathological lesions and their significance. In: Prusiner SB, Hadlow WJ (eds) Slow transmissible diseases of the nervous system, vol 1. Academic Press, New York, pp 253–270
110. Beck E, Daniel MP, Alpers MP et al (1969) Neuropathological comparisons of experimental kuru in Chimpanzees with human kuru. With a note on its relation to scrapie and spongiform encephalopathy. In Burdzy K, Kallos P (eds) Pathogenesis and etiology of demyelinating diseases. Proceedings of the Workshop on Contributions to the Pathogenesis and Etiology of Demylinating Conditions, Locarno, Switzerland, May 31–June 3, 1967. Additamentum to Int Arch Allergy Appl Immunol, vol 36. pp 553–562
111. Beck E, Daniel MP, Alpers MP et al (1966) Experimental kuru in chimpanzees. A pathological report. Lancet 2(7472):1056–1059
112. Beck E, Daniel MP, Asher MP et al (1973) Experimental kuru in chimpanzees. A neuropathological study. Brain 96(3):441–462
113. Beck E, Daniel MP, Gajdusek DC (1965) A comparison between the neuropathological changes in kuru and scrapie, system degeneration. In: Proceedings of the VIth International Congress Neuropathological, Zurich, pp 213–218
114. Beck E, Daniel MP (1965) Kuru and scrapie compared: are they examples of system degeneration? In: Gajdusek DC, Gibbs CJ Jr, Alpers MP (eds) Slow, latent, and temperate virus infections. US Department of Health, Education, Welfare, Washington, DC
115. Beck E, Daniel MP (1993) Prion diseases from a neuropathologist's perspective. In: Prusienr SB, Collinge J, Powell J, Anderton B (eds) Prion diseases of humans and animals. Ellis Horwood, New York
116. Beck E, Bak J, Christ JF et al (1975) Experimental kuru in the Spider monkey. Histopathological and Ultrastructural studies of the brain during early stages of incubation. Brain 98(4):595–612
117. Beasley A (2006) The promised medicine: fore reflections on the scientific investigation of kuru. Oceania 76:186–202
118. Alpers MP (1964) Kuru: age and duration studies. Mimeographed. Department of Medicine, University of Adelaide, Adelaide
119. Beasley A (2009) Frontier journals. Fore experiences on the kuru patrols. Oceania 79: 34–52
120. Beasley A (2004) Frontier science: the early investigation of kuru in Papua New Guinea. In: Dew K, Fitzgerald R (eds) Challenging science: issues in New Zealand. Dunmore Press, Palmerston North
121. Beasley A (2006) Kuru truths: obtaining Fore narratives. Field Methods 18:21–42
122. Beasley A (2008) Richard Hornabrook's first impressions of kuru and Okapa. Philos Trans R Soc Lond Ser B Biol Sci 363(1510): 3626–3627
123. Collinge CJ, Whitfield J, McKintosch E et al (2006) Kuru in the 21st century—an acquired human prion disease with very long incubation periods. Lancet 367(9528):2068–2074
124. Collinge J (2008) Lessons of kuru research: background to recent studies with some personal reflections. Philos Trans R Soc Lond Ser B Biol Sci 363(1510):3689–3696
125. Gajdusek DC, Sorenson ER, Meyer J (1970) A comprehensive cinema record of disappearing kuru. Brain 93(1):65–76
126. Prusiner SB, Gajdusek DC, Alpers MP (1982) Kuru with incubation periods exceeding two decades. Ann Neurol 12(1):1–9
127. Alpers MP (1964) Kuru: a clinical study, Mimeographed. US Department of Health, Education, Welfare, Washington, DC
128. Klatzo I, Gajusek DC, Zigas V (1959) Evaluation of pathological findings in twelve cases of kuru. In: Van Boagert L, Radermecker J, Hozay J, Lowenthal A (eds) Encephalitides. Elsevier, Amsterdam
129. Klatzo I, Gajdusek DC (1959) Pathology of kuru. Lab Investig 8(4):799–847
130. Biernat W, Liberski PP, Guiroy DC et al (1995) Proliferating cell nuclear antigen immunohistochemistry in astrocytes in experimental Creutzfeldt-Jakob disease and in human kuru, Creutzfeldt-Jakob disease and Gerstmann-Sträussler-Scheinker syndrome. Neurodegeneration 4(2):195–201
131. Fowler M, Robertson EG (1959) Observations on kuru. III: pathological features in five cases. Australas Ann Med 8(1): 16–26
132. Kakulas BA, Lecours A-R, Gajdusek DC (1967) Further observations on the pathology of kuru. J Neuropathol Exp Neurol 26(1):85–97

133. Piccardo P, Safar J, Ceroni M et al (1990) Immunohistochemical localization of prion protein in spongiform encephalopathies and normal brain tissue. Neurology 40(3): 518–522
134. Scrimgeour EM, Masters LC, Alpers MP et al (1983) A clinico-pathological study of case of kuru. J Neurol Sci 59(2):265–275
135. Scrimgeou EM (2008) Some recollections about kuru in a patient at Rabaul in 1978, and subsequent experiences with prion diseases. Philos Trans R Soc Lond Ser B Biol Sci 363(1510):3663–3664
136. Sikorska B, Liberski PP, Sobów T et al (2009) Ultrastructural study of florid plaques in variant Creutzfeldt-Jakob disease: a comparison with amyloid plaques in kuru, sporadic Creutzfeldt-Jakob disease and Gerstmann-Sträussler-Scheinker disease. Neuropathol Appl Neurobiol 35(1):46–59
137. Seitelberger F (1962) Eigenartige familiar-hereditare Krankheit des Zetralnervensystems in einer niederosterreichischen Sippe. Wien Klin Wochen 74:687–691
138. Neuman MA, Gajdusek DC, Zigas V (1964) Neuropathologic findings in exotic neurologic disorder among natives of the Highlands of New Guinea. J Neuropathol Exp Neurol 23:486–507
139. Masters CL, Gajdusek DC, Gibbs Jr CJ (1981) Creutzfeldt-Jakob disease virus isolations from the Gerstmann-Sträussler syndrome. With an analysis of the various forms of amyloid plaque deposition in the virus induced spongiform encephalopathies. Brain 104(3):559–588
140. Liberski PP, Sikorska B, Lindenbaum S et al (2012) Kuru: genes, cannibals and neuropathology. J Neuropathol Exp Neurol 71(2):92–103
141. Hainfellner J, Liberski PP, Guiroy DC et al (1997) Pathology and immunohistochemistry of a kuru brain. Brain Pathol 7(1):547–554
142. McLean CA (2008) The neuropathology of kuru and variant Creutzfeldt-Jakob disease. Philos Trans R Soc Lond Ser B Biol Sci 363(1510):3685–3687
143. McLean CA, Ironside JW, Alpers MP et al (1998) Comparative neuropathology of kuru with the new variant of Creutzfeldt-Jakob disease: evidence for strain of agent predominating over genotype of host. Brain Pathol 8(3):428–437
144. Peat A, Field EJ (1970) An unusual structure in kuru brain. Acta Neuropathol (Berl) 15(3):288–292
145. Liberski PP (1988) The occurrence of cytoplasmic lamellar bodies in scrapie infected and normal hamster brains. Neuropatol Pol 26:79–85
146. Field EJ, Mathews JD, Raine CS (1969) Electron microscopic observations on the cerebellar cortex in kuru. J Neurol Sci 8(2):209–224
147. Liberski PP, Yanagihara R, Gibbs Jr CJ, Gajdusek DC (1990) Re-evaluation of experimental Creutzfeldt-Jakob disease: serial studies of the Fujisaki strain of Creutzfeldt-Jakob disease virus in mice. Brain 113:121–137
148. Lampert PW, Earle KM, Gibbs Jr CJ, Gajdusek DC (1969) Experimental kuru encephalopathy in chimpanzees and spider monkey. J Neuropathol Exp Neurol 28(3): 353–370
149. Goodbrand IA, Ironside JW, Nicolson D, Bell JE (1995) Prion protein accumulations in the spinal cords of patients with sporadic and growth hormone-associated Creutzfeldt-Jakob disease. Neurosci Lett 183(1–2):127–130
150. Brandner S, Whitfield J, Boone K et al (2008) Central and peripheral pathology of kuru: pathological analysis of a recent case and comparison with other forms of human prion diseases. Philos Trans R Soc Lond Ser B Biol Sci 363(1510):3755–3763
151. Collinge J, Sidle KCL, Meads J et al (1996) Molecular analysis of prion strain variation and the etiology of "new variant" CJD. Nature 383(6602):685–670
152. Parchi P, Castellani R, Capellari S et al (1996) Molecular basis of phenotypic variability in sporadic Creutzfeldt-Jakob disease. Ann Neurol 39(6):767–778
153. Wadsworth JD, Joiner S, Hill AF et al (2001) Tissue distribution of protease resistant prion protein in variant Creutzfeldt-Jakob disease using a highly sensitive immunoblotting assay. Lancet 358(9277):171–180
154. Lee H-S, Brown P, Cervenakova L et al (2001) Increased susceptibility to kuru of carriers of the PRNP 129 methionine/methionine genotype. J Infect Dis 183(2):192–196
155. Mead S, Poulter M, Uphill J et al (2009) Genetic risk factors for variant Creutzfeldt-Jakob disease: a genome-wide association study. Lancet Neurol 8(1):57–66
156. Mead S, Stumpf MP, Whitfield J et al (2003) Balancing selection at the prion protein gene consistent with prehistoric kurulike epidemics. Science 300(5619):640–643
157. Mead S, Whitfield J, Poulter M et al (2008) Genetic susceptibility, evolution and the kuru epidemic. Philos Trans R Soc Lond Ser B Biol Sci 363(1510):3741–3746
158. Mead S (2006) Prion disease genetics. Eur J Hum Genet 14(3):1–9
159. Lantos B, Bhata K, Doey LJ et al (1997) Is the neuropathology of new variant Creutzfeldt-Jakob disease and kuru similar? Lancet 350(9072):187–188

160. Aguzzi A, Heikenwalder M (2003) Prion diseases: cannibals and garbage piles. Nature 423(6936):127–129
161. Brookfield JF (2003) Human evolution: a legacy of cannibalism in our genes? Curr Biol 13(15):R592–R593
162. Mead S, Whitfield J, Poulter M et al (2009) A novel protective prion protein variant that colocalizes with kuru exposure. N Engl J Med 361(21):2056–2065
163. Mead S, Uphill J, Beck J et al (2012) Genome-wide association study in multiple human prion diseases suggests genetic risk factors additional to PRNP. Hum Mol Genet 21(8):1897–1906
164. Marlar RA, Leonard BL, Billman BR et al (2000) Biochemical evidence of cannibalism at a prehistoric Puebloan site in southwestern Colorado. Nature 407(6800):25–26
165. Riel-Salvatore J (2008) Mad Neanderthal disease? Some comments on "A potential role for Transmissible Spongiform Encephalopathies in Neanderthal extinction". Med Hypotheses 71(3):473–474
166. Underdown S (2008) A potential role for transmissible spongiform encephalopathies in Neanderthal extinction. Med Hypotheses 71(1):4–7
167. Wadsworth JDF, Joiner S, Linehan JM et al (2008) The origin of the prion agent of kuru: molecular and biological strain typing. Philos Trans R Soc Lond Ser B Biol Sci 363(1510): 3747–3753
168. Wadsworth JDF, Joiner S, Linehan JM et al (2007) Kuru prions and sporadic Creutzfeldt-Jakob disease prions have equivalent transmission properties in transgenic and wild-type mice. Proc Natl Acad Sci U S A 105(10): 3885–3890
169. Hill AF, Desbruslais M, Joiner S et al (1997) The same prion strain causes vCJD and BSE. Nature 389(6650):448–450
170. Parchi P, Cescatti M, Notari S et al (2010) Agent strain variation in human prion disease: insights from a molecular and pathological review of the National Institutes of Health series of experimentally transmitted disease. Brain 133(10):3030–3042
171. Parchi P, Saverioni D (2012) Molecular pathology, classification, and diagnosis of sporadic human prion disease variants. Folia Neuropathol 50(1):20–45
172. Manuelidis L, Chakrabarty T, Miyazawa K et al (2009) The kuru infectious agent is a unique geographic isolate distinct from Creutzfeldt-Jakob disease and scrapie agents. Proc Natl Acad Sci U S A 106(32):13529–13534
173. Jaunmuktane Z, Mead S, Ellis M et al (2015) Evidence for human transmission of amyloid-β pathology and cerebral amyloid angiopathy. Nature 525(7568):247–250
174. Kovacs GG, Lutz MI, Ricken G et al (2016) Dura mater is a potential source of Aβ seeds. Acta Neuropathol 131(6):911–923
175. Frontzek K, Lutz MI, Aguzzi A et al (2016) Amyloid-β pathology and cerebral amyloid angiopathy are frequent in iatrogenic Creutzfeldt-Jakob disease after dural grafting. Swiss Med Wkly 146:w14287. doi:10.4414/smw.2016.14287. eCollection 2016

Chapter 2

Anthropological Methods Used in Kuru Research

Shirley Lindenbaum

Abstract

This historical account of the methods used by anthropologists studying kuru from 1961 to 2010 illustrates the identity of anthropology as both a humanist and natural science. To understand and analyze complex historical processes anthropologists employ both interpretive and explanatory research methods. This chapter documents the emergence of medical anthropology as a subfield in anthropology, changes that have taken place in the collaborative relations between anthropology and medicine, and importance of the political context in Papua New Guinea, all of which have had an impact on the research methods of anthropologists and medical investigators. Fore forms of health care have also changed as local therapists adopt some aspects of biomedicine while retaining a belief that sorcerers cause illness and death, a theory that supports their own methods of investigation.

Key words Anthropology, Humanism, Natural science, Medical anthropology, History, Kuru, Papua New Guinea

1 Introduction

In 1964 the distinguished anthropologist Eric Wolf described anthropology as the most scientific of the humanities, and the most humanist of the sciences, an assessment that holds true for a discipline that has emerged from a century of ethnographic and theoretical work. As a natural science, Wolf noted that anthropology is concerned with the organization and function of matter; as a humanistic discipline it is concerned with the organization and function of mind. Anthropologists thus mediate between human biology and ecology on the one hand, and the study of human understanding on the other. As both observers and participants in the internal dialogues of informants, anthropology is less subject matter than a bond between subject matters, and the anthropologist translates from one realm to the other [1].

Anthropologists who specialize in cultural anthropology, and who form the majority, are likely to have studied physical anthropology, archaeology, and linguistics, the hallmark of the "four-field"

Pawel P. Liberski (ed.), *Prion Diseases*, Neuromethods, vol. 129,
DOI 10.1007/978-1-4939-7211-1_2, © Springer Science+Business Media LLC 2017

approach in American anthropology. The breadth of training in human biology and social life, and the range of cultures past and present, equips them to make general statements about the human condition, as well as to understand human differences and human possibilities. Wary of Western "objective" research instruments, the anthropological search for meaning places them closer to history, which also explores phenomena as unique, and seeks to interpret them. The charge to generalize and to theorize, however, situates anthropology equally in the social sciences [2].

Anthropologists choose research methods appropriate for the topic and the question. Their methods include recording genealogies and mapping households and gardens. They collect data on particular issues using statistical samples, and provide more descriptive accounts on such matters as rituals, myths, gender, sexuality, health and illness, and use of plants. The most widely recognized form of anthropological research, called fieldwork, involves observing, documenting, and participating in day-to-day community life, to the degree that this is possible. This method of anthropological research applies equally to observing and examining the cultural beliefs and practices of the many actors encountered in the contemporary world. The archives provide a rich source for historical and comparative material.

Current research often requires understanding the worldwide networks of places linked to many field sites, although the anthropologist does not necessarily reside in all of them. Anthropologists record their observations and findings in notebooks and laptops. They use tape recorders, cameras, and video equipment, and keep a daily diary to record the sequence of events, and for personal reflections. With the help of interpreters they record and translate interviews, and ideally become fluent speakers of local languages. Those who work in Papua New Guinea speak Melanesian pidgin, allowing them to communicate across many linguistic boundaries. While living with people in their own communities for long periods of time, initial research usually lasts for at least a year, followed in many cases by one or more return visits lasting for shorter periods. Many anthropologists now communicate by cell phone with village friends and former research assistants, and sometimes by e-mail with those who have access to computers.

In recent years static accounts of social arrangements and cultural beliefs have been displaced by an appreciation of dynamic change. Anthropologists have moved away from an interest in the uniqueness and diversity of local cultures toward the study of regional organizations and historical processes. They look increasingly at the cultural formations of urban areas and nation states, and at transnational and global processes. The tribes and peasants of the world are seen to be responding to the variety of global institutions, ideas, and diseases that shape the experience of personal and collective life. Anthropologists often become advocates for indigenous people, a commitment that grows organically from an anthropology that gives voice to those whose opinions are rarely heard. As the world changes, analytical frameworks are updated, and most departments now teach a course in the history of anthropological theory. A number of subfields have also emerged. Ecological anthropologists, for example, study relationships among humans, animals, and their biophysical environments. Medical anthropologists study how health and illnesses are shaped, experienced, and understood in the context of global, historical, and political forces. They engage also with issues in biomedicine, once called Western medicine. As a product of the West, biomedicine, like all forms of medical knowledge and practice, takes form in a particular historical, social, and cultural context. Its assumed universalism is thus questioned.

Anthropologists are trained to interrogate their own cultural assumptions and biases. They are also aware that their own presence can influence the processes they seek to observe. The American Anthropological Association (www.aaanet.org) provides a Code of Ethics for the conduct expected of anthropologists as they undertake their research, as well as information about the range of subfields and associated journals sponsored by the Association.

The story of anthropological research methods that follows draws attention to two "phases" of research associated with kuru. The first describes the topics, questions, and methods that Robert Glasse and I employed from 1961 to 1963, and those I used during a number of short field trips between 1970 and 2008. The second describes the topics, questions, and methods used by the anthropologist Jerome Whitfield from 1996 to 2010.

2 Phase One

Early medical investigations of kuru were hampered by the lack of information about Fore social life, especially kinship. In 1961, Robert Glasse and I were sent to collect this data, supported by a grant from the Department of Genetics, Adelaide University. We chose to live in Wanitabe, a South Fore community in the Eastern Highlands, which had the highest reported incidence of the disease. Our focus on Fore kinship led us to question the genetic hypothesis current at the time. We had been charged to collect "pedigrees," not "genealogies," the latter an anthropological term that acknowledges notions of relatedness that are subject to historical and cultural construction. Our genealogies showed that many of the supposedly closely related kuru victims were not closely related biologically, but were kin in what we would call a social sense.

The process of acquiring and assembling genealogical information occupied much of our time during the first months in the field. We began by recording information from what we determined to be the largest locally recognized group of people at Wanitabe to form a distinct political and spatial entity, which we called a parish. This population of 350 people in 1962 was composed of 7 patrilineal lineages, which the Fore called *lounei*, or lines, and which were considered to be united by notions of patrilineal descent. The Fore lack of concern for strict descent reckoning, however, was manifest in the linguistic distinctions they made about Wanitabe co-residents, who were defined as *mago kina*, ground source people, or original residents (k/gina suffix = people), *tubagina*, people who gather, that is, immigrants, and *aguya gina*, those who have been beaten, or war refugees. Immigrants and refugees were welcomed as long as they demonstrated loyalty and observed their new social obligations. In time, the newcomers were said to possess "one blood" and a common ancestry, conveying the idea that those who reside and act together, and eat food grown on the same land, share bodily substance. Descent was a symbol of unity for a coalition of people with shared social and political interests.

In addition, childless couples could adopt an infant from another lineage family, and kinship could be "created" or "generated" in food-sharing ceremonies with people who had no recognized consanguineal ties. Held in the early afternoon, a time of day called the *kagi*-nei (nei suffix = the), the new kin were referred to as *kagi-sa-kina*, ceremonially created or "fictional" kin. Such flexibility in kinship definition meant that lines of cleavage could undermine unity, finding expression in accusations of sorcery and the emigration of lineage groups settling in distant locations. These two features, fictional kinship and population mobility, had relevance for the distribution of a disease thought to be based on a close breeding unit and a simple genetic factor.

Our genealogies recorded kinship two generations above the adult male being interviewed (the characteristic depth of Fore kinship memory), the informant's marriages, as well as those of his siblings and all their descendants, including infant deaths. We also interviewed women, but adult women often came from other local groups at the time of marriage, and although they had their own kin networks, they were unable to provide extensive knowledge of the relationships we were investigating. Robert and I first recorded the information in 1961 and 1962, and I brought them up to date in 1993 and 1996. Our genealogies recorded a wide range of data: male and female kuru deaths, male deaths from warfare, other deaths and their causes defined by the Fore, the exodus of women at the time of marriage and of men for employment, as well as religious affiliation (Open Bible, Seventh Day Adventist, Salvation

Army, or no affiliation). Additional information included the form of marriage from the male point of view: matrilateral cross cousins (mother's brother's daughter) the preferred form, more distant maternal cousins, and those said to be unrelated, as well as children's level of education. To make reading easier, the genealogies were color-coded.

The genealogy provided here (Fig. 1) is a simplified version of page 6 of 70 pages for Wanitabe parish. MANOVA, identified in line one as our main informant in this case was about 70 years old in 1961. The genealogy illustrates the gendered and generational impact of the epidemic. In-filled circles represent female kuru deaths, triangles represent men. Cannibalism ended in 1960, and no one born since that time has come down with the disease. The last case of kuru occurred in 2009, but as this genealogy shows, the epidemic was waning in 1996, and the population was increasing. Fewer men were dying during warfare, a significant cause of death still present in Manova's generation, with brothers inheriting the wives. A government clinic was now providing health care that included vaccinations for infants. All the children in the third generation were attending primary school, grades two to six, and only one child had died, one of Manova's great grand-daughters. The note in the right margin indicates that further information about this particular genealogy could be found on page 7.

Fore genealogical knowledge, like that of most Eastern Highlanders, is shallow. Hearing about my genealogical enquiries in 1996, some men from outside Wanitabe asked for their grandfather's names to bestow on their children. One interesting feature of our focus on where wives came from meant that we could provide a statistical estimate of the degree to which marriages were endogamous (marriage within Wanitabe) or exogamous (outside Wanitabe). Arrows in Fig. 1 show the exodus of women at marriage. The genealogy reveals the high frequency of endogamous marriage on the one hand, and the wide dispersion of exogamous marriage on the other, evidence of a high rate of gene flow within and between Wanitabe and other local groups, as well as a multitude of channels for interacting with many other groups [3].

The process of taking genealogies often attracted a small group of interested male participants (the women were busy working in the gardens). The men helped to document, and sometimes correct, the detailed data we were recording, which was enriched by spontaneous accounts of local politics and military victories over aggressive neighbors, as well as insight into local values, such as the way the hunting abilities of certain men made them attractive to women, or the psychodynamics of gender formation, when as youths they overcame the pain, terrors, and hardships of male initiation. These long genealogy-recording sessions proved to be a rich source of unsolicited ethnographic and historical information

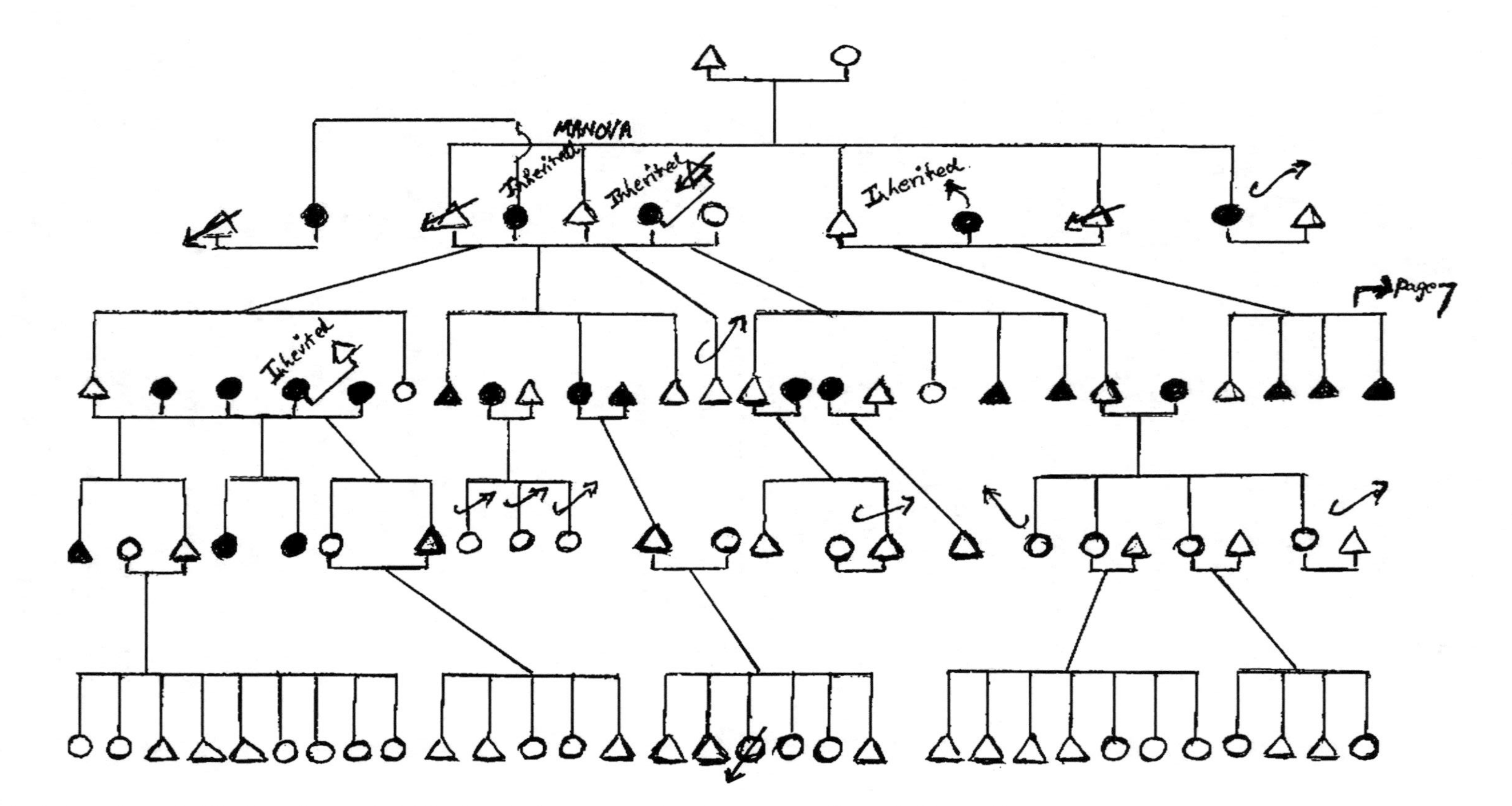

Fig. 1 Genealogy of a Wanitabe lineage in 1996 showing the gendered and generational impact of the epidemic

on many seemingly unrelated topics, and illustrate the often indirect anthropological method of collecting and interpreting data.

In addition to gaining an understanding of local concepts of kinship and marriage, our early investigations focused on kuru. We were surprised when people at Wanitabe said that kuru was a new disease, and that they could remember the first cases of the disease in their own and in neighboring communities. With several Fore research assistants we spent several weeks visiting Fore hamlets from the south to the north, and into the borderlands of the neighboring Keiagana, a different linguistic and cultural group, gathering accounts of "first sightings." We conducted interviews in Melanesian pidgin, our research assistants providing translations of Fore and other local languages. These oral histories showed that the Fore understood kuru to be of recent origin. Their stories depicted an epidemic that had spread within recent memory, emerging on the North Fore border around the turn of the century, and then spreading south along a traceable route until it reached Wanitabe and further south in the early 1930s [4]. This finding was a second challenge to the Mendelian model of genetics, which assumed that the disease was of remote evolutionary origin, and that it ought to have been in epidemiological equilibrium. However, as John Mathews, a medical investigator studying kuru, observed, the disease was too common and too fatal to be a purely genetic disorder, unless the hypothetical kuru gene was maintained at high frequency by a mechanism of balanced polymorphism. There was no evidence to support this latter suggestion [5]. Mathew's speculation would be shown to be relevant in the later identification of a human prion protein gene imposing strong balancing selection on the Fore population during the kuru epidemic.

The people we interviewed also said that those who died of kuru had been consumed. Before we began fieldwork we had read anthropological accounts of cannibalism in the Eastern Highlands, particularly among the Fore [6]. We collected more detailed information about the Fore practice and learned that all body parts were eaten, except the gallbladder, which was considered too bitter. Significantly, not all Fore were cannibals. Some elderly men rarely ate human flesh, but small children residing with their mothers ate what their mothers gave them. Youths who were initiated around the age of 9 or 10 moved to the men's house, where they began to observe the cultural practices and dietary regimes that defined masculinity. Consuming the dead was considered appropriate for adult women and small children, but not adult men, who feared the pollution and physical depletion associated with eating a corpse, especially that of a woman. Body parts were not distributed randomly, but were consumed by particular kin. We compiled tables to illustrate the different customary rights, defined by kinship, to consume a male or a female corpse [3].

In 1957, Carleton Gajdusek and Vincent Zigas had published two articles in which they referred to kuru as a disease new to Western medicine [7, 8]. The articles noted the high incidence of kuru in certain families and hamlets, its localization to the Fore and adjacent peoples with whom they intermarried, and the predilection for children and adult women. This epidemiological data seemed to match Fore rules for human consumption. Robert wrote about this in our 1962 and 1963 fieldwork reports to John Gunther, the Director of Public Health, the institute that had funded our second year of research [9, 10]. In 1967 Robert then presented the case for cannibalism to the Division of Anthropology at the New York Academy of Sciences [11].

Our reports were initially viewed with skepticism by many medical investigators and some anthropologists. John Mathews, however, embraced and further investigated our findings. His mathematical analysis of our genealogies [12] confirmed that kuru was of limited time depth, and that the genetic theory proposed by the geneticists was probably wrong. In 1968 Robert and I joined Mathews in a publication that would reach a wider medical audience [13]. This paper showed that the cannibalism hypothesis was supported by a wealth of epidemiological and ethnographic data consistent with Fore stories about named individuals who had taken part in mortuary feasts, and who had themselves died of kuru, and predicted what was likely to happen if the cannibalism theory was true.

A key focus of medical anthropology concerns the ways in which local populations experience, understand, and respond to illness and death, and the historical context in which this occurs. The late 1950s and early 1960s were crisis years for the Fore. The marital histories of Wanitabe men that we recorded in 1962 showed that they had contracted 76 marriages, or one-and-a-half per man. Two marriages had ended in divorce, and 45 with the death of the wife, 40 of them from kuru [4]. At that time, the motherless nuclear family was a common domestic unit.

When we first arrived in the South Fore in 1961, however, we were surprised by the absence of kuru victims. We learned that between April and August 1961, over 70 kuru victims had recently walked or were carried to Uvai, a community of Gimi-speaking peoples on the other side of the Yani river, where a Gimi therapist was said to be providing a cure for kuru. With our interpreters and a small team of Fore men carrying our equipment and food supplies, we followed the path of the pilgrims to Uvai. For over a week we observed and interviewed the patients and their relatives, as well as the Gimi therapist and his assistants, and then returned to Wanitabe traveling slowly through Fore villages where we interviewed returned patients who were still alive. In the following months, the Gimi therapist and his assistants visited the South Fore to provide follow-up treatments, which we witnessed and photographed. Forest medicines and bloodletting sessions were used again, but the therapist now alluded to the identity of the sorcerer, ambiguously suggesting his place of residence.

As kuru victims continued to die, however, the Fore were aware that they faced a demographic emergency, the dimensions of which they grasped clearly. From November 1962 to the end of March 1963, they held public meetings in different parishes to demand the outlawing of kuru sorcery, and to repair the sexual imbalance that was affecting daily life. Speakers declared that sorcerers had exceeded all moral boundaries, ignoring the limits they placed on their own behaviors during warfare. The ritualized meetings, a Fore method for investigating and stamping out the disease, provided a forum for reviewing mutual suspicions and hostilities, and a public arena for confession, expressions of self-denigration, and a concern that the colonial administration would consider them to be the guilty actors. To the often-expressed fear of extinction from the loss of women's reproductive powers, they now added the fear of internal disruption so great that the bonds that had long held them together were themselves endangered. Robert and I attended and recorded the meetings, and discussed them later with several speakers who came to our field house to register their own views [4].

In the 1960s anthropologists renewed an interest in the analysis of myths, rituals, and symbolism. Victor Turner's accounts of the ritual response to severe misfortune among the Ndembu in Africa provided comparative material for analyzing these Fore rituals in which common values were similarly stressed, as were the confessions of guilt by those who felt they had broken some norm governing the intercourse of the living with the dead, allowing individuals to purge themselves of rebellious wishes and emotions [14].

In short field trips between 1970 and 2008, I continued to document economic, political, and social changes taking place in the South Fore, particularly after Independence in 1975, showing how the changes had affected the way Fore now spoke about the epidemic and its causation. A particular challenge was how to explain the decline and apparent disappearance of the disease that had once endangered their survival. By the 1990s most people said that kuru had decreased with the arrival of Christianity, the school, and the market, a set of coherences that seemed causal, and that kuru would disappear when the last of the generation of elders had died, taking with them knowledge of the sorcery that had caused the epidemic. The Fore did not assume, as many scholars have supposed, that all belief in sorcery and witchcraft would end with "modernization" and modern science. This has not been the case in Papua New Guinea or elsewhere in the world [15]. The Fore still thought that kuru sorcerers caused kuru, but had turned to more profitable ventures, and they described a new form of sorcery coming to them from their Gimi neighbors. They were correct, however, in their historical observation that "modernity" had a place in ridding them of kuru. As a result of their encounters with the messengers of modernity, the missionaries and colonial administrators, who spoke out against cannibalism, which they viewed as a perversion and legal offense, the Fore gradually gave up the practice. At that time, the Fore, the missionaries, and government officers saw no relationship between kuru and the consumption of deceased relatives, but all three had unwittingly halted the transmission of the disease. New therapists had also begun to displace the local practitioners who had once provided cures for kuru. This new generation treated gonorrhea, syphilis, chlamydia, cancer, and AIDS, and they continued to use medicines from forest plants and trees, their identity no longer revealed in dreams, but in visions sent by God. Some of them took their medicines to be tested in biomedical institutes. In their view, the efficacy of their treatments did not depend on biomedical assessment, but on the foundations of Fore knowledge: visible proof, the power of spirits beings, and the authority of a Christian God.

As we see, the Fore now have access to two forms of medicine, which are not seen as incompatible, one based on their own understandings of the natural world, the anatomical self, and the world of social and spiritual relations, and the other on biomedical treatments provided by the local government clinic. Ethnographic accounts of health care elsewhere in the world provide similar reports of the integration of these two therapeutic modes.

3 Phase Two

The second phase of anthropological research takes place in a historical context in which the relationship between medicine and anthropology had changed. Kuru research in Phase Two is also carried out in a new political environment, following Papua New Guinea Independence in 1975.

Anthropological research in Phase One had begun during the 1960s, before medical anthropology was a recognized field of anthropological study. The Society for Medical Anthropology was formed in 1972, and its impact was not apparent until the 1980s, when the AIDS epidemic began to draw many anthropologists to study health and illness. In 1988, the Wenner Gren Foundation sponsored a conference to examine the historical development of medical anthropology, and to consider the directions in which it was headed [16]. By 2001, the Society for Medical Anthropology had become second largest Section (of 40 Sections) in the American

Anthropological Association. Medical anthropology is now taught in many departments of anthropology. Anthropologists also teach in medical schools, and some are members of multidisciplinary research teams, as kuru research Phase Two illustrates.

With the appearance of bovine spongiform encephalopathy (BSE) in the United Kingdom in 1986, and the identification of variant Creutzfeldt-Jakob disease, the kuru epidemic had acquired a new global relevance. The (MRC) Prion Institute in London established a research outpost at Waisa, a South Fore community, about an hour's walk from Wanitabe, where the Papua New Guinea Institute of Medical Research, under the direction of Michael Alpers, had a research base. The Institute's old research house at Waisa was refurbished and made liveable, equipped with solar power for lighting and recharging batteries. A small laboratory for the fractionation and freezing of blood samples was built and fully equipped with lighting, water supplies, a hand-powered centrifuge, and a −40 °C freezer. A specially adapted Land Rover to support the project was shipped from the United Kingdom, and a helicopter landing pad area was cleared for the use of charter helicopters when road conditions were hazardous or impractical. The Institute of Medicine in Goroka (IMR) provided the project with office space and administrative support. John Collinge, the Director of the London Prion Unit, made annual visits, and other members of the Unit occasionally visited the field site.

The Unit's scientific objectives were "to identify and study all the remaining kuru patients, and document the maximum inoculation periods; to provide further data for accurate epidemiological modeling of the kuru epidemic to estimate key epidemic parameters and traditional beliefs of the aetiology of kuru by interviewing surviving participants and other members of the Fore community; to study the clinical features of the current kuru patients and to compare clinical and other diagnostic features with other human prion diseases, notably iatrogenic and variant CJD; to investigate any evidence of maternal or other routes of kuru transmission; to identify genetic susceptibility factors to kuru by study of recent patients and archived samples, long-term survivors of multiple feast exposures and the normal Fore and adjacent (exposed and unexposed) populations; to study the peripheral pathogenesis of kuru and tissue distribution of infectivity; and investigate the possibility of sub-clinical prion infection by analysis of autopsy tissues from patients and elderly exposed, but clinically unaffected individuals" [17]. Jerome Whitfield, an anthropologist with a Diploma in Nursing and a background in emergency and tropical medicine, was recruited to assist in carrying out many of the program's social and scientific objectives.

Whitfield arrived in Papua New Guinea in 1996 to begin research on the anthropological and medical dimensions of this multidisciplinary project. With a local field staff of 14 young men, he lived and worked in the South Fore for 7 years, and then visited for a further seven. To update and keep track of the epidemic, ten "kuru reporters" in the field team were responsible for identifying kuru cases in their home locations. Whitfield and several team members then visited the family, and with the family's consent, Whitfield examined the patient, took a case history, and made a neurological examination, which was filmed and sent to Alpers and Collinge for verification. If they were available, Alpers or Collinge would visit the patient. This method of surveillance documented all cases of kuru from 1996 to 2010; one local reporter continued to provide information until 2012, when the project ended.

Six months into fieldwork, Whitfield began to collect blood samples. Sometimes assisted by IMR nursing staff, Jerome and the team collected blood samples from 3322 individuals in the kuru-affected region, and another 984 from surrounding linguistic groups (Gimi, Keiagana, Kanite, Yagaria, Yate, Kamano, Auyana, Awa, and Usurufa). Where blood sampling was of little benefit to the health of participants, they introduced malaria and anemia screening, and with the results in hand, Whitfield and the team returned to the village to treat those with malaria and anemia. Genealogical information associated with each blood sample, such as language group, place of birth, and ancestral group names, was recorded. In some cases the genealogies resulted only in the names of parents and children; in others, the information included the names of grandparents or occasionally great-grandparents and their children. Adoptions were recorded when this was common knowledge, as well as the names of second and third wives, or husbands. Fore age estimates were calculated based on a chronology of local events; non-Fore age estimates for other populations in the Eastern Highlands, taken from government census data, were less complete. Genealogical data associated with blood samples in populations not affected by kuru consisted mostly of father's name, mother's name, village, and language group. Initially logged in field books, the data are now digitalized, allowing the information to be linked readily to the blood donor. The field books were also copied and entered into the Prion Unit database [18].

This body of data contributed to a number of important medical findings. In 2003, some elderly women known to have consumed deceased kin were shown to have a distinct genetic type of prion protein gene (heterozygous at position 129), which provided protection against kuru as a result of exposure to the prion infectious agent, in keeping with balancing selection. Global patterns of diversity in the same gene suggest that European populations show similar, but older, evidence of selection and survival advantage as a result of presumed endocannibalism and prion-related epidemics in the past [19]. In 2009, with data from a larger sample of the

Eastern Highlands' material, the medical investigators located a novel variant of the prion protein gene (at position 127), again in elderly women survivors of the epidemic, which was also a powerful acquired prion-disease resistance factor [20]. And in June 2015, the Unit reported that the amino acid change that occurs at codon 127 (127 V), replacing a glycine with a valine, had a different and more powerful effect than the substitution at codon 129 (129 M). Transgenic mice with the codon 127 mutation were completely resistant to kuru and Creutzfeldt-Jakob disease (CJD). The selection of a single genetic change which occurred during the kuru epidemic is said to provide a striking example of Darwinian evolution in humans. While the collapse of the Fore population was prevented by the cessation of endocannibalism in the late 1950s, this new data suggested that if transmission had continued, the epicenter of the affected region in the South Fore might have been repopulated with kuru-resistant individuals in a population genetic response to the epidemic [21].

With the earlier anthropological hypothesis that kuru was transmitted by cannibalism now accepted, Whitfield was charged with providing further information about Fore mortuary practices that might have relevance for the changing epidemiological patterns of the disease. This ethnographic research had one novel methodological addition: two Fore members of the research team initially interviewed their own family members about the beliefs and practices of mortuary cannibalism in order to test the feasibility of the study. With this established, unstructured interviews were followed by larger group discussions with many of the team members' elders. The members of the team had not undergone initiation, where this information would have been transmitted, and thus had no knowledge of beliefs and practices that had ended in the 1960s, and like their elders, they now considered themselves to be Christians. As the elders confirmed, information about mortuary cannibalism had been kept secret from the younger generation and from most outsiders. Conducted in the Fore language, the interviews were translated into pidgin and back again into Fore, to ensure the accuracy of translation.

Spreading out from Waisa, Whitfield and several team members interviewed 65 men and 11 women in 20 different villages, the gender imbalance reflecting the death of women as a result of kuru. Although adult males did not participate in eating the dead, the elders had participated in, or had witnessed, mortuary feasts as children. The interviews focused on concepts and behaviors associated with body, the soul, and the mourners, the three dimensions identified in Robert Hertz' legendary contribution to the sociological study of death, first published in 1907 [22]. This provided a template for Whitfield's analysis of traditional mortuary ceremonies among the Fore and in nearby Eastern Highland populations where cannibalism had once been practiced [23].

Postcolonial Papua New Guinea in the late 1990s, with its heightened sense of nationalism, presented a difficult political environment for the Prion Unit's investigations. Sensitive to these conditions, Whitfield had been charged with establishing and maintaining contact with local communities, continuing a legacy of the reciprocal interactions and obligations established earlier by anthropologists and medical investigators, which had consisted mainly of gifts and support to individuals and families. (My fieldwork obligations during the 1990s also expanded to include contributions to an education fund to pay school fees for South Fore children, and toward construction of a new elementary school at Wanitabe.)

With his medical qualifications Whitfield could assist in providing health care to communities in the project. This support, of necessity, differed from the "community medicine" adopted elsewhere in the Eastern Highlands, in which epidemiological studies could identify the causative factors in illness and death, such as the consumption of contaminated pork during local festivals that resulted in a form of food poisoning called *enteritis necroticans* (pig-bel in pidgin). The epidemiologist could then work to alter dietary behaviors to prevent occurrence of the disease, and to introduce economic development (such as smoke houses) to sustain the change [24]. The behaviors causing kuru, however, had already changed, the epidemic was ending, and the Prion Unit wished to undertake research that could promise no apparent future benefit. Instead, responding to local requests for assistance, the Unit provided the local health center with medicines when clinic supplies ran out, evacuated patients by car or helicopter to the hospital in Goroka, and, in collaboration with Save the Children Fund, initiated a village birth attendant program. The field team, led by Whitfield, also held medical clinics in remote areas, and provided malaria and anemia screening and treatment in low-lying villages affected by malaria, assisted at times by nursing staff from the Institute of Medicine. Outside funding was gained to provide water supplies for the local marketplace, and in a number of villages, for the construction of primary schools. When the last kuru autopsy took place in 2003 (carried out by an Eastern Highlander), and with the agreement and cooperation of the extended family, a compensation payment was made for the future education of the family's children. The "End of Kuru" conference, held in London in 2007, was attended by 15 Papua New Guineans, 12 of them Fore. Several of them provided testimonies of their own memories and experience [25].

4 Discussion

This historical account of anthropological research methods associated with kuru can be read in several ways. It demonstrates the similarity and variety of the research methods anthropologists deployed in Phases One and Two: recording customary rules associated with kinship, marriage and the consumption of the dead; a focus on ritual behaviors, observed or recalled; maintaining a field base at a specific location and gathering data from a wide area; and working with research assistants to interpret Fore and other languages to enhance an understanding of local histories, concepts, and social events. In both phases, anthropologists compared Fore data with examples in the anthropological record.

In Phase Two, however, the relationship between anthropology and medicine had changed, as medical anthropology became institutionalized in academia. As the story of kuru research illustrates, anthropologists may maintain a collaborative relationship with medical investigators, and some may become part of a multidisciplinary research team. Changes in the political environment also had an impact on the relationships among anthropologists, medical investigators, and the Fore, reflecting a general shift in sensibilities. Anthropologists and medical investigators began to focus more on the community services that the Fore expected. Medical investigators would no longer carry out autopsies. Not to be overlooked, Fore methods for investigating and preventing kuru also changed. The ritual events, seen as a solution in the 1960s, were no longer held. With the decline in kuru cases, their long collaboration with a variety of investigators, and some familiarity with biomedical clinics and hospitals, the Fore began to fold biomedical practices into their own forms of inquiry and health care. To study the complexity of this interacting community, anthropologists selected from a range of procedures, both interpretive and explanatory, for a discipline that, as Wolf observed [1], is part history, part literature, part natural science, and part social science.

References

1. Wolf ER (1974/1964) Anthropology. W.W. Norton, New York
2. Keesing RM (1981) Cultural anthropology. Holt, Rinehart and Winston, New York
3. Glasse R, Lindenbaum S (1976) Kuru at Wanitabe. In: Hornabrook RW (ed) Essays on kuru. E.W. Classey, Faringdon, Berks
4. Lindenbaum S (2013) Kuru sorcery. Disease and danger in the New Guinea Highlands, 2nd edn. Paradigm Publishers, Boulder
5. Mathews JD (1971) Kuru. A puzzle in culture and environmental medicine. Dissertation, University of Melbourne
6. Berndt RM (1952) A cargo movement in the Central Eastern Highlands of New Guinea. Oceania 23:40–65
7. Gajdusek DC, Zigas V (1957) Degenerative diseases of the central nervous system in New Guinea. N Engl J Med 257(20): 974–978

8. Zigas V, Gajdusek DC (1957) Kuru: clinical study of a new syndrome resembling paralysis agitans in natives of the Eastern Highlands of Australian New Guinea. Med J Aust 44(21):745–754
9. Glasse R (1962) South fore cannibalism and kuru. Territory Papua New Guinea, Department of Public Health
10. Glasse R (1963) Report of fieldwork by R.M. Glasse and S. Lindenbaum. Territory Papua New Guinea, Department of Public Health
11. Glasse R (1967) Cannibalism in the kuru region of New Guinea. Trans N Y Acad Sci 29(6):748–754
12. Mathews JD (1965) The changing face of kuru: an analysis of pedigrees collected by R.M.Glasse and Shirley Glasse and of recent census data. Lancet 1(7396):1139–1142
13. Mathews JD, Glasse R, Lindenbaum S (1968) Kuru and cannibalism. Lancet 2(7565): 449–452
14. Turner V (1957) Schism and continuity in an African Society: a study of Ndembu village life. Manchester University Press, Manchester
15. Geschiere P (1997) The modernity of witchcraft. University of Virginia, Charlottesville
16. Lindenbaum S, Lock M (eds) (1993) Knowledge, power & practice. The anthropology of medicine and daily life. University of California Press, Berkley
17. Collinge J (2008) Lessons of kuru research: background to recent studies with some personal reflections. Philos Trans R Soc Lond Ser B Biol Sci 363(1510):3689–3696
18. Whitfield J (2015) Personal Communication
19. Mead S, Stumf MP, Whitfield J et al (2003) Balancing selection at the prion protein gene consistent with prehistoric kurulike epidemics. Science 300(5619):640–643
20. Mead S, Whitfield J, Poulter M et al (2009) A novel protective prion protein variant that colocalizes with kuru exposure. N Engl J Med 361(21):2056–2065
21. Asante EA, Smidak M, Grimshaw A et al (2015) A naturally occurring variant of the human prion protein completely prevents prion disease. Nature 522(7557):478–481
22. Hertz R (1960/1907]) A contribution to the study of collective representation of death. Free Press, Glencoe
23. Whitfield J, Pako WH, Alpers M (2015) Metaphysical personhood and traditional South Fore mortuary rites. J Soc Océan 141:303–321
24. Brener M (2015) Infectious personalities: the public health legacy of three Australian Doctors in Papua New Guinea. Health Hist 17(1):73–24
25. Tarr PI (2008) The late 1970s: a lull in the action on kuru. Philos Trans R Soc Lond Ser B Biol Sci 363(1510):3668–3670

Chapter 3

Nonhuman Primates in Research on Transmissible Spongiform Encephalopathies

David M. Asher, Pedro Piccardo, and Luisa Gregori

Abstract

Nonhuman primates played an important role in early studies of the human transmissible spongiform encephalopathies (TSEs): kuru, Creutzfeldt-Jakob disease (CJD), and similar degenerative diseases of the nervous system. The high cost of procuring and maintaining nonhuman primates and growing public resistance have discouraged their use for TSE research in recent years. Invasive research with chimpanzees has effectively ended in the USA. A few situations remain, however, in which laboratory primates offer advantages for biomedical research on TSEs: (1) transmission attempts when etiology and host range of a human disease are not yet known or when more accessible tests to detect an infectious agent are either unavailable or not feasible; (2) studies of pathogenesis using monkeys instead of rodents may be more directly relevant to a human TSE; and (3) preparation of certain biological reference materials, such as blood or tissues infected with a human-derived TSE agent, might require a species of animal genetically close to humans because of antigenically similar prion proteins.

Key words Transmissible spongiform encephalopathy (TSE), Prion, Creutzfeldt-Jakob disease (CJD), Monkey, Macaque, Chimpanzee, Squirrel monkey, Pathogenesis, Biological reference material

1 Use of Nonhuman Primates to Demonstrate Transmissibility of Human Diseases

William Hadlow presciently observed that the human progressive neurological disease kuru [1] resembled ovine scrapie [2]—a contagious disease with a very long silent incubation period, at the time experimentally transmitted only from sick to healthy sheep and to a closely related species, goats. Hadlow's observation immediately suggested two important requirements for future attempts to transmit kuru experimentally: human materials should be

Disclaimer: The authors' contributions to this chapter are informal communications representing our own best judgment that do not bind or obligate the US FDA.

Pawel P. Liberski (ed.), *Prion Diseases*, Neuromethods, vol. 129,
DOI 10.1007/978-1-4939-7211-1_3, © Springer Science+Business Media LLC 2017

injected into animals as closely related to humans as possible and long incubation periods were to be expected. Stimulated by Hadlow's observation, D. Carleton Gajdusek and colleagues injected nonhuman primates with brain suspensions from kuru patients in 1963; several inoculated chimpanzees became ill 2 years later [3] with a relentlessly progressive illness recapitulating many clinical and histopathological features of kuru [4]. Primates of other species later proved susceptible to kuru as well [5]. The successful experimental transmission of kuru immediately suggested that other human neurological diseases might also be transmissible. Within 2 years, brain suspension from a case of sporadic CJD had transmitted a similar spongiform encephalopathy to chimpanzees [6], later to monkeys, and eventually to nonprimate animals [7]. From that time until now, only kuru, CJD, and its variants and similar spongiform encephalopathies, like the Gerstmann-Sträussler-Scheinker syndrome, have been experimentally transmitted to nonhuman primates [8, 9].

It might be argued in retrospect that the infectious nature of the transmissible spongiform encephalopathies (TSEs) could have been demonstrated without recourse to terminal studies in primates, because other animals were later found to be susceptible, albeit less consistently [8, 9]; however, as early as the first third of the twentieth century, chimpanzees and monkeys were used to establish an infectious etiology and to study pathogenesis of several transmissible illnesses of humans—yellow fever [10–12] and poliomyelitis [13] come to mind—so that by the mid-twentieth century nonhuman primate models had long been considered a normal part of the scientific armamentarium to investigate human infectious diseases. In any case, the remarkable similarities between clinical and histopathological manifestations of human kuru and CJD to the diseases transmitted to chimpanzees quickly quelled doubts that the human spongiform encephalopathies were infections and accelerated research on the diseases and their mysterious self-replicating agents, and that research eventually led to derivation of transgenic mice susceptible to human TSE agents [14, 15] that have generally replaced primates in the study of TSEs. (Certain cell lines susceptible to infection with a few strains of TSE agent have also been used to study characteristics of the agents [16] but have not yet detected the agents in tissues of naturally infected humans or animals.) Nonetheless, until better models are developed, attempts to transmit infection to primates might be justified under some circumstances to study TSEs and other human diseases of unknown etiology.

Another appropriate use of nonhuman primates to detect human TSE agents might be to investigate some situations in which iatrogenic infection is suspected, because primates accommodate the large volumes of inoculum often needed to detect very small amounts of infectivity present in body fluids or tissues. As an example, detecting CJD agent in only one of 76 lots of reconstituted lyophilized cadaveric human pituitary growth hormone employed more than 200 squirrel monkeys inoculated by the intracranial (IC) and by other routes and then maintained for more than 5 years as an infectivity assay [17]; the study would have required a huge number of CJD-susceptible transgenic mice (not yet available at the time) simply to accommodate the volumes of hormone tested without assurance that even susceptible mice would have survived long enough to express recognizable disease after infection with the tiny amount of CJD agent present.

Primates might also be used to test special inocula otherwise difficult to assay in small laboratory animals or in TSE-susceptible cell cultures (still lacking for most human TSE agents). For example, the experimental transmission of CJD by IC implantation into a chimpanzee of an electrode contaminated during epilepsy surgery [18] would have been very difficult in smaller animals. Another recent example of the unique contribution to research by TSE infectivity assays in primates was an effort to determine a dose-response relationship for transfusion-transmitted CJD (the probable amount of infectivity in human blood needed to transmit an infection to a recipient exposed by transfusion [19]); that estimate was derived from reanalysis of data in a much earlier summary report of nonhuman primates' inoculated IC with suspensions of brain tissue from patients with CJD [8].

Regarding the species of nonhuman primate most suitable to detect human TSE agents, although chimpanzees were the earliest species studied [3, 6] and were more consistently susceptible to infection than other species (Table 1), squirrel monkeys proved almost as sensitive to infection with both kuru and CJD agents as chimpanzees (Table 2) [8, 9] and remain the most accessible nonhuman primate to study TSEs, now that chimpanzees are no longer to be used for invasive research. (Indeed, after 1985 no new chimpanzees were introduced into TSE research by the NIH.)

However, the advantages of squirrel monkeys for detecting TSE agents transmissible to humans should not be overstated; for example, squirrel monkeys were less sensitive to infection with the agent of bovine spongiform encephalopathy (BSE) and had much longer incubation periods than did TgBo transgenic mice that express bovine prion protein (PrP) [20].

There are also regulatory impediments to TSE research with nonhuman primates. Squirrel monkeys (and their "parts and derivatives") are currently listed in Appendix II of the Convention on International Trade in Endangered Species of Wild Fauna and Fauna [21]; CITES Appendix-II animals are " … not necessarily now threatened with extinction … [but] may become so unless trade is closely controlled." CITES permits the international shipment of such animals but only with an export certificate. National authorities determine the criteria for granting export certificates for CITES Appendix-II species and may impose even stricter

Table 1
Comparative transmission rates, incubation periods, and durations of illness in seven species of nonhuman primate (382 animals) inoculated with 10% suspensions of brain tissue from 213 cases of confirmed sporadic CJD (adapted from [8])

	Anthropoid apes	New world monkeys			Old world monkeys		
	Chimpanzee	Squirrel	Spider	Capuchin	Vervet	Rhesus	Cynomolgus
Animals with TSE/ animals inoculated (% of transmission)	28/29 (97)	196/211 (93)	30/31 (97)	36/45 (80)	13/15 (87)	19/28 (68)	6/23 (22)
Average incubation ± SD (range) in months	17 ± 7 (11–36)	25 ± 5 (11–37)	32 ± 8 (15–50)	40 ± 9 (29–73)	49 ± 5 (46–58)	64 ± 7 (44–73)	61 ± 10 (53–74)
Average duration of illness ± SD (range) in months	1.7 ± 1.6 (0.5–6)	1.3 ± 1.6 (0–11)	1.6 ± 1.6 (0–8)	2.4 ± 2.4 (0–11)	4.4 ± 3.6 (0–12)	3.2 ± 2.9 (0–10)	2.1 ± 2.3 (0.5–7)

SD standard deviation. Calculated averages, SDs, and ranges intentionally excluded seven animals with exceptionally long incubation periods (four chimpanzees: 51, 66, 71, and 75 months; one squirrel monkey: 160 months; two capuchin monkeys: 160 and 168 months). Chimpanzee, *Pan* sp.; squirrel monkey, *Saimiri* sp.; spider monkey, *Ateles* sp.; capuchin monkey, *Cebus* sp.; vervet monkey, *Chlorocebus* sp.; rhesus and cynomolgus monkeys, *Macaca* sp.

Table 2
Comparison of transmissions from brain tissues of patients with three histopathologically confirmed TSEs to two nonhuman primate species (adapted from [9])

	Disease	Cases	Animals	AST (SD)	Range of survival
Chimpanzee	CJD	31	36	27 (21)	11–76
	GSS	1	1	17	
	Kuru	10	17	33 (15)	20–84
Squirrel monkey	CJD	168	223	28 (16)	11–166
	GSS	2	3	23 (1)	20–22
	Kuru	10	15	25 (9)	8–38

The comparison was limited to animals inoculated IC with 1–20% suspensions of brain tissue from confirmed cases of disease. All forms of CJD were included, though most had sporadic CJD. AST, average survival time in months (mo); SD, standard deviation; CJD, Creutzfeldt-Jakob disease; GSS, Gerstmann-Sträussler-Scheinker syndrome

requirements (for example, the US Department of the Interior currently prohibits the export of their tissues and even extracts of tissues). Chimpanzees, seriously threatened with extinction in the wild, are listed in CITES Appendix-I, prohibiting international trade under almost all circumstances.

2 Use of Nonhuman Primates to Study the Pathogenesis of TSEs

Humans are phylogenetically closer to other primates than to rodents. Pathogenesis of TSEs in conventional or transgenic rodents, although informative, does not always faithfully recapitulate all aspects of human infections with similar TSE agents; consequently, there are situations in which the similarity of nonhuman primates to humans allows experimental studies more directly relevant to pathogenesis of human TSEs than studies in rodents would be. For example, the feeding of infected brains to squirrel monkeys transmitted typical kuru and CJD [22], confirming the plausibility of an oral route of exposure by which humans acquired kuru and most vCJD infections. The same study also transmitted scrapie—never implicated as likely source of a human TSE—to squirrel monkeys, serving to warn that, at least under some experimental circumstances, the oral route was effective and a "species barrier" failed to protect a primate against scrapie infection. Recent studies by Comoy et al. confirmed transmission of natural sheep scrapie to a cynomolgus macaque that became ill 10 years after intracerebral inoculation of brain extract [23]. Marsh et al. showed that squirrel monkeys infected by the IC route with brain homogenate from a mule deer with chronic wasting disease (CWD) developed TSE approximately 30 months post-infection [24]. Race and colleagues [14] tested two primate species and two routes of inoculation (IC and oral) to study the susceptibility of primates to CWD, finding that most squirrel monkeys developed TSE but only following long incubation periods. However, cynomolgus macaques (Old World monkeys evolutionarily closer to humans than are New World squirrel monkeys) showed no signs of disease 6 years post-inoculation. (Results of a recent study–unpublished but publicly presented [Czub S et al. Prion 2017, Edinburgh, May 2017]–showed that macaques were infected by the oral route with muscle of overtly healthy deer incubating CWD). These studies highlight the importance of using animal models with long life spans but also indicate that experimental susceptibility of monkeys to a TSE does not necessarily predict human susceptibility. It is important to note that, in spite of the fact that food-borne exposures to scrapie and CWD agents transmitted disease to monkeys, no epidemiological evidence suggests that humans have actually been infected with either agent [25–27].

Recent comparative studies of TSEs in different species of monkey help to understand variable and constant features of TSE histopathology—including variability in properties of the abnormal form of PrP associated with TSEs, variously designated scrapie-type prion protein (PrP^{Sc}), protease-resistant PrP (PrP^{res}), disease-associated PrP (PrP^{D}), or TSE-associated PrP (PrP^{TSE}) [28, 29]—the term used here. There can be no doubt that animals must express normal "cellular" PrP (PrP^{C}) to be infected with TSE agents [30] and that mutations in the PrP-encoding gene are linked to the autosomal dominant expression of familial human TSEs [31]. However, the role of PrP^{TSE} in pathogenesis of TSEs is not yet fully understood and its role in etiology—proposed by the widely accepted "prion hypothesis" to be the self-replicating pathogenic agent itself [32]—remains controversial [16].

Studies in monkeys have helped to elucidate some aspects of the role that PrP^{TSE} may play in pathogenesis of TSEs. For example, brains of cynomolgus monkeys infected with the BSE agent had amyloid plaques composed of PrP^{TSE} resembling those in human vCJD [33], while brains of squirrel monkeys with experimental BSE did not [34, 35]. Brains of both simian species contained diffuse deposits of abnormal PrP^{TSE}. Those differences suggest that, because they were not a constant feature of TSEs in monkeys, amyloid plaques cannot be essential to pathogenesis of BSE. (Plaques have also not been described in brains of cattle with "classical" BSE [36]). On the other hand, because diffuse parenchymal deposits composed of PrP^{TSE} are consistently present in the brain in TSEs, they can be considered to be an important feature of disease.

Differences in the molecular mass and relative abundance of isoforms of PrP^{TSE} (resulting from differences in amounts of diglycosylated, monoglycosylated, and nonglycosylated PrP remaining

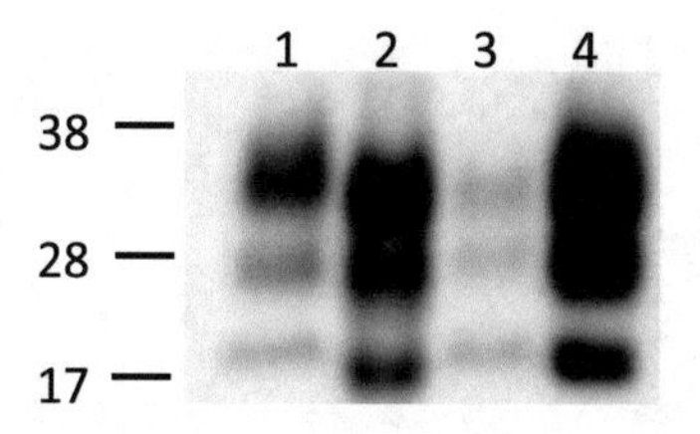

Fig. 1 Comparison of PrP^{TSE} extracted from brains with TSEs originating from BSE. Western blots with brain homogenates from a bovine infected with BSE (lane 1), a human with vCJD (lane 2), a cynomolgus macaque experimentally infected with macaque-adapted vCJD agent (lane 3), and a squirrel monkey experimentally infected with BSE agent from an affected cow (lane 4). Brain homogenates were treated with 100 μg/mL proteinase K for 1 h at 37 °C. Molecular masses in kDa are indicated on the left. The blot was probed with a mixture of two monoclonal antibodies to PrP: 6D11 and 3F4. (Antibody 6D11 stained bovine and macaque PrP- and 3F4 stained human and squirrel monkey PrP.) The non-glycosylated band (lowest mass) of PrP^{TSE} extracted from a case of human vCJD in lane 2 appears to have a mass smaller than those of the PrP^{TSE} in the other three lanes. While PrP in lanes 1, 2, and 4 shows an increased abundance of the diglycosylated form often associated with BSE-derived infections [37], the PrP in lane 3 (macaque with vCJD) does not

after PK digestion) are sometimes asserted to be stable properties of infections with the BSE agent [37], suggesting that the origin of a TSE can be reliably deduced from its PrP "glycoform." However, glycoforms of PrP^{TSE} extracted from brains of humans and animals with TSEs presumed to have originated from BSE may be more variable than commonly thought (Fig. 1) [20, 33, 35]. Indeed, results of studies in transgenic mice expressing mutant PrP generated by targeted genes suggested that some PrP^{TSE} caused no TSE at all, even though the mice were highly susceptible to infection with several TSE agent strains [38]. Interestingly, other researchers found substantial amounts of TSE infectivity in brains lacking any detectable PrP^{TSE} [39, 40]. Recent studies using in vitro protein amplification systems to enhance detection of PrP^{TSE} showed infectivity in peripheral tissues of cattle terminally ill with natural BSE but no detectable PrP^{TSE} [41]. Data obtained from the inoculations of brain tissue suspensions from 300 human cases into nonhuman primates showed the highest transmission rates from iatrogenic and sporadic cases and lower transmission rates from familial TSEs [8].

Recent studies showed that squirrel monkeys infected with BSE agent developed a typical spongiform encephalopathy resembling vCJD with severe spongiform degeneration, gliosis, and large amounts of diffuse PrP^{TSE} deposited in multiple areas of the brain [35]; affected animals also accumulated large amounts of other proteins thought to be important in neurodegeneration (i.e., hyperphosphorylated tau protein [p-tau] and α-synuclein) suggesting a dose-dependent toxicity triggered by PrP^{TSE} [34] (Fig. 2). However, none of the squirrel monkeys

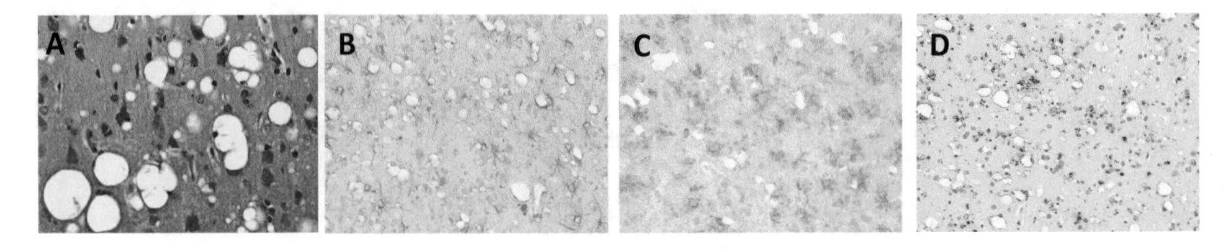

A – D BSE: Squirrel Monkey. Cerebral cortex. 40x magnification.

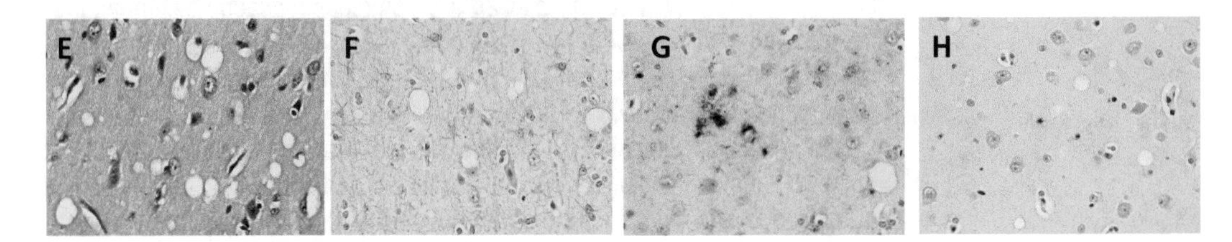

E – H vCJD: Cynomolgus Monkey. Cerebral cortex. 40x magnification.

A,E – HE stain **B,F** – anti-GFAP (rabbit polyclonal) **C,G** – anti-PrP (6H4 mAb) **D,H** – anti-phosphorylated tau protein (AT8 mAb)

Fig. 2 Comparison of histopathology and immunohistochemistry in brains from monkeys of two species with two experimental TSEs. *Upper row*: cerebral cortex of a squirrel monkey inoculated intracerebrally with a suspension of brain from a cow with the classical form of bovine spongiform encephalopathy 34 months earlier shows (**a**) severe spongiform degeneration, (**b**) astrogliosis, (**c**) widespread accumulation of PrP in the neuropil, and (**d**) hyperphophorylated tau protein. Lower row: cerebral cortex of a cynomolgus macaque inoculated intravenously and intraperitoneally with a macaque-adapted vCJD brain suspension 30 months earlier shows (**e**) spongiform degeneration, (**f**) astrogliosis, (**g**) coarse and pericellular PrP deposits in the neuropil, and (**h**) isolated deposits of hyperphosphorylated tau. (**a**, **e**) hematoxylin and eosin stain; (**b**, **f**) immunostain for glial-fibrillary acidic protein (rabbit polyclonal antibody to GFAP); (**c**, **g**) immunostain for prion protein (mouse monoclonal antibody 6H4); (**d**, **h**) immunostain for hyperphosphorylated tau protein (mouse monoclonal antibody AT8). The original magnification of all images was 40×

with TSE and complex protein aggregates contained the amyloid-ß (A-ß) or apolipoprotein-E typical of Alzheimer disease, so the phenomenon does not involve indiscriminately all proteins important in neurodegeneration. The colocalization with PrP^{TSE} of other proteins found in neurodegenerative suggests that misfolded PrP probably induces selective posttranslational modifications in p-tau and α-synuclein. This hypothesis is strengthened by reports that the amino-terminal and repeat regions of tau protein interact with the octapeptide-repeat region of PrP to generate heterologous protein aggregates [42], and that α-synuclein induces formation of tau fibrils—each molecule synergistically increasing polymerization of the other [43]. In contrast, cynomolgus monkeys infected with the agent of vCJD, while developing cortical vacuolation, gliosis, and abundant accumulations of PrP^{TSE} in brain, accumulated only small amounts of p-tau. Thus, the role of "secondary" proteins in the pathogenesis of TSEs and replication of the infectious agents remains to be determined.

The extremely long incubation periods in some experimental TSEs of primates [14, 23, 24] resemble those of naturally acquired human TSEs, which occasionally exceed 35 years in iatrogenic CJD [44] and 50 years in kuru [45]. At first glance, primate models of TSEs with very long incubation periods appear too cumbersome to warrant further investigation; however, because of their similarity to human TSEs, they might provide an opportunity to improve understanding of pathogenesis occurring during the silent incubation period of a TSE and even—eventually—to test candidate prophylactic and therapeutic interventions.

Taken together, these results suggest that, although the central importance of PrP in susceptibility to TSEs and some features of illness is clear, the role of PrP^{TSE} in neurodegeneration, replication of the essential infectious molecule, and transmission of TSEs between individuals is not yet fully understood. Studies with nonhuman primates may contribute to solving some puzzles remaining in pathogenesis of TSEs.

3 Primates as a Source of Biological Reference Materials (Candidate Biological Standards)

Nonhuman primates might serve another purpose in applied TSE research: providing well-characterized biological reference materials with " … defined biological activity … ensuring the reliability of in vitro biological diagnostic procedures used for diagnosis of diseases and treatment monitoring" [46]. WHO and national authorities sometimes develop biological reference materials or standards for infectious diseases, assemblages of many identical aliquots of the same pedigreed, well-characterized and carefully stored materials offered as measurement standards to calibrate working stocks and to compare performance characteristics (sensitivity, limits of detection, and predictive values) for different diagnostic and donor screening tests, among other uses. Such reference materials, when used to validate and evaluate tests for "conventional" human infections, have most often been derived from human blood or other tissues containing the infectious agents of concern or antibodies to them. For TSEs, the WHO supported the preparation and worldwide distribution of samples of brain suspensions from patients with well-documented sporadic or variant CJD as candidate biological reference materials suitable to evaluate tests detecting infectivity and PrP^{TSE} [47]; reference brain suspensions, however, are mainly useful to evaluate assays intended for use with brain-derived materials—postmortem or biopsy samples.

Brain suspensions do not fill an important need: a reference material for naturally TSE-infected human blood. First, blood is the biological sample most commonly used in clinical practice to diagnose systemic human infections. Second, blood is of special importance for improving the control of human TSEs, because transfusions have transmitted vCJD. Red blood cell concentrates from three blood donors latently infected with vCJD transmitted infection to four transfusion recipients in the UK [48]; three developed clinical vCJD [49], while a fourth—dying of unrelated causes—had accumulations of PrP^{TSE} in spleen and a lymph node [50]. A coagulation factor derived from pooled human plasma, containing at least one donation from a donor who later developed vCJD, was implicated in a fifth iatrogenic transmission, also in the UK [51]. Although there is a theoretical risk that sporadic CJD might also be transmitted by transfusion, surveillance over many years failed to find evidence of TSE in any recipient of components from blood of US donors who later became ill with sporadic CJD [52]. A recent report that two recipients of plasma-derived coagulation factors were diagnosed with sporadic CJD, while concerning, may represent a chance occurrence [53].

The US FDA and other national authorities that evaluate various diagnostic and donor screening tests have long encouraged development of antemortem TSE tests and have even proposed schemes for eventually validating such tests [53, 54]. Useful assays must reliably discriminate between samples from known infected and uninfected individuals. While vCJD brain preparations have been used to "spike" blood as a crude surrogate blood reference material—spiked blood employed at an early stage to validate the performance of candidate blood-based tests for TSEs—there is a concern that properties of infectivity and PrP^{TSE} in brain might differ from those in blood. Review of UK transfusion-transmitted vCJD case reports and experimental studies with blood from sheep with BSE and scrapie transfused into normal sheep [55] suggest that concentrations of infectivity in blood of humans and animals with TSEs, while sufficient to transmit infection, are likely to be very low [56]—perhaps a million times lower than in brain. For those reasons, it would be desirable to have reference materials composed of blood from naturally infected humans in addition to human brain reference materials.

Blood from patients with well-documented CJD, including vCJD, has not been collected in volumes suitable to serve as reference materials, and blood will probably never become available from overtly healthy people silently incubating vCJD—the group for which a blood-based antemortem test would be most useful. Therefore, as noted, blood of TSE-infected animals has been proposed as a surrogate reference material. To date, all research tests to identify animals or humans silently incubating TSEs have attempted to detect the small amounts of PrP^{TSE} thought to be present in blood. Blood of TSE-infected rodents can provide "proof of principle" by demonstrating that a candidate antemortem blood-based test method shows promise; however, rodent PrP^TSE differs antigenically from human PrP^TSE and may not provide a suitable surrogate control analyte for a test that detects PrP^TSE in human blood. The PrP in monkey blood, on the other hand, is (like many other proteins) antigenically very close to its human analog, and the reagents used to test human blood are likely to react with PrP in monkey blood.

To assist test developers and prepare for eventual evaluation of antemortem blood-based tests, FDA investigators recently prepared large volumes of blood components from cynomolgus macaques during both the asymptomatic incubation period and overt illness of experimental vCJD (Fig. 3) [57]. Earlier studies found both infectivity and PrP^TSE (detected by the sensitive albeit laborious protein misfolding cyclic amplification [PMCA] technique [58]) in blood of cynomolgus macaques infected with vCJD agent [59], providing proof of concept.

Macaque monkeys were inoculated intravenously and intraperitoneally with a suspension of vCJD-infected macaque brain (a gift from J-P Deslys and colleagues [60]) (Fig. 3); the monkeys became ill with typical signs of TSE about 2 years later (range 23–27 months). Blood was collected from infected monkeys monthly throughout the incubation period and illness, and from control uninfected monkeys less often. Blood was anticoagulated and separated into components: plasma, buffy coat (nucleated cells), and red blood cells (RBC). Multiple aliquots were stored

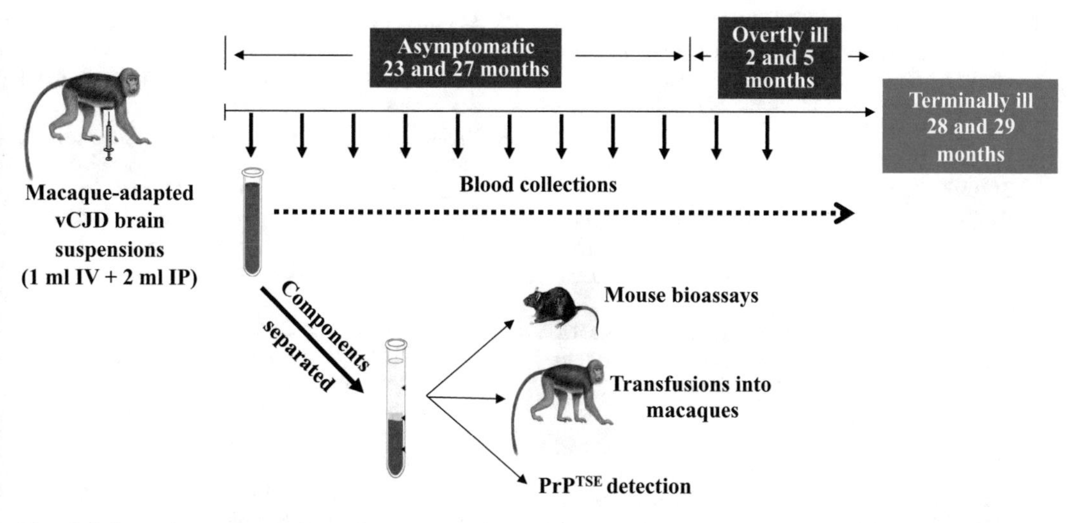

Fig. 3 Schematic representation of a protocol for developing biological reference materials from blood of macaques with experimental vCJD. Blood was collected from control and vCJD-infected macaques at intervals after infection and separated into components. Samples of buffy coat and plasma were assayed for infectivity by injection into vCJD-susceptible mice and aliquots of whole blood were transfused into macaques (in progress). Plasma was used in tests to detect PrP^TSE [57]

frozen at −80 °C. The materials have been provided to qualified assay developers and might eventually serve as reference panels to help validate and control the quality of TSE diagnostic tests and screening tests for donors of blood or tissues.

4 Conclusions

While small animals and in vitro assays have reduced the reliance on nonhuman primates in the study of TSEs, we remain convinced that a need remains for a limited number of monkeys in certain critical studies and to generate reference materials relevant to human diseases.

Acknowledgements

The authors are grateful to Lewis Shankel, Steven Harbaugh, Anthony Cook, F. Salih Muhammad, Jill Ascher, and John Dennis for expert care of monkeys, and Oksana Yakovleva and Juraj Cervenak for carefully preparing the Western immunoblot and histological sections. Arifa Khan read the manuscript critically and offered helpful suggestions. We owe special thanks to the late Kitty L. Pomeroy for her many years of dedicated technical assistance.

References

1. Gajdusek DC, Zigas V (1957) Degenerative disease of the central nervous system in New Guinea; the endemic occurrence of kuru in the native population. N Engl J Med 257(20):974–978
2. Hadlow WJ (1959) Scrapie and kuru. Lancet 2(7097):289–290
3. Gajdusek DC, Gibbs CJ Jr, Alpers M (1966) Experimental transmission of a kuru-like syndrome in chimpanzees. Nature 209:794–796
4. Beck E, Daniel PM, Alpers M, Gajdusek DC, Gibbs CJ Jr (1966) Experimental "kuru" in chimpanzees. A pathological report. Lancet 2(7472):1056–1059
5. Gajdusek DC, Gibbs CJ Jr, Asher DM, David E (1968) Transmission of experimental kuru to the spider monkey (Ateles geoffreyi). Science 162(3854):693–694
6. Gibbs CJ Jr, Gajdusek DC, Asher DM, Alpers MP, Beck E, Daniel PM, Matthews WB (1968) Creutzfeldt-Jakob disease (spongiform encephalopathy): transmission to the chimpanzee. Science 161(3839):388–389
7. Manuelidis EE, Kim J, Angelo JN, Manuelidis L (1976) Serial propagation of Creutzfeldt-Jakob disease in guinea pigs. Proc Natl Acad Sci U S A 73(1):223–227
8. Brown P, Gibbs CJ Jr, Rodgers-Johnson P, Asher DM, Sulima MP, Bacote A, Goldfarb LG, Gajdusek DC (1994) Human spongiform encephalopathy: the National Institutes of Health series of 300 cases of experimentally transmitted disease. Ann Neurol 35(5):513–529
9. Asher DM, Gibbs CJ Jr, Sulima MP, Bacote A, Amyx H, Gajdusek DC (1993) Transmission of human spongiform encephalopathies to experimental animals: comparison of the chimpanzee and squirrel monkey. Dev Biol Stand 80:9–13
10. Anonymous (1927) Obituary. Adrian Stokes. Br Med J 2(3482):615–618
11. Stokes A, Bauer J, Hudson N (1928) The transmission of yellow fever to Macacus rhesus. J Am Med Assoc 90(4):253–254
12. Stokes A, Bauer JH, Hudson NP (1928) Experimental transmission of yellow fever

to laboratory animals. Am J Trop Med 8(2):103–164

13. Howe HA, Bodian D (1942) Neural mechanisms in poliomyelitis. The Commonwealth Fund, New York
14. Race B, Meade-White KD, Miller MW, Barbian KD, Rubenstein R, LaFauci G, Cervenakova L, Favara C, Gardner D, Long D, Parnell M, Striebel J, Priola SA, Ward A, Williams ES, Race R, Chesebro B (2009) Susceptibilities of nonhuman primates to chronic wasting disease. Emerg Infect Dis 15(9):1366–1376
15. Telling GC, Scott M, Hsiao KK, Foster D, Yang SL, Torchia M, Sidle KC, Collinge J, DeArmond SJ, Prusiner SB (1994) Transmission of Creutzfeldt-Jakob disease from humans to transgenic mice expressing chimeric human-mouse prion protein. Proc Natl Acad Sci U S A 91(21):9936–9940
16. Botsios S, Manuelidis L (2016) CJD and scrapie require agent-sssociated nucleic acids for infection. J Cell Biochem 117(8):1947–1958. doi:10.1002/jcb.25495
17. Gibbs CJ Jr, Asher DM, Brown PW, Fradkin JE, Gajdusek DC (1993) Creutzfeldt-Jakob disease infectivity of growth hormone derived from human pituitary glands. N Engl J Med 328(5):358–359
18. Gibbs CJ Jr, Asher DM, Kobrine A, Amyx HL, Sulima MP, Gajdusek DC (1994) Transmission of Creutzfeldt-Jakob disease to a chimpanzee by electrodes contaminated during neurosurgery. J Neurol Neurosurg Psychiatry 57(6):757–758
19. Huang Y, Gregori L, Anderson SA, Asher DM, Yang H (2014) Development of dose-response models of Creutzfeldt-Jakob disease infection in nonhuman primates for assessing the risk of transfusion-transmitted variant Creutzfeldt-Jakob disease. J Virol 88(23):13732–13736
20. Piccardo P, Cervenakova L, Vasilyeva I, Yakovleva O, Bacik I, Cervenak J, McKenzie C, Kurillova L, Gregori L, Pomeroy K, Asher DM (2011) Candidate cell substrates, vaccine production, and transmissible spongiform encephalopathies. Emerg Infect Dis 17(12):2262–2269
21. Convention on International Trade in Endangered Species of Wild Fauna and Flora (CITES) (2015) The CITES Appendices. https://www.cites.org/eng/app/index.php. Accessed 17 Dec 2015
22. Gibbs CJ Jr, Amyx HL, Bacote A, Masters CL, Gajdusek DC (1980) Oral transmission of kuru, Creutzfeldt-Jakob disease, and scrapie to nonhuman primates. J Infect Dis 142(2):205–208
23. Comoy EE, Mikol J, Luccantoni-Freire S, Correia E, Lescoutra-Etchegaray N, Durand V, Dehen C, Andreoletti O, Casalone C, Richt JA, Greenlee JJ, Baron T, Benestad SL, Brown P, Deslys JP (2015) Transmission of scrapie prions to primate after an extended silent incubation period. Sci Rep 5:11573. doi:10.1038/srep11573
24. Marsh RF, Kincaid AE, Bessen RA, Bartz JC (2005) Interspecies transmission of chronic wasting disease prions to squirrel monkeys (Saimiri sciureus). J Virol 79(21):13794–13796
25. Belay ED, Maddox RA, Williams ES, Miller MW, Gambetti P, Schonberger LB (2004) Chronic wasting disease and potential transmission to humans. Emerg Infect Dis 10(6):977–984
26. Mawhinney S, Pape WJ, Forster JE, Anderson CA, Bosque P, Miller MW (2006) Human prion disease and relative risk associated with chronic wasting disease. Emerg Infect Dis 12(10):1527–1535
27. Anderson CA, Bosque P, Filley CM, Arciniegas DB, Kleinschmidt-Demasters BK, Pape WJ, Tyler KL (2007) Colorado surveillance program for chronic wasting disease transmission to humans: lessons from 2 highly suspicious but negative cases. Arch Neurol 64(3):439–441
28. World Health Organization (2006) WHO guidelines on tissue infectivity distribution in transmissible spongiform encephalopathies. In: Padilla A, Asher DM (eds). Geneva. http://www.who.int/bloodproducts/TSEREPORT-LoRes.pdf. Accessed 18 Apr 2016
29. World Health Organization (2010) WHO tables on tissue infectivity distribution in transmissible spongiform encephalopathies. Updated. In: Padilla A, Asher DM (eds). Geneva. http://www.who.int/bloodproducts/tablestissueinfectivity.pdf. Accessed 14 Apr 2016
30. Bueler H, Aguzzi A, Sailer A, Greiner RA, Autenried P, Aguet M, Weissmann C (1993) Mice devoid of PrP are resistant to scrapie. Cell 73(7):1339–1347
31. Owen F, Poulter M, Lofthouse R, Collinge J, Crow TJ, Risby D, Baker HF, Ridley RM, Hsiao K, Prusiner SB (1989) Insertion in prion protein gene in familial Creutzfeldt-Jakob disease. Lancet 1(8628):51–52
32. Prusiner SB (1982) Novel proteinaceous infectious particles cause scrapie. Science 216:136–144
33. Lasmezas CI, Comoy E, Hawkins S, Herzog C, Mouthon F, Konold T, Auvre F, Correia E, Lescoutra-Etchegaray N, Sales N, Wells G, Brown P, Deslys JP (2005) Risk of oral infec-

tion with bovine spongiform encephalopathy agent in primates. Lancet 365(9461):781–783
34. Piccardo P, Cervenak J, Bu M, Miller L, Asher DM (2014) Complex proteinopathy with accumulations of prion protein, hyperphosphorylated tau, alpha-synuclein and ubiquitin in experimental bovine spongiform encephalopathy of monkeys. J Gen Virol 95(Pt 7):1612–1618
35. Piccardo P, Cervenak J, Yakovleva O, Gregori L, Pomeroy K, Cook A, Muhammad FS, Seuberlich T, Cervenakova L, Asher DM (2012) Squirrel monkeys (Saimiri sciureus) infected with the agent of bovine spongiform encephalopathy develop tau pathology. J Comp Pathol 147(1):84–93
36. Wells GA, Hancock RD, Cooley WA, Richards MS, Higgins RJ, David GP (1989) Bovine spongiform encephalopathy: diagnostic significance of vacuolar changes in selected nuclei of the medulla oblongata. Vet Rec 125(21):521–524
37. Collinge J, Sidle KC, Meads J, Ironside J, Hill AF (1996) Molecular analysis of prion strain variation and the aetiology of 'new variant' CJD. Nature 383(6602):685–690
38. Piccardo P, Manson JC, King D, Ghetti B, Barron RM (2007) Accumulation of prion protein in the brain that is not associated with transmissible disease. Proc Natl Acad Sci U S A 104(11):4712–4717
39. Barron RM, Campbell SL, King D, Bellon A, Chapman KE, Williamson RA, Manson JC (2007) High titers of transmissible spongiform encephalopathy infectivity associated with extremely low levels of PrPSc in vivo. J Biol Chem 282(49):35878–35886
40. Manuelidis L (2013) Infectious particles, stress, and induced prion amyloids: a unifying perspective. Virulence 4(5):373–383
41. Balkema-Buschmann A, Eiden M, Hoffmann C, Kaatz M, Ziegler U, Keller M, Groschup MH (2011) BSE infectivity in the absence of detectable PrP(Sc) accumulation in the tongue and nasal mucosa of terminally diseased cattle. J Gen Virol 92(Pt 2):467–476
42. Wang XF, Dong CF, Zhang J, Wan YZ, Li F, Huang YX, Han L, Shan B, Gao C, Han J, Dong XP (2008) Human tau protein forms complex with PrP and some GSS- and fCJD-related PrP mutants possess stronger binding activities with tau in vitro. Mol Cell Biochem 310(1–2):49–55
43. Giasson BI, Forman MS, Higuchi M, Golbe LI, Graves CL, Kotzbauer PT, Trojanowski JQ, Lee VM (2003) Initiation and synergistic fibrillization of tau and alpha-synuclein. Science 300(5619):636–640
44. Croes EA, Roks G, Jansen GH, Nijssen PC, van Duijn CM (2002) Creutzfeldt-Jakob disease 38 years after diagnostic use of human growth hormone. J Neurol Neurosurg Psychiatry 72(6):792–793
45. Collinge J, Whitfield J, McKintosh E, Beck J, Mead S, Thomas DJ, Alpers MP (2006) Kuru in the 21st century—an acquired human prion disease with very long incubation periods. Lancet 367(9528):2068–2074
46. World Health Organization (2015) Blood products and related biologicals. International references. http://www.who.int/bloodproducts/ref_materials/en/. Accessed 18 Apr 2016
47. Minor P, Newham J, Jones N, Bergeron C, Gregori L, Asher D, van Engelenburg F, Stroebel T, Vey M, Barnard G, Head M, Working Group on International Reference Materials for the Diagnosis and Study of Transmissible Spongiform, Encephalopathies (2004) Standards for the assay of Creutzfeldt-Jakob disease specimens. J Gen Virol 85(Pt 6):1777–1784
48. Editorial team (2007) Fourth case of transfusion-associated vCJD infection in the United Kingdom. Euro Surveill 12:1–2
49. Llewelyn CA, Hewitt PE, Knight RS, Amar K, Cousens S, Mackenzie J, Will RG (2004) Possible transmission of variant Creutzfeldt-Jakob disease by blood transfusion. Lancet 363(9407):417–421
50. Peden AH, Head MW, Ritchie DL, Bell JE, Ironside JW (2004) Preclinical vCJD after blood transfusion in a PRNP codon 129 heterozygous patient. Lancet 364(9433):527–529
51. Peden A, McCardle L, Head MW, Love S, Ward HJ, Cousens SN, Keeling DM, Millar CM, Hill FG, Ironside JW (2010) Variant CJD infection in the spleen of a neurologically asymptomatic UK adult patient with haemophilia. Haemophilia 16(2):296–304
52. Crowder LA, Schonberger LB, Dodd RY, Steele WR (2017) Creutzfeldt-Jakob disease lookback study: 21 years of surveillance for transfusion transmission risk. Transfusion (April). do:i10.1111/trf.14145
53. Urwin P, Thanigaikumar K, Ironside JW, Molesworth A, Knight KS, Hewitt PE, Llewelyn C, Mackenzie J, Will RG (2017) Sporadic Creutzfeldt-Jakob disease in 2 plasma product recipients, United Kingdom. Emerg Infect Dis 23(6):893–897
54. United States Food and Drug Administration (2006) Potential screening assays to detect blood and plasma donors infected with agents

of transmissible spongiform encephalopathies (TSE agents or prions). Issue Summary; Transmissible Spongiform Encephalopathies Advisory Committee Meeting, September 19, 2006. http://www.fda.gov/ohrms/dockets/ac/06/briefing/2006-4240B1_2.pdf. Accessed 18 Apr 2016

55. Houston F, McCutcheon S, Goldmann W, Chong A, Foster J, Siso S, Gonzalez L, Jeffrey M, Hunter N (2008) Prion diseases are efficiently transmitted by blood transfusion in sheep. Blood 112(12):4739–4745

56. Gregori L, Yang H, Anderson S (2011) Estimation of variant Creutzfeldt-Jakob disease infectivity titers in human blood. Transfusion 51(12):2596–2602

57. McDowell KL, Nag N, Franco Z, Bu M, Piccardo P, Cervenak J, Deslys JP, Comoy E, Asher DM, Gregori L (2015) Blood reference materials from macaques infected with variant Creutzfeldt-Jakob disease agent. Transfusion 55(2):405–412

58. Saborio GP, Permanne B, Soto C (2001) Sensitive detection of pathological prion protein by cyclic amplification of protein misfolding. Nature 411(6839):810–813

59. Lescoutra-Etchegaray N, Jaffre N, Sumian C, Durand V, Correia E, Mikol J, Luccantoni-Freire S, Culeux A, Deslys JP, Comoy EE (2015) Evaluation of the protection of primates transfused with variant Creutzfeldt-Jakob disease-infected blood products filtered with prion removal devices: a 5-year update. Transfusion 55(6):1231–1241

60. Herzog C, Riviere J, Lescoutra-Etchegaray N, Charbonnier A, Leblanc V, Sales N, Deslys JP, Lasmezas CI (2005) PrP^{TSE} distribution in a primate model of variant, sporadic, and iatrogenic Creutzfeldt-Jakob disease. J Virol 79(22):14339–14345

Chapter 4

Clinical Features and Diagnosis of Human Prion Diseases

Richard Knight

Abstract

Human prion diseases are rare fatal neurodegenerative conditions that occur as sporadic, inherited, and acquired disorders. Despite clinico-pathological variations, there are common clinical features and characteristic neuropathological changes accompanied by the accumulation of a disease-associated form of the prion protein in the brain (and, in variant, CJD also in lymphoid tissues). While neuropathological examination of the brain is required for an absolutely definite diagnosis, probable or highly probable clinical diagnosis is possible in the majority of cases.

In this chapter, the clinical assessment of human prion disease is reviewed with emphasis on the role of clinical features and relevant clinical investigations in their diagnosis.

Key words Human prion disease, CJD, Clinical features, Clinical diagnosis, Diagnostic tests, MRI, CSF proteins: EEG, Genetic tests

1 Introduction

Prion diseases (or transmissible spongiform encephalopathies, TSEs) are a group of rare, animal and human, neurodegenerative diseases affecting the central nervous system (CNS) (listed elsewhere in this book). While varying in many characteristics, these diseases have common neuropathological changes and a common molecular underpinning involving a post-translational conformational change in a normal host-encoded protein: prion protein (described in the neuropathology chapter). This chapter is concerned with the human prion diseases and these are generally classified according to causation: sporadic (of unknown cause), iatrogenic, and acquired; there are further subdivisions based on aetiology and other factors (Table 1).

Sporadic Creutzfeldt-Jakob disease (sCJD) is the commonest form of human prion disease. It has been subdivided according to clinicopathological features, molecular characteristics, and genetic analysis. Iatrogenic CJD (iCJD) has resulted most commonly from the surgical use of cadaveric derived dura mater tissue and human pituitary hormones [2]. Kuru was an important human disease but limited to a specific

Pawel P. Liberski (ed.), *Prion Diseases*, Neuromethods, vol. 129,
DOI 10.1007/978-1-4939-7211-1_4, © Springer Science+Business Media LLC 2017

Table 1
Human prion diseases

Main classification		
Major type	**Subtypes**	**Cause**
Sporadic	Sporadic CJD	Unknown
	VPSPr	
Genetic	Genetic CJD	*PRNP* mutation
	GSS syndrome	*PRNP* mutation
	FFI	*PRNP* mutation
Iatrogenic	Cadaveric derived pituitary hormones	As indicated
	Cadaveric derived dura mater grafts	As indicated
	Neurosurgery/EEG depth electrodes	As indicated
	Corneal transplant	As indicated
Acquired	Kuru	Ritual mourning cannibalism
	Primary variant CJD	BSE dietary contamination
	Secondary variant CJD	Blood and blood products
Subclassification of sporadic CJD		
sCJD Subtype **129 genotype, Protein type** **C: Cortical; T: Thalamic**	**% Of cases (Approx)**	**Some general clinical features** **Note: There are also pathological features that characterize each subtype—not given here**
MM I	67	Cognitive, ataxic, visual, ~4-month duration, periodic EEG common
MV I	3	Broadly similar to MM I
VV I	3	No visual problems, ~15-month duration, no periodic EEG
MM II C	2	Progressive dementia. Duration ~16 months. No periodic EEG
MM II T	2	Usually insomnia, psychomotor hyperactivity. Duration ~15 months. No periodic EEG
MV II	9	Duration ~17 months. Usually no periodic EEG
VV II	14	Duration ~15 months. Usually no periodic EEG

Data as reported in Parchi et al. 1999 [1]

geographical area in Papua New Guinea and related to a particular transmission via ritualistic mourning cannibalism [3]. Variant CJD (vCJD) arose as a zoonosis via the contamination of human food with material from BSE (bovine spongiform encephalopathy) cattle and has been transmitted secondarily (human to human) via blood and blood products [4–6]. Genetic prion disease (gPD) arises from pathogenic mutations of *PRNP*, the human prion protein gene [7]. The terminological archaeology of these diseases is still visible in the modern classification system. The original terms were essentially eponymous (Creutzfeldt-Jakob disease; Gerstmann-Sträussler-Scheinker syndrome), then descriptive (fatal familial insomnia, transmissible spongiform encephalopathies), and then molecular (prion disease) with aetiological qualifiers (sporadic; acquired; genetic). In recent years, other prion diseases have been described, such as variably protease-sensitive prionopathy (VPSPr), that have uncertain relationships to the existing categories, but which may essentially be part of the sCJD spectrum [8, 9]. Variant CJD has become, thankfully, a very rare disease indeed with, at December 2015, no known living cases in the world and only one new case of vCJD identified within the UK during the preceding 4 years (2011–2015) [10].

However, a recent study examining large numbers of archived appendix samples removed between the years of 2000 and 2012 estimated the prevalence of subclinical vCJD infection at 1 in 2000 of the UK population [11]. In addition, there are uncertainties over the range of incubation periods in human dietary-related vCJD and the effect of the *PRNP* codon-129 polymorphism on this. Further clinical cases of vCJD are, therefore, expected, although estimates suggest a relatively low number [12].

While absolutely definitive diagnosis requires neuropathological examination of the brain (usually at autopsy but occasionally via brain biopsy), a probable or even highly probable clinical diagnosis is possible in the majority of cases.

The clinical diagnostic approach is reviewed below.

2 Methods

The diagnosis of prion disease uses the standard methods of neurological clinical practice. There are validated diagnostic criteria that are widely used by many disease surveillance systems throughout the world (Table 2).

Clinical diagnosis rests on three broad steps:

1. Recognition of the possibility of a prion disease.
2. Exclusion of other possible diagnoses.
3. Diagnostic test results supportive of the diagnosis

Table 2
Human prion diseases: diagnosis criteria

a SPORADIC CJD

DEFINITE:

Neuropathologically/ immunocytochemically confirmed

PROBABLE:

1.2.1 I + 2 of II + III

OR

1.2.2 I + 2 of II + IV

OR

1.2.3 Possible + positive 14-3-3

POSSIBLE:

I + 2 of II + duration < 2 years

I Rapidly progressive dementia

II A Myoclonus

B Visual or cerebellar problems

C Pyramidal or extrapyramidal features

D Akinetic mutism

III Typical EEG

IV High signal in caudate/putamen on MRI brain scan

b IATROGENIC CJD

DEFINITE

Definite CJD with a recognised iatrogenic risk factor

PROBABLE

Progressive predominant cerebellar syndrome in human pituitary hormone recipients

Probable CJD with recognised iatrogenic risk factor (see box)

POSSIBLE

Possible CJD with a recognised risk factor

RELEVANT EXPOSURE RISKS FOR THE CLASSIFICATION AS IATROGENIC CJD

The relevance of any exposure to disease causation must take into account the timing of the exposure in relation to disease onset

- Treatment with human pituitary growth hormone, human pituitary gonadotrophin or human dura mater graft.
- Corneal graft in which the corneal donor has been classified as definite or probable human prion disease.
- Exposure to neurosurgical instruments previously used in a case of definite or probable human prion disease.

This list is provisional as previously unrecognised mechanisms of human prion disease may occur

(continued)

Table 2 (continued)

c GENETIC PRION DISEASE

DEFINITE

Definite TSE + definite or probable TSE in 1st degree relative

Definite TSE with a recognised pathogenic *PRNP* mutation

PROBABLE

Progressive neuropsychiatric disorder + definite or probable TSE in 1st degree relative

Progressive neuropsychiatric disorder + recognisedv pathogenic *PRNP* mutation

d vCJD

DEFINITE

IA **and** neuropathological confirmation of vCJD

PROBABLE

4.2.1 I **and** 4/5 of II **and** IIIA **and** IIIB

4.2.2 I **and** IV A

POSSIBLE

I **and** 4/5 of II **and** III A

I A Progressive neuropsychiatric disorder
B Duration of illness > 6 months
C Routine invesitgations do not suggest an alternative diagnosis
D No history of potential iatrogenic exposure
E No evidence of a familial form of TSE

II A Early psychiatric symptoms[a]
B Persistent painful sensory symptoms[b]
C Ataxia
D Myoclonus or chorea or dystonia
E Dementia

III A EEG does not show the typical appearance of sporadic CJD[c] in the early stages of illness
B Bilateral pulvinar high signal on MRI scan

IV A Positive tonsil biopsy

2.1 Recognition of the Possibility of a Prion Disease

Recognition is based on two considerations: any knowledge of a relevant particular risk factor for prion disease and the clinical features of the case.

2.1.1 Risk Factors

The important factors that indicate a particular risk of developing prion disease are the following:

1. A family history of prion disease (gPD): Most specifically, the known possession of a recognized *PRNP* pathogenic mutation following preclinical testing of an individual.

2. A past exposure to a recognized risk factor (such as a cadaveric derived human dura mater graft or cadaveric derived human pituitary hormone).
3. Receipt of a blood transfusion or blood product from a vCJD blood donor.

There may be other particular risk factors (such as a known accidental exposure to potentially contaminated surgical instruments) but these are rare [2]. Some known risk factors such as age (for sCJD), living in the UK (for vCJD), or *PRNP* codon 129 genotype are too general to be of particular relevance in suspecting a prion disease in general (though they may be helpful in considering the likelihood of a particular diagnosis).

2.1.2 Clinical Features

Although there is significant clinical heterogeneity in human prion disease, there are some core features. Firstly, these diseases are primarily brain diseases that present with cerebral symptoms and signs with a progressive (indeed ultimately fatal) course. Secondly, cognitive dysfunction, cerebellar ataxia, and involuntary movements (especially myoclonus) are common features [13].

In addition, in prion diseases, there are generally no systemic abnormalities (such as pyrexia) unless these relate to secondary complications (such as pneumonia).

It is important, however, to recognize that the clinical presentation of prion diseases is generally nonspecific and, in the initial stages, the differential diagnosis may well be wide. The particular presentation and overall clinical course vary according to the particular prion disease; a general summary is given in Table 3.

The clinical features of vCJD are relatively uniform as are those of hGH-related iCJD; those of sCJD and gPD are potentially quite varied [7, 14–16]. It is important to note that the clinical presentation of some gPD can be very similar to, or even indistinguishable from, that of sCJD and a family history may be absent in up to 40% of gPD cases [7].

A detailed review of the clinical features of gPD is beyond the scope of this chapter but is available in published studies [7].

A significant majority of cases of sCJD present as a rapidly progressive, multifocal, encephalopathy, with dementia, cerebellar ataxia, and myoclonus as prominent features [16]. This typical, rapidly progressive, form generally results in an akinetic mute state and death, after an illness duration of usually only a few weeks to a few months. A significant minority of individuals present with an isolated progressive impairment (typically visual or cerebellar) before progressing on to a more general encephalopathic illness [17, 18]. There are other variations and the clinic-pathological heterogeneity of sCJD has been studied in some detail with attempts made to subclassify the disease in terms of correlations between the clinic-pathological features, the prion protein type, and the *PRNP*-codon 129

Table 3
Human prion diseases: presenting & overall clinical features

Disease	Typical presentation	General features
Sporadic CJD	Rapidly progressive encephalopathy (cognitive and cerebellar features)	Dementia, ataxia, myoclonus, terminal akinetic mutism
Variant CJD	Psychiatric/behavioral features +/− unpleasant/painful sensory symptoms	Psychiatric features, dementia, sensory symptoms, involuntary movements (chorea, dystonia, myoclonus)
Iatrogenic human growth hormone CJD	Typically cerebellar ataxia. Tremor may be present	Progressive cerebellar syndrome, tremor, myoclonus, pyramidal signs. Cognitive impairment usually in later stages
Other iatrogenic CJD	Essentially as sCJD	Essentially as sCJD
Genetic prion disease	Variable, dependent at least in part on relevant *PRNP* mutation. May present in similar manner to sCJD	Dementia and ataxia common features. Variable, dependent at least in part on relevant *PRNP* mutation

genotype (Table 1) [1]. These subclassifications have utility in considering the diagnosis in particular cases, but the distinctions are not always clear-cut and there are individual cases that do not altogether fit these defined categories. The clinico-pathological-molecular subclassification of sCJD, with its partial dependence on pathological prion protein type, is further complicated by the fact that more than one prion protein type can be found in one brain. An attempt to include this molecular phenomenon has led to a more complex classification system that is of uncertain practical clinical diagnostic utility [19]. It is uncertain as to whether diseases such as VPSPr are distinct entities or essentially part of the sporadic CJD spectrum. There are relatively few data on the clinical characteristics of VPSPr and, as data have accumulated, the phenotypic spectrum has broadened in a way that makes a clear-cut distinction from sCJD difficult [8, 9].

Variant CJD tends to present with psychiatric or behavioral disturbances and the emergence of more obviously neurological features often delayed for a few months (a median of 6 months) [20]. The presentation may, therefore, be very nonspecific; a presentation of an agitated depression in a young person is clearly not an uncommon clinical event and only exceptionally rarely due to vCJD.

2.2 Exclusion of Other Possible Diagnoses

As mentioned above, the presentation of prion disease is often fairly nonspecific. For example, a combination of ataxia and cognitive impairment in a 55-year-old or a depressive illness in a 25-year-old has much likelier causes than a prion disease. In addition, as prion diseases are currently always fatal without disease-modifying treatments, it is vitally important to consider and exclude other possible diagnoses, especially potentially treatable ones. The general differential diagnosis of prion disease is outside the scope of this chapter; however the broad principles and methodology can be stated.

1. Various inflammatory and infectious diseases may present in a similar way. These are generally excluded by the absence of any inflammatory markers or specific infection features (pyrexia, raised peripheral white blood cell count, raised ESR or CRP, CSF pleocytosis, certain MRI brain abnormalities, etc). Specific infections may need consideration such as HIV. Immune encephalopathies (nonmetastatic malignancy based or idiopathic) also need particular consideration as does cerebral vasculitis.
2. Structural disease (including cerebral malignancy—primary or secondary) needs consideration via cerebral imaging.
3. A number of other neurodegenerative diseases can present in similar ways to prion diseases and one should always consider these, including considering genetic testing for appropriate possible genetic diseases.

In general, therefore, most cases of suspect prion disease will require fairly comprehensive investigation including the following:

1. Cerebral imaging: Typically cerebral CT initially and then cerebral MRI.
2. A wide-ranging set of blood tests looking for possible inflammatory, infective, metabolic, and other causes.
3. Consideration of tests for non-CNS primary tumors (related to secondary CNS tumors or nonmetastatic encephalopathy).
4. A lumbar puncture to obtain CSF to look for evidence of inflammatory causes

2.3 Diagnostic Test Results Supportive of the Diagnosis

There are several investigations that my provide support for a diagnosis of a prion disease: the cerebral MRI, the EEG, CSF nonspecific protein testing, CSF RT-QuIC testing, direct detection assay on blood, tonsil biopsy, and *PRNP* genetic testing [21–27]. Recent reports suggest a role for other tests: urine testing and nasal brushing [28, 29]. The detailed methodologies of these tests are given in other chapters.

Brain biopsy is the most definitive clinical test in life for prion disease but carries obvious risk and, if confirmatory for prion disease, does not lead to any specific therapy. It is, therefore, arguable that it should be limited to those cases in which there is important diagnostic uncertainty and where there is a real possibility of an alternative, treatable, diagnosis thereby being made.

2.3.1 The Cerebral MRI

The cerebral MRI has two important roles: the exclusion of other possible diagnoses and providing support for a diagnosis of prion disease; here, the latter is being considered. The basic points about the cerebral MRI in prion disease are as follows:

1. In general, the relevant MR sequences used can be listed in the following descending order in relation to sensitivity for prion disease features: DWI, FLAIR, PD, and T2 [21, 30]. If the most sensitive sequences have not been performed on an initial scan that did not show any relevant abnormality, a repeat scan should be performed with additional sequences included.
2. If the relevant changes are not seen on the initial scan, it is worth repeating the MRI as changes may develop with illness progression. It is difficult to give precise guidance on the intervals between scans as most scans have been undertaken in the course of clinical investigation, rather than in planned studies. However, the tempo of the illness should be taken into account; a repeat scan after a week or two might be reasonable in rapidly progressing illnesses.
3. The relevant MR abnormalities are the following:
   (a) Signal change in the putamen/caudate region, Fig. 1a.
   (b) Signal change in the thalamic region, Fig. 1b.
   (c) Signal change in the cerebral cortex, Fig. 1c.
4. None of the MR abnormalities seen in prion diseases are absolutely specific: they may be seen in other diseases and their

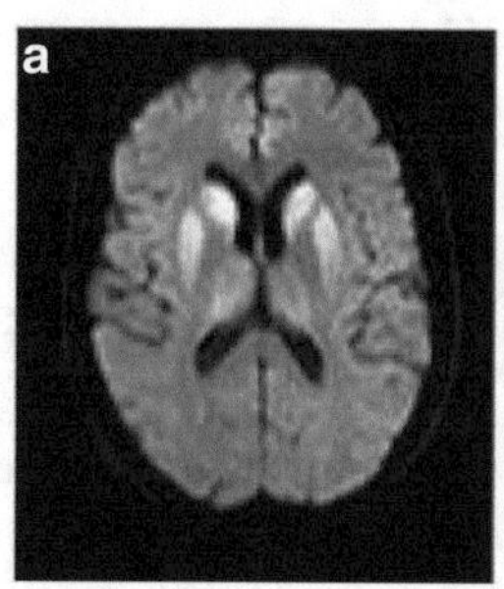

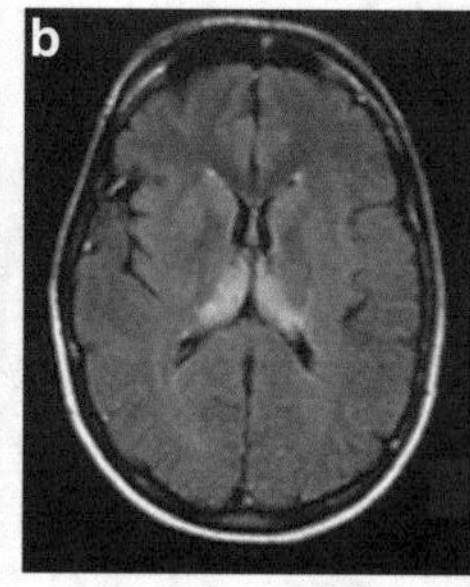

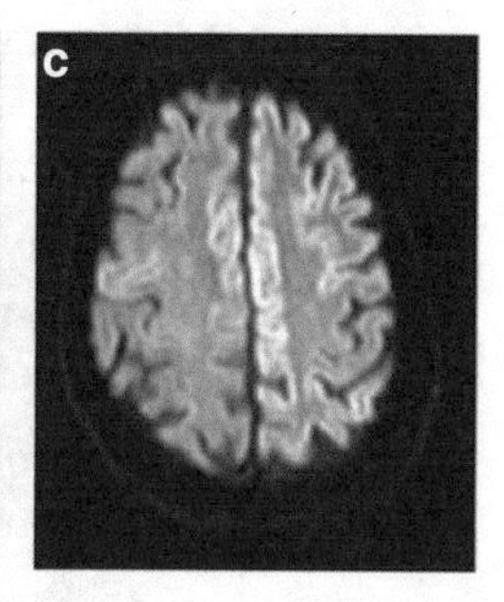

Fig. 1 MR images in human prion disease. (**a**) MR image (DWI) showing putamen/caudate hyperintensity in sCJD. (**b**) MR image (FLAIR) showing posterior thalamic hyperintensity in vCJD (the "pulvinar sign"). (**c**) MR image (DWI) showing areas of cortical hyperintensity in sCJD. *MR image courtesy of Dr. David Summers, Neuroradiology, Edinburgh*

significance always needs to be determined within the overall clinical context.

5. The relevant MR abnormalities vary somewhat with different diseases and disease subtypes [30].

This last point needs amplification. Firstly, there are key MRI differences between sCJD and vCJD. Secondly, the frequency and pattern of abnormality have been studied in the different subtypes of sCJD.

1. In vCJD, the characteristic abnormality is the "pulvinar sign" consisting of signal change in the pulvinar area of the thalamus. This area can show signal change in other forms of CJD such as sCJD, but the pulvinar sign is defined by the relative magnitude of signal change: in vCJD, it is greater in the pulvinar region than elsewhere [22].
2. In sporadic CJD, the frequency of abnormality and its distribution pattern do vary across the described clinico-pathological-molecular subtypes, as has been described in detail by Meissner et al. [30]. For example, in MMI sCJD, basal ganglia signal hyperintensity is frequently seen with widespread cerebral cortex involvement in around half of cases but with no hippocampal and thalamic hyperintensity. In contradistinction, VVI cases typically show cerebral cortex signal changes but without basal ganglia or thalamic hyperintensity.

2.3.2 The EEG

The EEG played an important role in the diagnosis of CJD before the use of the cerebral MRI and CSF protein tests. However, it can still be helpful despite the widespread introduction of these other tests. The four key points are the following:

1. In many, but not all, cases of sCJD, the EEG shows the loss of normal background rhythms and the appearance of generalized, periodic, bi-, or triphasic complexes, Fig. 2 [31].
2. These may appear early in the disease course. If they are not present initially, and the EEG is playing an important role in the diagnostic process, then it is worth repeating the test at around weekly intervals.
3. The typical sCJD EEG appearance is, however, not specific: it can be seen in a variety of clinical circumstances and its finding needs to be considered in the overall clinical context.
4. The typical periodic pattern is generally not seen in vCJD although it has been reported in two cases at the terminal stages of disease [32, 33].

2.3.3 CSF Nonspecific Protein Tests

The two proteins that have been most widely used in the diagnosis of CJD are 14-3-3 and S100b [24]. Tau has also been used [24, 34].

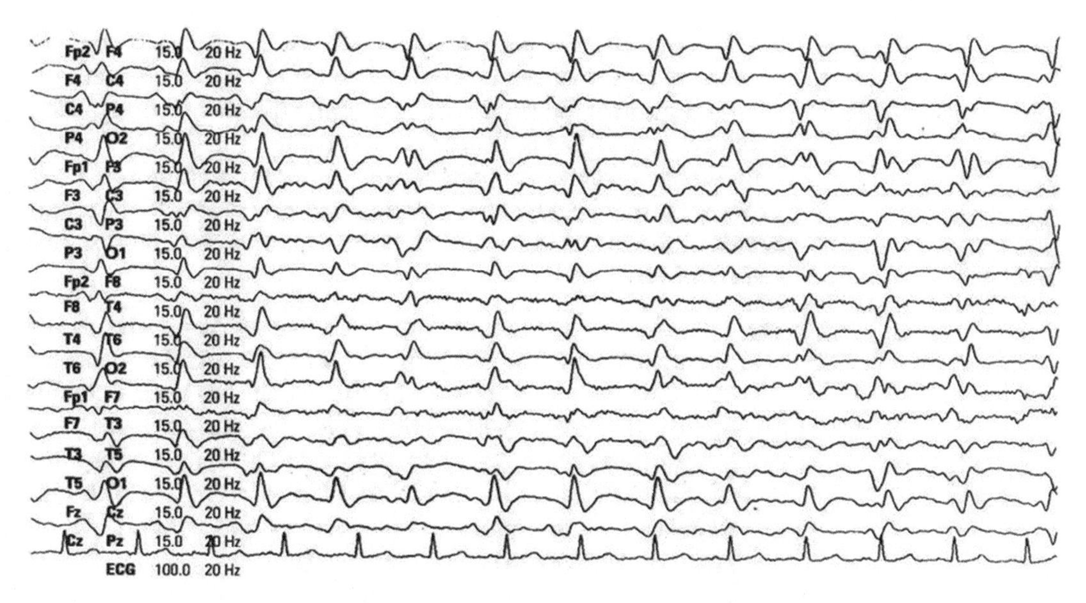

Fig. 2 Characteristic periodic EEG discharges seen in sCJD

2.3.4 CSF RT-QuIC Test

The CSF RT-QuIC test is of particular significance since it is based on the detection of abnormal prion protein and, therefore, related to the fundamental disease process, rather than being incidentally useful (as is so for 14-3-3 and S100b). Interestingly, in its current form, it is a highly sensitive test for sCJD but not for vCJD. The reported sensitivity and specificity for sCJD are 87% and 100%, respectively [35].

2.3.5 DDA Blood Test

This is another test based on the fundamental disease process, with the aim of detecting abnormal prion protein in the blood [26]. The principal use of this test is in the diagnosis of vCJD. Its reported sensitivity and specificity for a diagnosis of vCJD are 71% and 100%, respectively.

2.3.6 Tonsil Biopsy

Since, in vCJD, abnormal prion protein is found in lymphoid tissue as well as in the CNS, tonsil biopsy has been used as a method of confirming a diagnosis of vCJD [27]. Clearly, this is a method of detecting the disease-related prion protein in tissue and so is the most specific clinical test other than brain biopsy. However, it is of utility in only vCJD and is a procedure that has potential morbidity. Arguably, its principal use is in confirming the diagnosis of vCJD in cases where the clinical picture is not entirely typical and/ or other tests, such as the cerebral MRI, have not helped.

2.3.7 PRNP Genetic Testing

PRNP genetic analysis has two roles: the detection of pathogenic mutations and the characterization of potentially important polymorphisms.

The detection of pathogenic mutations is a relatively straightforward process that can be undertaken using DNA extracted from any relevant material (in the clinical context, usually blood). It is a definitive test in the sense that it is capable of detecting relevant mutations; however, the presence of a mutation does not, in itself, prove that the individual has a prion disease: it proves that they are at risk of the disease and clinical evaluation of their clinical state is still required.

1. The characterization of any *PRNP* polymorphisms is not directly diagnostically helpful, but it can be useful in considering which type of prion disease the patient has or in weighing up the diagnostic possibilities when the clinical features or other investigation results are atypical [1]. The polymorphism of real significance is the 129 M/V polymorphism (although an E219K polymorphism is of some importance for CJD in the Japanese population) [35].

2.3.8 Urine Testing

A recent publication has reported positive results in testing urine from patients with vCJD but not sCJD (based on the protein misfolding cyclic amplification, PMCA, technique) [28].

2.3.9 Nasal Brushing

Olfactory neurons are accessible through the nose and nasal brushing can obtain neurons that are then analyzable for the presence of abnormal prion protein. This has, therefore, been proposed as a test for prion disease and one that directly confirms the disease process, with disease-related prion protein being detected in neuronal tissue [29].

3 Notes

1. While the clinical features of each type of human prion disease are relatively uniform in most cases, there is significant variation in a minority of cases, especially in sCJD and genetic prion disease. In unusual, progressive neuropsychiatric illness, prion disease should be considered at some point.
2. In genetic prion disease, there may be no family history and a variable clinical illness: *PRNP* genetic testing should always be considered in progressive neuropsychiatric illness with no alternative diagnosis and in apparently sporadic CJD.
3. Clinical diagnostic tests in human prion disease can be considered in two groups: (a) tests which are essentially nonspecific and which need interpretation within the clinical context (cerebral MRI, EEG, CSF 14-3-3) and (b) tests whose abnormalities relate to the fundamental prion pathology and thus should be very highly specific (tonsil biopsy, direct detection assay blood test, nasal brushing, PMCA urine testing, CSF RT-QuIC).

References

1. Parchi P, Giese A, Capellari S et al (1999) Classification of sporadic Creutzfeldt–Jakob disease based on molecular and phenotypic analysis of 300 subjects. Ann Neurol 46(2):224–233
2. Brown P, Brandel J-P, Sato T et al (2012) Iatrogenic Creutzfeldt-Jakob disease, final assessment. Emerg Infect Dis 18(6):901–907
3. Liberski PP (2013) Kuru: a journey back in time from Papua New Guinea to the Neanderthals' extinction. Pathogens 2(3):472–505
4. Bruce ME, Will RG, Ironside JW, McConnell I, Drummond D, Suttie A, McCardle L, Chree A, Hope J, Birkett C et al (1997) Transmissions to mice indicate that 'new variant' CJD is caused by the BSE agent. Nature 389(6650):498–501
5. Hewitt PE, Llewelyn CA, Mackenzie J et al (2006) Creutzfeldt-Jakob disease and blood transfusion: results of the UK Transfusion Medicine Epidemiology Review study. Vox Sang 91(3):221–230
6. Peden A, McCardle L, Head MW, Love S, Ward HJT, Cousens SN, Keeling DM, Millar CM, Hill FGH, Ironside JW (2010) Variant CJD infection in the spleen of a neurologically asymptomatic UK adult patient with haemophilia. Haemophilia 16(2):296–304
7. Kovács G, Puopolo M, Ladogana A, Pocchiari M, Budka H et al (2005) Genetic prion disease: the EUROCJD experience. Hum Genet 118(2):166–174
8. Gambetti P, Puoti G, Zou WG (2011) Variably protease-sensitive prionopathy: a novel disease of the prion protein. J Mol Neurosci 45(3):422–424
9. Head MW, Yull HM, Ritchie DL, Ironside JW et al (2013) Variably protease-sensitive prionopathy in the UK: a retrospective review 1991–2008. Brain 136(4):1102–1115
10. www.cjd.ed.ac.uk
11. Gill ON, Spencer Y, Richard-Loendt A et al (2013) Prevalent abnormal prion protein in human appendixes after bovine spongiform encephalopathy epizootic: large scale survey. BMJ 347:f5675
12. Garske T, Ghani AC (2010) Uncertainty in the tail of the variant Creutzfeldt-Jakob disease epidemic in the UK. PLoS One 5(12):e15626
13. Knight R (2008) Clinical features and diagnosis of human prion diseases. Future Neurol 3(4):473–481
14. Heath CA, Cooper SA, Murray K, Knight RSG, Will RG et al (2010) Validation of diagnostic criteria for variant CJD. Ann Neurol 67(6):761–770
15. Rudge P et al (2015) Iatrogenic CJD due to pituitary-derived growth hormone with genetically determined incubation times of up to 40 years. Brain 138(11):3386–3399
16. Will RG, Alpers MP, Dormont D, Schonberger LB (2004) In: Prusiner SB (ed) Infectious and sporadic prion diseases. Prion biology and diseases. Cold Spring Harbor Laboratory Press, New York, pp 629–671
17. Cooper SA, Murray KL, Heath CA, Will RG, Knight RSG (2006) Sporadic Creutzfeldt–Jakob disease with cerebellar ataxia at onset in the UK. J Neurol Neurosurg Psychiatry 77(11):1273–1275. doi:10.1136/jnnp.2006.088930
18. Cooper SA, Murray KL, Heath CA, Will RG, Knight RSG (2005) Isolated visual symptoms at onset in sporadic Creutzfeldt-Jakob disease: the clinical phenotype of the "Heidenhain variant". Br J Ophthalmol 89(10):1341–1342. doi:10.1136/bjo.2005.074856
19. Parchi P, Strammiello R, Notari S et al (2009) Incidence and spectrum of sporadic Creutzfeldt–Jakob disease variants with mixed phenotype and co-occurrence of PrPSc types: an updated classification. Acta Neuropathol 118(5):659–671. doi:10.1007/s00401-009-0585-1
20. Spencer MD, Knight RSG, Will RG (2002) First hundred cases of variant Creutzfeldt-Jakob disease: retrospective case note review of early psychiatric and neurological features. BMJ 324(7352):1479–1482
21. Collie DA, Sellar RJ, Zeidler M, Colchester A, Knight RSG, Will RG (2001) MRI of Creutzfeldt±Jakob disease: imaging features and recommended MRI protocol. Clin Radiol 56(9):726–739
22. Collie DA, Summers DM, Sellar RJ et al (2003) Diagnosing variant Creutzfeldt-Jakob disease with the pulvinar sign: MR imaging findings in 86 neuropathlogically confirmed cases. Am J Neuroradiol 24(8):1560–1569
23. Zerr I, Pocchiari M, Collins S, Brandel J-P, de Pedro Cuesta J, Knight RS et al (2000) Analysis of EEG and CSF 14-3-3 proteins as aids to the diagnosis of Creutzfeldt-Jakob disease. Neurology 55(6):811–815
24. G Chohan G, Pennington C, Mackenzie JM, Andrews M, Everington D, Will RG, Knight RSG, Green AJE (2010) The role of cerebrospinal fluid 14-3-3 and other proteins in the diagnosis of sporadic Creutzfeldt-Jakob disease in the UK: a 10-year review. J Neurol Neurosurg Psychiatry 81(11):1243–1248
25. McGuire L, Peden AH, Orru CD (2012) Real time quaking-induced conversion analysis of

cerebrospinal fluid in sporadic Creutzfeldt–Jakob disease. Ann Neurol 72(2):278–285
26. Edgeworth JA, Farmer M, Sicilia A (2011) Detection of prion infection in variant Creutzfeldt-Jakob disease: a blood-based assay. Lancet 377(9764):487–493
27. Hill AF, Zeidler M, Ironside J, Collinge J (1997) Diagnosis of new variant Creutzfeldt-Jakob disease by tonsil biopsy. Lancet 349(9045):99–100
28. Moda F, Gambetti P, Notari S, Concha-Marambio L, Catania M, Park KW, Maderna E, Suardi S, Haïk S, Brandel JP, Ironside J, Knight R, Tagliavini F, Soto C (2014) Prions in the urine of patients with variant Creutzfeldt-Jakob disease. N Engl J Med 371(6):530–539
29. Orru C et al (2014) A test for Creutzfeldt-Jakob disease using nasal brushings. New Engl J Med 371(6):519–529. doi:10.1056/NEJMoa1315200
30. Meissner B, Kallenberg K, Sanchez-Juan P, Collie D, Summers DM et al (2009) MRI lesion profiles in sporadic Creutzfeldt_Jakob disease. Neurology 72(23):1994–2001
31. Steinhoff BJ, Racker S, Herrendorf G, Poser S, Grosche S et al (1996) Accuracy and reliability of periodic sharp wave complexes in Creutzfeldt-Jakob disease. Arch Neurol 53(2):162–166
32. Yamada M, Variant CJD Working Group (2006) The first Japanese case of variant Creutzfeldt-Jakob disease showing periodic electroencephalogram. Lancet 367 (9513):874
33. Binelli S, Agazzi P, Giaccone G, Will RG, Bugiani O, Franceschetti S, Tagliavin F (2006) Periodic electroencephalogram complexes in a patient with variant Creutzfeldt-Jakob disease. Ann Neurol 59(2):423–427
34. Sanchez-Juan P, Green A, Ladogana A et al (2006) CSF tests in the differential diagnosis of Creutzfeldt-Jakob disease. Neurology 67(4):637–643
35. Shibuya S, Higuchi J, Shin RW, Tateishi J, Kitamoto T (1998) Codon 219 Lys allele of *PRNP* is not found in sporadic Creutzfeldt-Jakob disease. Ann Neurol 43(6):826–828

Chapter 5

Neuropathology, Immunohistochemistry, and Biochemistry in Human Prion Diseases

Diane L. Ritchie and James W. Ironside

Abstract

Human prion diseases are rare fatal neurodegenerative conditions that occur as sporadic, inherited, and acquired disorders. All are defined by characteristic neuropathological changes and by the accumulation of a disease-associated form of the prion protein in the brain. In variant CJD, prion protein also accumulates in lymphoid tissues and in the peripheral nervous system. Examination of the brain in human prion diseases is essential for a definite diagnosis and for disease surveillance.

In this chapter we bring over 20 years' experience of research and diagnosis in human prion diseases to provide detailed information on how to handle brain tissue from a case of human prion disease safely in a laboratory. We also describe how to make the diagnosis of a human prion disease using histological and biochemical techniques for the detection of disease-associated prion protein in the brain, and provide helpful notes for practical guidance and troubleshooting in the laboratory.

Key words Neuropathology, Brain, CJD, prion protein, Immunohistochemistry, Western blot, PET blot

1 Introduction

Prion diseases, formerly referred to as transmissible spongiform encephalopathies (TSEs), are a group of rare and inevitably fatal degenerative diseases of the central nervous system (CNS) that occur in humans as well as a number of animal species (Table 1). These diseases are characterized by their neuropathological changes, comprising varying degrees of spongiform change (vacuoles that occur in the grey matter of the brain), neuronal loss, reactive proliferation of astrocytes and microglia, and, in certain forms of the disease, formation of amyloid plaques (Fig. 1a–d). The central feature of prion diseases involves the conformational change of a normal host-encoded cellular protein, the prion protein (PrP^{C}), into a misfolded, highly infectious and disease-specific form, termed PrP^{Sc} [1]. The accumulation and detection of PrP^{Sc} within CNS tissues has become an important diagnostic marker for prion diseases (Fig. 1e, f).

Pawel P. Liberski (ed.), *Prion Diseases*, Neuromethods, vol. 129,
DOI 10.1007/978-1-4939-7211-1_5, © Springer Science+Business Media LLC 2017

Table 1
Prion diseases in animals and humans

Prion disease	Natural host	Etiology
• Sporadic Creutzfeldt-Jakob disease (sCJD)	Human	Idiopathic
• Sporadic fatal insomnia	Human	
• Variably protease-sensitive prionopathy (VPSPr)	Human	
• Familial CJD	Human	Familial
• Gerstmann-Sträussler-Scheinker	Human	
• Fatal familial insomnia	Human	
• Kuru	Human	Acquired
• Iatrogenic CJD (iCJD)	Human	
• Variant CJD (vCJD)	Human	
• Bovine spongiform encephalopathy (BSE)	Cattle	
• Feline spongiform encephalopathy	Cats (domestic and large)	
• Exotic ungulate encephalopathy	Nyala, Kudu etc., cats	
• Scrapie	Sheep and goats	Probably acquired
Classical		
Atypical or Nor98 scrapie		
• Atypical BSE (H or L type)	Cattle	
• Transmissible mink encephalopathy	Mink	
• Chronic wasting disease	Mule deer, white-tailed deer, and elk	

Sporadic Creutzfeldt-Jakob disease (CJD) is the commonest form of human prion disease, but rarer inherited forms and acquired human prion diseases also occur [2, 3]. The infectious nature of prion diseases distinguishes this group of disorders from other more common protein misfolding neurodegenerative diseases, such as Alzheimer's disease and Parkinson's disease. Although rare, the transmissibility and in some forms the zoonotic potential of prion diseases present a recognizable and continued risk to public health, since no prophylaxis or treatment is available to date. This is best demonstrated with the identification of variant Creutzfeldt-Jakob disease (vCJD) resulting from human exposure to the bovine spongiform encephalopathy (BSE) agent [4] and the subsequent secondary human-to-human transmission of vCJD infectivity via blood transfusion [5]. Although vCJD appears to be in decline in the UK, with only one new case of vCJD identified

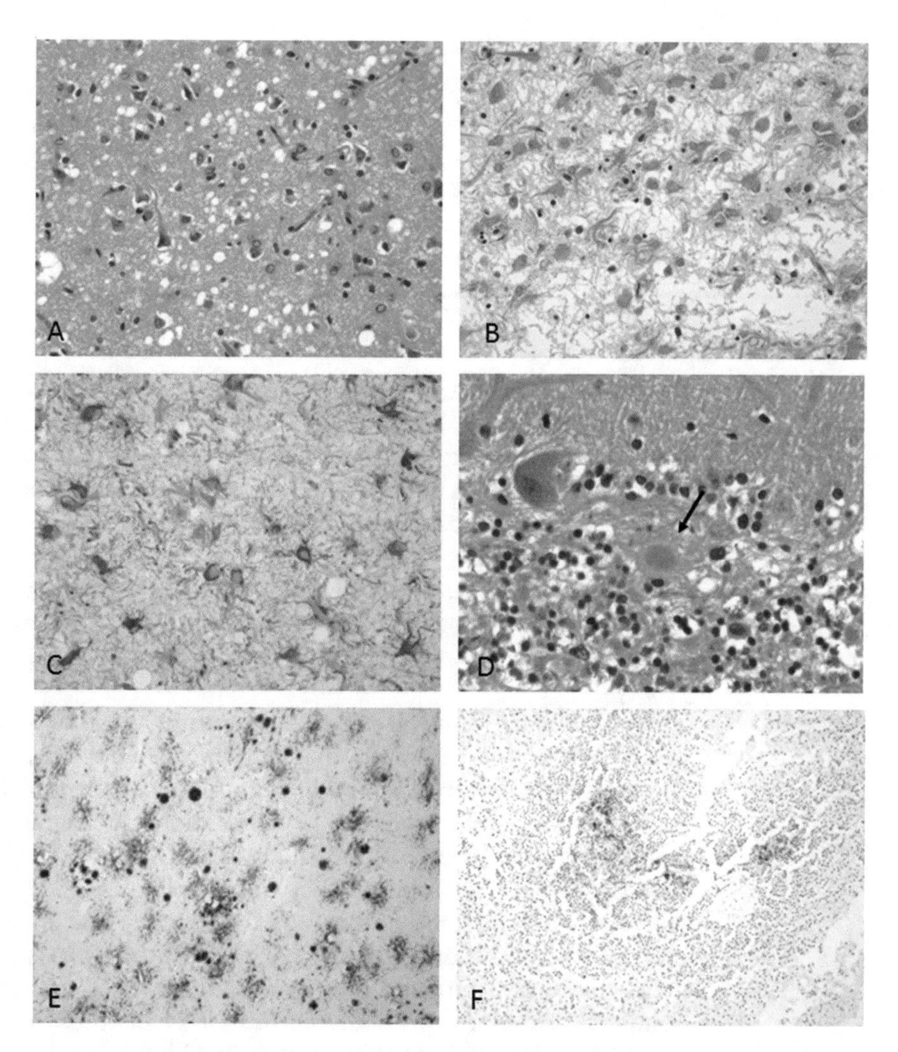

Fig. 1 Pathological changes in human prion diseases. (**a**) microvacuolar degeneration in the frontal cortex of a sporadic CJD case (H&E stain). (**b**) Status spongiosis accompanied by extensive neuronal loss and widespread gliosis in a case of panencephalopathic CJD (H&E stain). (**c**) Immunohistochemistry demonstrating intense gliosis within the basal ganglia in a vCJD patient GFAP antibody. (**d**) Kuru-type amyloid plaques in the cerebellum in a sporadic MV2 case. Immunohistochemistry for the prion protein in the (**e**) brain and (**f**) spleen of a vCJD patient. 12F10 anti-PrP antibody

within the UK during the past 4 years [6] a recent study examining large numbers of archived appendix samples removed between the years of 2000 and 2012 estimated the prevalence of subclinical vCJD infection at 1 in 2000 of the UK population [7].

With such estimates of subclinical vCJD infection in the UK population and the continued identification of new forms of human prion disease, further developments in the specificity and sensitivity of techniques in the identification and diagnosis of human prion diseases remain of paramount importance, in order to limit further accidental exposure to the infectious agent. A range of investigatory techniques are available for the confirmation of a diagnosis of human prion disease, the vast majority of which rely on the detection and characterization of PrP^{Sc} in brain tissue. This chapter outlines in detail protocols used for the immunohistochemical and Western blot analysis of PrP^{Sc}, which are used routinely in the diagnosis of human prion diseases, and discusses the caveats of each technique.

2 Methods

Definitive diagnosis of a human prion disease requires histological, immunohistochemical, and biochemical examination of the brain. This places a high priority on the need to preserve and investigate fixed brain tissue and unfixed frozen brain tissue from autopsies carried out on suspected cases of prion disease. Similar methods can be used on brain biopsy specimens, but brain biopsy carries a risk of contamination of neurosurgical instruments and is therefore usually performed only if a potentially treatable alternative diagnosis has to be excluded.

2.1 Macroscopic Pathology

2.1.1 Protocol 1: Postmortem Examination

The postmortem removal and examination of the brain on a suspected case of prion disease require a number of important health and safety considerations. The infectious agent in prion diseases shows a remarkable resilience to conventional decontamination and disinfection protocols that successfully inactivate viruses and bacteria [8, 9]. Therefore, autopsy and sampling of infected tissue represent a potential risk of exposure to mortuary staff and pathologists, with unfixed brain tissue carrying the highest levels of infectivity. Robust protocols and guidance are in place for the examination of suspected cases of prion disease in order to limit the risks of exposure and contamination [10, 11].

1. Where possible, autopsies are performed in a dedicated "high-risk" autopsy room or in a general autopsy room only when no other autopsies are being carried out.
2. The use of personal protective equipment including cut-resistant gloves, disposable aprons and gowns, protective footwear, and a visor or helmet is essential when carrying out an autopsy on a suspected case of prion disease in order to prevent accidental exposure to prion-infected tissue through cuts or through mucous membranes.
3. Single-use instruments should be used wherever possible and disposed by incineration. However, some mortuary equipment is not disposable, so establishing a set of dedicated instruments for use only on suspected cases of prion disease is a suitable alternative.
4. Equipment should be cleaned and decontaminated with appropriate chemicals and physical reagents that reduce prion infectivity.

These include gravity-displacement autoclaving at 132 °C for 1 h, porous load autoclaving at 134–138 °C for 18–60 min, and if suitable, treatment with 1 M sodium hydroxide (NaOH) for 1 h at room temperature [8, 9]. The use of sodium hypochlorite solution at 20,000 ppm is also an effective method of prion decontamination, but this must be used in a well-ventilated space.

2.1.2 Protocol 2: Macroscopic Examination and Tissue Sampling

Macroscopic pathology of the brain in cases of prion disease, particularly in cases of sporadic CJD (sCJD), is often unremarkable, with the brain showing only age-related changes. However, there are certain disease subtypes where macroscopic pathology is apparent, such as those cases with extended disease durations where severe cerebral and cerebellar atrophy may be observed. In the most extreme cases, severe cerebral and cerebellar atrophy is accompanied by secondary degeneration of the white matter of the cerebral cortex, cerebellum, and brain stem and is sometimes referred to as "panencephalopathic" CJD. As macroscopic examination of the brain often shows no obvious abnormalities, extensive sampling of the brain is recommended for subsequent examination of microscopic pathology.

1. Macroscopic examination and tissue sampling of unfixed prion-infected tissue should be carried out in a Class 1 safety cabinet in order to minimize potential laboratory contamination from unfixed tissues containing high levels of prion infectivity.
2. For tissue sampling, the use of disposable gowns, cut-resistant gloves, and eye protection is essential.
3. Consistent with autopsy (*protocol 1*), single-use instruments (scalpel blades) are recommended. Following sampling, instruments are sprayed with 1 M NaOH and placed in a sharp safe for incineration.
4. Extensive sampling of the brain is recommended for neuropathological examination as the severity and distribution of microscopic pathology are often variable. As a minimum, samples should be taken from all regions of the cerebral cortex, the hippocampus, basal ganglia, thalamus, cerebellum, and brain stem.
5. For biochemical detection of PrP^{Sc} by Western blot analysis, an approximate 5 mg sample of frontal cortex is a minimum requirement, but the freezing of samples (store at −80 °C) from multiple regions of the cerebral and cerebellar hemisphere is desirable.

2.1.3 Protocol 3: Tissue Fixation and Processing

Like most histological samples, formaldehydes (usually 15% unbuffered formalin) are suitable for the fixation of tissue samples from cases of suspected prion disease. The infectious agent in prion diseases can survive formaldehyde fixation, so autopsy tissue blocks for processing into paraffin wax should be fixed in formalin followed by immersing in 96% formic acid for 1 h, a step introduced to reduce the levels of prion infectivity [8]. This fixation protocol is adapted for biopsy samples from patients with suspected prion disease by reducing the treatment in formic acid to 30 min. Tissues are reimmersed in fresh fixative prior to tissue processing. Formic acid treatment of tissue samples has been shown to have no detrimental effect on tissue morphology or on antigen preservation. Although infectivity is not completely abolished, this fixation protocol allows titers of infectivity to be sufficiently lowered that formic acid-treated paraffin-embedded blocks may be cut in an open laboratory, preferably using disposable microtome blades.

2.2 Microscopic Pathology

Prion diseases are characterized by their neuropathological changes comprising spongiform change, neuronal loss, reactive proliferation of astrocytes and microglia, and, in certain forms of the disease, formation of amyloid plaques [12]. All of these neuropathological changes are observed using conventional histological staining techniques, usually hematoxylin and eosin staining (H&E) (Fig. 1a–c) [13]. While none of these pathological features alone are specific for prion disease, their occurrence in defined regions of the brain is characteristic of a prion disease. The distribution, pattern, and severity of the pathological changes in human prion disease are often variable between different forms of the disease, but can also show variability within a single brain. Spongiform vacuoles can vary in size from 2 to 20 μm in diameter, consistent with the microvacuolar type, to larger and often confluent vacuoles (Fig. 2a, b). The formation of amyloid plaques is a feature of only some forms of human prion disease (*see* **Note 1**).

2.3 PrP^{Sc} Immuno histochemistry

Immunohistochemical detection of the prion protein continues to be an indispensable tool in the diagnosis of human prion diseases that allows the study of the pathogenesis of prion diseases at the tissue level. In addition, the identification of different PrP^{Sc} immunostaining patterns in the brain, from the subtle synaptic/granular and peri-neuronal deposits to the more intensely labeled perivacuolar and plaque-like accumulation, has contributed considerably to the subclassification of human prion disease [12, 14] as well as assisting in the identification of new prion diseases (Fig. 2c, d).

2.3.1 Protocol 5: Pretreatment Protocols

The inclusion of a number of pretreatment steps is an essential part in the immunohistochemical detection of the prion protein (PrP). Like many antigens following fixation in formalin, a pretreatment protocol is essential for the unmasking and exposing of PrP epitopes (*see* **Note 2**). An added complication in PrP immunostaining is that normal (PrP^{C}) and disease-specific (PrP^{Sc}) forms of the protein share the same primary structure. As a consequence, the majority of anti-PrP antibodies are unable to discriminate between PrP^{C} and PrP^{Sc}. Therefore, a number of pretreatment steps have been introduced to optimize the detection of PrP^{Sc} while minimizing or eliminating the detection of PrP^{C}.

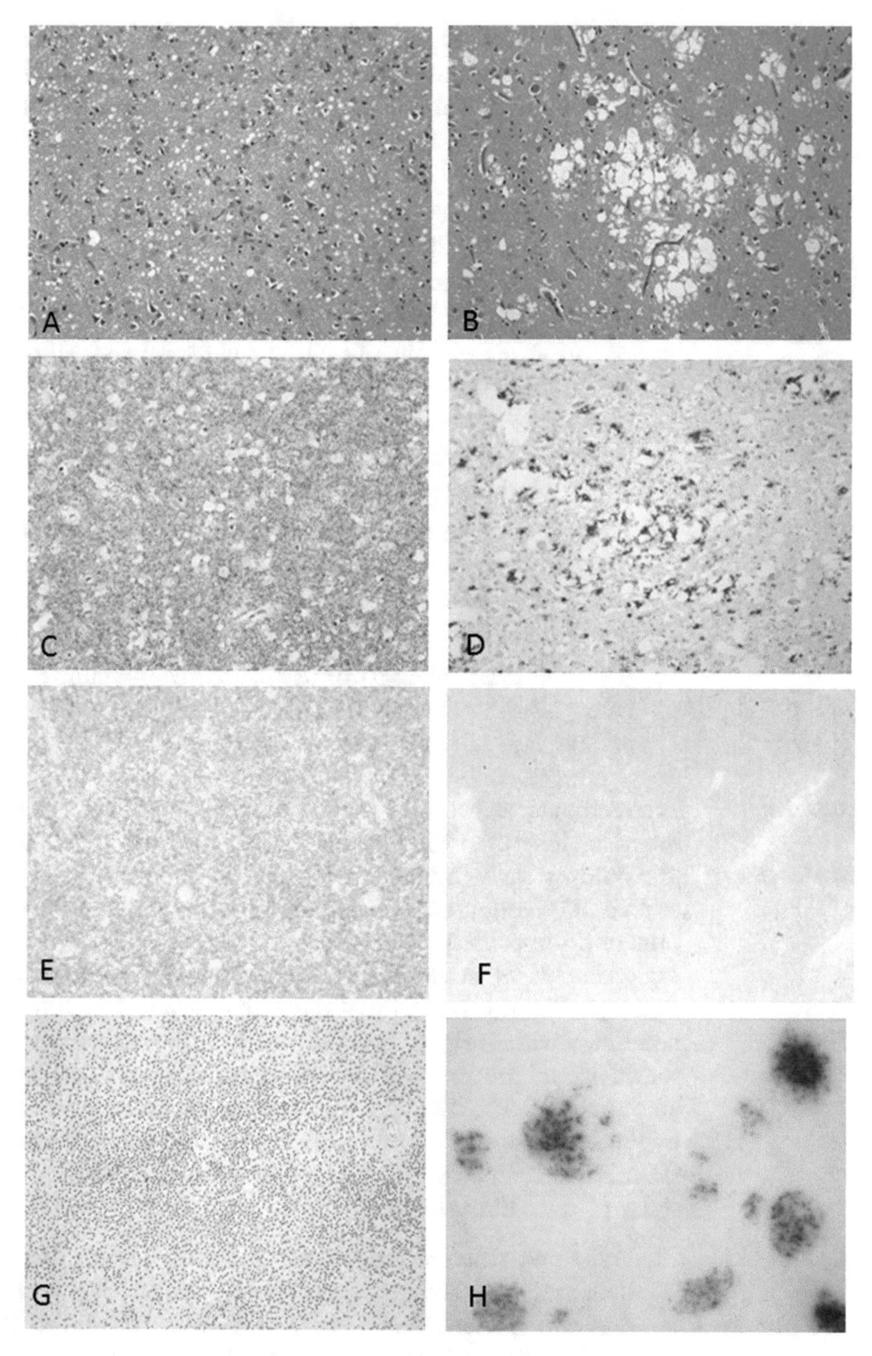

Fig. 2 (**a**, **b**) Microscopic pathology in human prion diseases. (**a**) Microvacuolar spongiform change in the frontal cortex in sporadic CJD MM1 subtype. (**b**) Confluent spongiform change in the frontal cortex in the sCJD MM2 subtype. (**c**, **d**) Patterns of PrP accumulation following immunohistochemistry. (**e**) Granular and synaptic-like pattern, which occurs as small aggregates in a relatively regular distribution in the neuropil. (**d**) Intense perivacuolar accumulation around areas of confluent spongiform change. (**e**) Widespread deposits of PrP^{C} in the frontal cortex in Lewy body disease following PrP immunohistochemistry. (**f**) Negative PET blot from a corresponding tissue section in Lewy body disease. (**g**, **h**) Labeling of peripheral tissues in vCJD. (**g**) Faint labeling of PrP^{Sc} within a lymphoid follicle in the tonsil following immunohistochemistry. (**h**) Intense labeling of PrP^{Sc} in several lymphoid follicles of the tonsil following PET blot analysis

1. Like all histological and immunohistochemical techniques, 5 μm paraffin-embedded brain sections are deparaffinized in xylene for 5 min followed by stepwise rehydration (5-min incubations) in absolute alcohol, 74% industrial methylated spirits, and 70% IMS prior to use.
2. An initial step of heat-induced epitope retrieval is an essential component in all PrP immunohistochemical protocols. The most effective and widely used method is hydrated autoclaving at 121 °C for 10 min [15, 16] but hydrolytic autoclaving [17] microwaving in citric acid buffer and pressure cooking of tissue sections are also effective [18].
3. Epitope retrieval is commonly followed by immersion of tissue sections in 96% formic acid for 5 min, a step that enhances PrP immunostaining [15, 16, 19].
4. Treatment of tissue sections with proteinase K solution, at a concentration of ~5 μg/mL for 5 min at room temperature, has been included in the pretreatment protocol as a valuable step in the digestion of PrP^{C} [20] (*see* **Note 3**).
5. A further denaturing treatment with 4 M guanidine isothiocyanate (4 °C for 2 h) is also used, although less frequently, in protocols for the detection of PrP^{Sc} [16].

2.3.2 Protocol 6: PrP Antibodies and Immunohistochemistry

Pretreatment protocols are followed by incubation with antibodies raised against the prion protein. A large number of PrP antibodies are commercially available for use on formalin-fixed paraffin-embedded tissue, each recognizing different epitopes on the human prion protein (*see* **Note 4**). A variety of PrP immunolabeling protocols are available using a wide range of commercial detection kits. The vast majority of these kits use amplification methods combined with antigen retrieval protocols in order to increase the sensitivity of the immunohistochemical assay and detect minute amounts of PrP^{Sc}. The PrP immunohistochemistry protocol used by the National CJD Research & Surveillance Unit in the diagnosis of human prion disease uses a Novolink max polymer detection system (Leica Biosystems, UK).

1. Following the pretreatment protocol (a combination of hydrated autoclaving, formic acid treatment, and proteinase K digestion), tissue sections are washed in Tris-buffered saline (TBS) (50 mM Tris; 150 mM NaCL; pH 7.6).
2. Tissue sections are incubated in the primary monoclonal anti-PrP antibodies 3F4, KG9, and 12F10 for 1 h at room temperature. It is recommended that the primary antibodies are diluted in the primary antibody diluent (Leica Biosystems) which can help reduce any background and nonspecific labeling issues.
3. Tissues are washed in TBS and immunolabeling completed with the two-part Novolink max polymer detection system.

The first step is the incubation of tissue sections in postprimary block for 30 min at room temperature. Following washing in TBS, tissue sections are incubated in the Novolink polymer for 30 min at room temperature and further washed in TBS.

4. PrP immunolabeling is visualized with 3,3′-diaminobenzidine (DAB) (*see* **Note 5**).

2.4 Paraffin-Embedded Tissue (PET) Blot

The PET blot is a staining technique with the ability to discriminate between PrP^C and PrP^{Sc} while improving on the sensitivity of immunohistochemistry in the detection of PrP^{Sc} [21]. Therefore, the PET blot proves particularly beneficial for use on tissues where upregulation of normal PrP^C or nonspecific labeling following immunohistochemistry was problematic to the interpretation of staining (Fig. 2e, f). In addition, the superior sensitivity of the PET blot in the detection of PrP^{Sc} is particularly useful on brain biopsy tissues, where there are often limited morphological changes and levels of PrP^{Sc} may be lower than those of autopsy tissue. The PET blot has also been useful in the detection of PrP^{Sc} in peripheral tissues from patients with vCJD, where levels in lymphoreticular tissues are much lower than those of CNS tissue (Fig. 2g, h). The PET blot combines several of the technical aspects of immunohistochemistry with that of Western blot analysis.

2.4.1 Protocol 7: PET Blotting

1. 5 μm Formalin-fixed, paraffin-embedded tissue sections are blotted onto prewetted nitrocellulose membrane and dried at 60 °C overnight.
2. Nitrocellulose membranes/tissue sections are deparaffinized in xylene for 5 min followed by stepwise rehydration (5-min incubations) in propan-1-ol (absolute, 95%, 85%, and 70%). Sections are air-dried before immersing in Tris-buffered saline, 0.05% Tween 20 (TBST) (10 mM Tris–HCL pH 7.8, 100 mM NaCL, 0.05% Tween 20).
3. The superior adherence of tissue sections to the nitrocellulose membrane allows a rigorous treatment with proteinase K solution. Proteinase K is used at a concentration of ~25 μg/mL in digestion buffer (10 mM Tris–HCL pH 7.8, 100 mM NaCL, 0.1% Brij) for ~18 h at 60 °C, thus ensuring the complete digestion of PrP^C within tissues leaving only the disease-associated form, PrP^{Sc}.
4. The sensitivity of the PET blot is further enhanced by a denaturing treatment of tissue sections with 3 M guanidine thiocyanate for 10 min at room temperature.
5. Tissue sections are washed in TBST, and blocked in casein for 30 min before immersion in the anti-PrP antibodies for 2 h at room temperature. Antibodies are diluted in casein (*see* **Note 6**).

6. Following the washing of tissue sections in TBST, immunolabeling is completed using a Vectastain ABC amplification detection system (Vector Laboratories).
7. Labeling is visualized using NBT/BCIP (5-bromo-4-chloro-3-indolyl phosphate/nitroblue tetrazolium) and sections are observed on a stereoscopic microscope.

Although a sensitive and specific technique in the detection of PrP^{Sc}, PET blot preparations lack the cellular detail obtained with conventional immunohistochemistry techniques. This, combined with the large volumes of antibodies required for immunostaining and a longer protocol, has discouraged the use of the PET blot as a routine diagnostic tool for human prion diseases. Furthermore, in the examination of postmortem brain tissue from patients with prion disease, levels of PrP^{Sc} in the brain are normally sufficiently high that any differences in the staining between immunohistochemistry and the PET blot are difficult to determine.

2.5 Biochemistry of the Human Prion Protein

The pathogenesis of prion diseases is associated with the conversion of the normal cellular form of the protein (PrP^C) into the abnormal and disease-associated form (PrP^{Sc}). The mechanism or mechanisms underlying this protein conversion are yet to be fully resolved, but have been attributed to a seeded aggregation process, during which the direct interaction of PrP^C and PrP^{Sc} results in the refolding of the α-helical and coil structure of PrP^C into the β-pleated sheet-rich complex that characterizes PrP^{Sc} [22]. This change in conformation alters the physicochemical properties of the prion protein. PrP^{Sc} has a decreased solubility in nondenaturing detergents and an increased resistance to protease treatment with proteinase K when compared to PrP^C [23]. When tissue homogenates are digested with proteinase K, a proportion of the PrP^{Sc} present is truncated at the N-terminus (and, in some forms of human prion disease, the C-terminus), producing a protease-resistant core fragment commonly referred to as PrP^{res}. The resistance of PrP^{Sc} to protease digestion forms the basis for its biochemical detection by Western blot analysis. When tissue homogenates are examined by Western blot using antibodies to PrP, samples containing PrP^{Sc} will be detected on the blot, even after proteinase K digestion, while PrP^C is completely digested into peptides that are not resolved on the blot.

2.6 Western Blot Analysis

Western blot detection of PrP^{res} in CNS or peripheral tissues obtained at autopsy or, more rarely, from biopsy tissue has become an essential component in the diagnosis of human prion diseases. The identification of several distinct types of PrP^{res} following Western blot analysis, which are associated with different disease phenotypes, is also valuable in the subclassification of human prion diseases and forms the basis of molecular subtyping [24]. A variety of Western blot protocols are available [24–26], but a critical element shared by all comprises tissue extraction, protease digestion, and denaturation.

2.6.1 Protocol 8: Tissue Extraction of PrP^{res} from Frozen Brain Tissue

1. Tissue samples (~100 mg) are homogenized in sufficient extraction buffer (20 mM Tris pH 7.4; 0.5% Nonident P-40; 0.5% sodium deoxycholate) to give a 10% (wt/vol) brain homogenate.
2. Insoluble material is pelleted by spinning at 2000 rpm for 5 min as 4 °C. The supernatant is removed and the remaining pellet frozen at −20 °C.
3. PrP^C is digested by adding proteinase K to the supernatant to a final concentration of 50 μg/mL and incubates at 37°C for 1 h. The digestion is stopped with the addition of Pefabloc to 1 mM. Proteinase K-treated extracts can be stored at −20 °C until required.

2.6.2 Protocol 9: Sodium Dodecyl Sulfate-Polyacrylamide gel Electrophoresis (SDS-PAGE) and Western Transfer

1. Proteinase K-treated samples are given a short vortex before removing a 5 μL volume and adding to 5 μL of 4× NuPAGE LDS sample buffer, supplemented with 10 μL of 1× extraction buffer (*see* **Note 7**). After a brief vortex, samples are denatured by boiling at 100 °C for 10 min before a brief spin at 14,000 rpm. Protein molecular weight markers are included in each run (1 μL Magic Marker™ XP, 5 μL Benchmark, 9 μL of 1× extraction buffer, 5 μL of NuPAGE LDS sample buffer) (*see* **Note 8**).
2. Centrifugal concentration is an optional step used for increased sensitivity in the Western blot detection of PrP^{res}. Following proteinase K treatment, the sample is given a brief vortex before centrifugation at 14,000 rpm for 1 h at 4 °C (*see* **Note 9**). The supernatant is removed and the pellet resuspended in 5 μL of 4× NuPAGE LDS sample buffer (Invitrogen) supplemented with 15 μL of 1× extraction buffer (20 mM Tris pH 7.4; 0.5% NP-40; 0.5% sodium deoxycholate). After a short vortex, the sample is boiled at 100 °C for 10 min and stored at −20 °C until required.
3. Proteins are separated by running at 200 V (constant current) for 55 min in 1× NuPAGE MES running buffer (*see* **Note 10**).
4. Proteins separated by SDS-PAGE gel electrophoresis are transferred onto polyvinylidene difluoride (PVDF) membrane at 30 V (constant current) for 1 h using the NuPAGE transfer system. Transfers are carried out in 1× NuPAGE transfer buffer.

2.6.3 Protocol 10: Immunodetection of PrP Bound to PDVF Membrane

1. Membranes are rinsed in Tris-buffered saline and Tween 20 (TBST) (20 mM Tris HCL pH 7.4, 0.9% NaCL, 0.01% Tween 20) before incubating in 5% powdered milk/TBST for 45 min at room temperature or overnight at 4 °C.

2. Membranes are washed (3 × 5 min) in TBST prior to incubating in the primary anti-PrP antibody, most commonly 3F4, diluted in TBST for 1 h at room temperature.
3. After a further wash, immunolabeling is completed by incubating the membrane in an anti-mouse IgG-linked horseradish peroxidise-conjugated secondary for 1 h at room temperature.
4. Immunodetection is carried out by incubating the membrane with the chemiluminescent substrate ECL plus for 5 min (GE Healthcare, UK).
5. Membranes are then exposed to Hyperfilm ECL for 30 s, 3 min, and 30 min before the films are developed in an automatic film hyperprocessor. The glycoform ratio and mobility of the unglycosylated band of samples under investigation are directly compared to the diagnostic reference standards (*see* **Note 8**).

The detection of PrP^{res} in tissue homogenates results in a characteristic pattern of three bands following Western blotting that correlate to the variable occupancy of two N-linked glycosylation sites (asparagine 181 and 187) such that the N-terminally truncated PrP^{res} exists in three resolvable glycoforms (non-, mono-, and diglycosylated fragments) in the 18–30 kDa molecular mass range. Several distinct banding patterns of PrP^{res} have been identified in human prion diseases, based on differences observed in the electrophoretic mobility (fragment size) and in the glycosylation ratio (the relative abundance of the glycoforms). In relation to variability in fragment size, two major cleavage sites have been identified following proteinase K digestion resulting in differently sized nonglycosylated protease-resistant core fragments; truncation to glycine 82 yields a 21 kDa nonglycosylated fragment (termed type 1) whereas truncation at glycine 97 yields a 19 kDa nonglycosylated fragment (termed type 2). For variation in the glycoform ration, the suffix A is given to cases in which the mono- or nonglycosylated fragment predominates such as those found in sCJD cases, with the suffix B given to cases such as vCJD where the diglycosylated band predominates. The most common typing protocol used in the diagnosis of human prion diseases and that used by the NCJDRSU is based on that of Parchi et al. [27]. This protocol has evolved and continues to evolve to encompass all forms of human prion disease and is shown in diagrammatic form in Fig. 3a.

2.6.4 Protocol 10: Sodium Phosphotungstic Acid (NaPTA) Precipitation

PrP^{Sc} accumulates in peripheral tissues in some forms of human prion disease, most notably in lymphoreticular tissues of vCJD patients. However, levels of PrP^{Sc} within peripheral tissues are much lower than those detected in CNS tissues [28]. An increase in Western blot sensitivity can be obtained by utilizing the semiselective precipitation of PrP^{Sc} from sarkosyl-solubilized tissue by

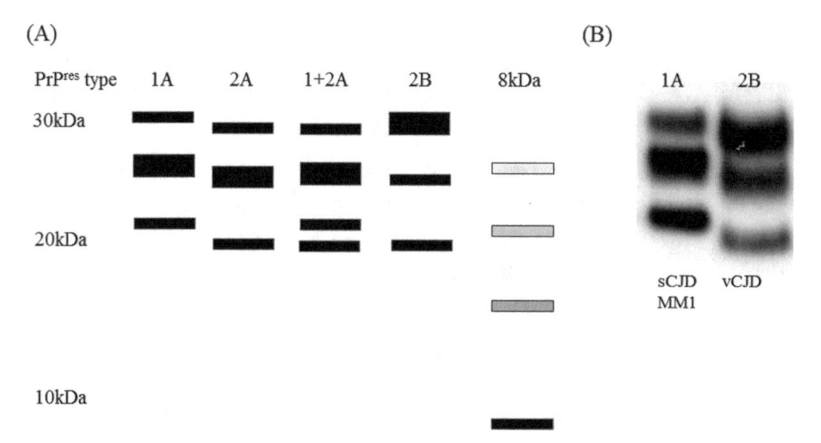

Fig. 3 Western blot analysis of PrPres types in human prion disease. (**a**) Diagrammatic representation of the major protease-resistant prion protein (PrPres) types found in human prion disease after proteinase K treatment and Western blot analysis. (**b**) Western blot analysis of PrPres in brain homogenate from a sCJD MM1 and vCJD patient. These two different human prion diseases are routinely included in Western blot analysis of brain tissue from cases with suspected human prion disease, serving as appropriate reference control for both fragment mobility and glycoform ratio

sodium phosphotungstic acid (NaPTA) precipitation prior to proteinase K treatment and Western blotting [29].

1. Tissue samples (~60–70 mg) are placed in 2 mL Lysing matrix D screw-capped tubes (Qbiogen) with sufficient lysis buffer (50 mM Tris pH 7.4; 5 mM $MgCl_2$; 2 mM CaCl2; 100 mM NaCl; 0.5% N-laurosarcosine; 2% Nonident P-40) added to give a 10% (wt/vol) tissue homogenate (*see* **Note 11**).
2. Samples are placed in a Fastprep instrument and run for 45 s at a speed of 65 ms^{-1}. Samples are removed and placed at 4 °C for 5 min before a further two more runs in the Fastprep using this procedure.
3. Samples are then centrifuged at 7000 rpm for 5 min at 4 °C. 500 μL of the supernatant is removed for NaPTA analysis. Any excess supernatant is stored at −80 °C.
4. To the 500 μL sample add 500 μL of 2% (w/v) sarkosyl in phosphate-buffered saline (PBS) pH 7.4 and incubate at 37 °C for 10–30 min.
5. Add 5 μL of 10 U/μL benzonase and 5 μL of 0.2 M magnesium chloride ($MgCl_2$) to the sample. After a brief vortex, incubate the sample for a further 30 min at 37 °C.
6. To the sample, add 81 μL of 4% (w/v) NaPTA stock solution (4% NaPTA, 170 mM $MgCl_2$) to give a final concentration of 0.32% (w/v) NaPTA. Vortex the sample and incubate at 37 °C for 30 min.

7. Centrifuge the sample at 14,000 rpm for 30 min at 37 °C and discard the supernatant. Resuspended the pellet in 20 μL of 0.1% sarkosyl in PBS pH 7.4.
8. Protease digest the sample by adding 2 μL of 550 μg/mL proteinase K solution and incubating at 37 °C for 30 min. Proteinase K digestion is stopped by adding 1 μL of 23 mM Pefabloc.
9. Add 8 μL of 4× NuPAGE sample buffer to the sample and boil for 10 min at 100 °C.
10. Western blot analysis is completed using the electrophoresis (*protocol 9*), transfer (*protocol 9*), and immunolabeling (*protocol 10*) protocols described in Sect. 2.6.

2.7 Ancillary Techniques in the Detection of Human Prion Disease

The majority of protocols discussed thus far detect PrP^{Sc} based on its protease resistance in relation to PrP^{C}. However, an alternative approach has been developed to detect PrP^{Sc} based on the conformational difference between PrP^{C} and PrP^{Sc}, known as conformation-dependant immunoassay (CDI) [30]. In CDI, PrP^{Sc} is detected on the basis of the binding of the anti-PrP antibody 3F4 to a PrP epitope that is concealed in PrP^{Sc} but made accessible by denaturation with the chaotrope guanidine hydrochloride. As yet, CDI is not commonly used in the diagnosis of human prion diseases.

Immunohistochemistry and Western blot analysis continue to be the primary diagnostic tools in the diagnosis of human prion disease. Although highly sensitive, these techniques are restricted to use on brain tissue sampled at autopsy, or more rarely from biopsy tissue in patients showing clinical signs of disease where levels of prion infectivity are high. Current research in prion diseases focuses on the development and application of in vitro amplification assays with the aim of detecting very low levels of PrP^{Sc} from biological fluids such as blood, urine, and cerebrospinal fluid (CSF) in patients with suspected prion disease [31]. The principle behind the in vitro assays is to mimic the replication of the misfolded, disease-associated protein by increasing the concentration of PrP^{Sc} in an accelerated manner prior to detection. Protein misfolding cyclic amplification (PMCA) was the first amplification assay to be described and its methodology has been modified and forms the basis of other in vitro assays [32]. In PMCA, a sample containing a seed (PrP^{Sc}) is diluted into a substrate containing excess PrP^{C}. The seeded aggregation process associated with replication of PrP^{Sc} is then accelerated by successive rounds of incubation and sonication prior to detection by Western blotting. In a more recent development, the real-time quaking-induced conversion (RT-QuIC) method monitors the conversion of recombinant PrP (used as the seed) in real time using the fluorescent dye thioflavin T which binds to β-sheet structures [33]. In contrast to

PMCA, RT-QuIC substitutes shaking instead of sonication. Although both PMCA and RT-QuIC are not fully established in the diagnosis of human prion diseases, RT-QuIC in particular has shown considerable promise as a clinical CSF diagnostic test [34].

2.8 Conclusion

Human prion diseases remain a rare group of neurodegenerative disorders, but continue to have considerable public health implications as a result of their capacity for human-to-human transmission [5]. Therefore, the accurate identification and laboratory diagnosis of human prion diseases are essential. Although diagnosis of a human prion disease can be made with a high degree of accuracy based on the characteristic clinical phenotype, our extensive experience in the laboratory diagnosis of a human prion disease indicates that a combined histological and biochemical approach is essential, combined with adequate clinical and genetic data for clinicopathological correlation. Therefore, examination of the brain in known and suspected cases of human prion disease is necessary. The unique nature of the infectious agent in prion disease, specifically the extreme resistance to conventional decontamination and disinfection methods, requires a number of robust protocols for the safe laboratory handling of brain and other CNS tissues from cases of suspected prion disease. These protocols are described in this chapter and are based on the principles of containment and decontamination. Most of the widely used techniques to diagnose human prion diseases reply on the identification of protease-resistant prion protein (PrP^{Sc}), allowing it to be distinguished from the normal cellular from of the prion protein (PrP^{C}), which is protease sensitive. However, PrP^{C} can be upregulated in certain disease states in the human brain, e.g., hypoxia, so the use of the techniques we describe to distinguish upregulated PrP^{C} from PrP^{C} is essential.

The development of a wide range of laboratory techniques to specifically identify and amplify PrP^{Sc} are in current research use; these are beginning to be applied to human CSF samples and brain tissue samples for diagnosis and we anticipate that these methods will become more widespread in diagnostic use in the near future.

3 Notes

1. Several distinct morphologies of plaques have been described in human prion diseases, the majority of which can be identified with tinctorial amyloid stains such as Congo red and thioflavin T in addition to H&E staining [35]. However, as all plaques are composed of the aggregated and abnormal form of the prion protein they are most easily identified following immunohistochemistry for PrP.

2. Fixation helps to preserve cellular architecture and composition of cells in the tissue to allow them to withstand subsequent processing. However, formalin fixation results in the cross-linking of protein molecules within tissues that can "mask" protein epitopes corresponding to the targets of antibodies used in immunohistochemistry. The PrP antigen is masked during formalin fixation requiring a number of denaturing pretreatment protocols to expose the epitope [15, 16].
3. Treatment of tissue sections with proteinase K solution has proved to be particularly useful in the digestion of endogenous PrP^{C}. However, concentrations and time of exposure must be kept to a minimum to preserve tissue morphology. As a result, the differentiation between PrP^{C} and PrP^{Sc} can prove problematic.
4. In the immunohistochemical diagnosis of prion disease, it is advisable to use a panel of anti-PrP antibodies, as the use of antibodies with different epitopes within PrP can often result in differences in the pattern and intensity of staining. This is best demonstrated in cases of variable protease-sensitive prionopathy (VPSPr) where differential staining has been observed between the monoclonal PrP antibodies 3F4, 12F10, KG9, and 12F10, four of the most commonly used antibodies used in the diagnosis of human prion diseases [36].
5. Interpretation of PrP immunolabeling patterns can prove problematic, most commonly due to the inability of the primary antibodies to discriminate between PrP^{C} and PrP^{Sc}. PrP^{C} is often upregulated in other neurodegenerative conditions such as Alzheimer's disease and cerebral hypoxia [37, 38]. Also, levels of PrP^{C} are likely to be higher in rapidly fixed brain biopsy specimens than autopsy brain samples, as there is less opportunity for degradation. It is unlikely that the concentration of proteinase K used in immunohistochemistry is sufficient to degrade these high levels of PrP^{C}; therefore, a diffuse labeling of PrP^{C} may be present in these tissues. Under these circumstances, it is crucial that the patterns of apparently positive immunolabeling should be differentiated from PrP^{Sc}.
6. Consistent with immunohistochemistry, it is advisable to use a panel of anti-PrP antibodies in the PET blot analysis of PrP^{Sc}. It is important to include a tissue section in which the primary antibody is omitted and replaced with blocking serum as an appropriate negative control.
7. The abundance of PrP^{res} can vary among samples. Loading the sample under investigation at a range of volumes is recommended, and including a lane of the sample prior to proteinase digestion is helpful.

8. In the Western blot analysis of brain homogenates from suspected cases of human prion disease, it is important to include appropriate reference control for both fragment mobility and glycoform ratio. The diagnostic reference standards included in the Western blot diagnostic protocol at the NCJDRSU are from a sCJD MM1 patient and a vCJD (MM2B) patient (Fig. 3b).
9. Centrifugal concentration is recommended for the analysis of peripheral tissues and brain biopsy samples from suspected cases of prion disease, where levels of PrP^{res} are lower than those of brain samples taken at autopsy. For a diagnostic brain biopsy sample concentrate 100 μL in the first instance. For peripheral tissues, concentrate 200 μL.
10. Some forms of human prion disease, such as Gerstmann-Sträussler-Scheinker (GSS) disease and VPSPr, are characterized by the presence of a low-molecular-weight band at ~8 kDa following proteinase K digestion and Western blot analysis [37, 39]. To detect these bands it is recommended to reduce the time of the electrophoresis to 40 min. The reader should be aware that this will slightly reduce the separation of the bands in the 18–30 kDa range.
11. The inclusion of appropriate control samples is an important component in the detection of PrP^{Sc} by NaPTA precipitation. A negative control sample from a patient considered for a diagnosis of CJD but given an alternative final diagnosis (non-CJD) should be included to allow better interpretation of any background signal that may become apparent in the test samples. Background signal following NaPTA precipitation is a particular problem associated with the analysis of peripheral tissues. The inclusion of a positive control is also important and this should comprise a second sample from the non-CJD patient but this should be spiked with CJD brain (~1 μL) to provide a positive control.

References

1. Prusiner SB (1982) Novel proteinaceous infectious particles cause scrapie. Science 216(4542): 136–144
2. Gambetti P, Kong Q, Zou W et al (2003) Sporadic and familial CJD: classification and characterisation. Br Med Bull 66(1): 213–239
3. Prusiner SB, DeArmond SJ (1994) Prion diseases and neurodegeneration. Annu Rev Neurosci 17(1):311–339
4. Will RG, Ironside JW, Zeidler M et al (1996) A new variant of Creutzfeldt-Jakob disease in the UK. Lancet 347(9006):921–925
5. Peden AH, Ironside JW, Head MW (2013) Risk of transmission of Creutzfeldt-Jakob disease by blood transfusion. In: Zou WQ, Gambetti P (eds) Prions and diseases. Springer, New York
6. National CJD Research & Surveillance Unit (1990–2015) Monthly surveillance figures. http://www.cjd.ed.ac.uk/documents/figs.pdf. Accessed 21 Sept 2015
7. Gill ON, Spencer Y, Richard-Loendt A et al (2013) Prevalent abnormal prion protein in human appendixes after bovine spongiform encephalopathy epizootic: large scale survey. Br Med J 347:f5675

8. Taylor DM (2000) Inactivation of transmissible degenerative encephalopathy agents. A review. Vet J 159(1):10–17
9. Taylor DM (2003) Preventing accidental transmission of human transmissible spongiform encephalopathy. Br Med Bull 66(1):293–303
10. Advisory committee on Dangerous Pathogens (2013) Guidance: Minimise Transmission Risks of CJD and vCJD in Healthcare settings. http://www.gov.uk/government/publications/guidance-from-the-acdp-tse-risk-management-subgroup-formerly-tse-working--group. Accessed 21 Sept 2015
11. World Health Organization (2003) Infection control guidelines for transmissible spongiform encephalopathies. http://who.int/csr/resources/publications/bse/WHO_CDS_APH_2000_3/en/. Accessed 21 Sept 2015
12. Head MW, Ironside JW, Ghetti B et al (2015) Prion diseases. In: Love S, Budka H, Ironside JW, Perry A (eds) Greenfields neuropathology, 9th edn. CRC Press/Taylor & Francis Group, Boca Raton
13. Suvarna KS, Layton C, Bancroft JD (2013) The hematoxylins and eosin. In: Suvarna KS, Layton C, Bancroft JD (eds) Bancroft's theory and practice of histological techniques, 7th edn. Churchill Livingston/Elsevier, London
14. Ironside JW, Ritchie DL, Head MW (2005) Phenotypic variability in human prion diseases. Neuropathol Appl Neurobiol 21:565–579
15. Bell JE, Gentleman SM, Ironside JW et al (1997) Prion protein immunocytochemistry-UK five centre consensus report. Neuropathol Appl Neurobiol 23(1):26–35
16. Kovacs G, Head MW, Hegyi I et al (2002) Immunohistochemistry for the prion protein: a comparison of different monoclonal antibodies in human prion disease subtypes. Brain Pathol 12(1):1–11
17. Kitamoto T, Shin RW, Doh-ura K et al (1992) Abnormal isoform of prion protein accumulates in the synaptic structures of the central nervous system in patients with Creutzfeldt-Jakob disease. Am J Pathol 140(6):1285–1294
18. Liberski PP, Yanagihara R, Brown P et al (1996) Microwave treatment enhances the immunostaining of amyloid deposits in both the transmissible and non-transmissible amyloidoses. Neurodegeneration 5(1):95–99
19. Kitamoto T, Ogomori K, Tateishi J et al (1987) Formic acid pretreatments enhances immunostaining of cerebral and systemic amyloids. Lab Investig 57(2):230–236
20. Hilton DA, Ghani AC, Conyers L et al (2004) Prevalance of lymphoreticular prion protein accumulation in UK tissue samples. J Pathol 203(3):733–739
21. Ritchie DL, Head MW, Ironside JW (2004) Advances in the detection of prion protein in peripheral tissues of variant Creutzfelt-Jakob disease patients using paraffin-embedded tissue blotting. Neuropathol Appl Neurobiol 30(4):360–368
22. Jarrett JT, Landsbury PT Jr (1993) Seeding "one dimensional crystallization" of amyloid: a pathogenic mechanism in Alzheimers disease and scrapie? Cell 73(6):1055–1058
23. Prusiner SB, Scott MR, DeArmond SJ (1998) Prion protein biology. Cell 93(3):337–348
24. Parchi P, Castellani R, Capellari S et al (1996) Molecular basis of phenotypic variability in sporadic Creutzfeldt-Jakob disease. Ann Neurol 39(6):767–778
25. Collinge J, Sidle KCL, Meads J et al (1996) Molecular analysis of prion strain variation and the aetiology of "new variant" CJD. Nature 383(6602):685–690
26. Head MW, Bunn TJ, Bishop MT et al (2004) Prion protein heterogeneity in sporadic but not variant Creutzfeldt-Jakob disease UK cases 1991–2002. Ann Neurol 55(6):851–859
27. Parchi S, Capellari S, Chen SG et al (1997) Typing prion isoforms. Nature 386(6622):232–233
28. Bruce ME, McConnell I, Will RG et al (2001) Detection of variant Creutzfeldt-Jakob disease infectivity in extraneural tissues. Lancet 358(9277):208–209
29. Wadsworth JD, Joiner S, Hill AF et al (2001) Tissue distribution of protease resistant prion protein in variant Creutzfeldt-Jakob disease using a highly sensitive immunoblotting assay. Lancet 358(9277):171–180
30. Safar J, Geschwind MD, Deering C et al (2005) Diagnosis of human prion diseases. Proc Natl Acad Sci U S A 102(9):3501–3506
31. Orru CD, Caughey B (2011) Prion seeded conversion and amplification assays. Top Curr Chem 305:121–133
32. Saborio GP, Permanne B, Soto C (2001) Sensitive detection of pathological prion protein by cyclic amplification of protein misfolding. Nature 411(6839):810–813
33. Atarashi R, Moore A, Sim VL et al (2007) Ultrasensitive detection of scrapie prion protein using seeded conversion of recombinant prion protein. Nat Methods 4(8):645–650
34. McGuire LI, Peden AH, Orru CD et al (2012) Real time quaking-induced conversion analysis of cerebrospinal fluid in sporadic Creutzfelt-Jakob disease. Ann Neurol 72(2):278–285
35. Suvarna KS, Layton C, Bancroft JD (2013) Amyloid. In: Suvarna KS, Layton C, Bancroft

JD (eds) Bancroft's theory and practice of histological techniques, 7th edn. Churchill Livingston/Elsevier, London

36. Head MW, Yull HM, Ritchie DL et al (2013) Variably protease-sensitivity prionopathy in the UK: a retrospective review 1991–2008. Brain 136(4):1102–1115
37. Kovacs GG, Zerbi P, Voigtlander T et al (2002) The prion protein in human neurodegenerative diseases. Neurosci Lett 329(3):269–272
38. McLennan NF, Brennan PM, McNeil A et al (2004) Prion protein accumulation in hypoxic brain damage. Am J Pathol 165(1): 227–235
39. Parchi P, Giese A, Capellari S et al (1998) Different patterns of truncated prion protein fragments correlate with distinct phenotypes in P102L Gerstmann-Sträusler-Scheinker disease. Proc Natl Acad Sci U S A 95(14): 8322–8327

Chapter 6

Subcellular and Molecular Changes Associated with Abnormal PrP Accumulation in Brain and Viscera of Classical and Atypical Prion Diseases

Martin Jeffrey and Gillian McGovern

Abstract

Immunogold electron microscopy has shown that the primary toxic effects on cells of disease-associated prion protein (PrP^d) accumulation are to be found on plasma membranes. The nature of the membrane change is tissue and cell type specific. PrP^d accumulation on neurites is associated with bizarre and unique twisted or branched membrane invaginations that are sometimes coated and sometimes have a spiral twist. Neuronal PrP^d elicits submembrane ubiquitination by interaction with a presumptive transmembrane ligand and this protein complex undergoes clathrin-mediated endocytosis. PrP^d accumulation on plasma membranes of glial cells causes membrane ruffling but on follicular dendritic cells of the lymphoid system it causes dendritic process hyperplasia, while on adrenal chromaffin cells it causes the membrane to be rearranged into linear palisades. PrP^d interacts with different molecular partners on different cells suggesting that these different morphological defects are caused by variable molecular interactions related to a scaffolding function of normal membrane PrP. The presence of a glycophosphoinositol (GPI) anchor is necessary for membrane binding and pathology. The absence of a GPI anchor promotes release of PrP^d into interstitial fluid and, on interaction with extracellular matrix binding proteins, its accumulation in basement membranes of blood vessels to form cerebral amyloid angiopathy. Membrane PrP^d may be released to the extracellular space to form amyloid fibrils or be internalized into the endolysosomal systems where it is initially N-terminally truncated prior to complete degradation. These three subcellular locations of PrP^d, plasma membranes, endolysosomes, and extracellular aggregates of PrP^d, are common to all classical prion diseases and prion disease sources in man, sheep, cattle, and deer. However, atypical scrapie PrP^d may be found on membranes of myelinated axons and in the inner mesaxon of oligodendroglial cells suggesting that differing trafficking pathways for PrP^d occur in atypical scrapie relative to classical prion diseases. While PrP^d is consistently associated with cellular lesions that most likely contribute to neurological deficits, it is unlikely that PrP^d colocalized lesions are, alone, the proximate cause of clinical disease. Other lesions such as apoptosis that do not colocalize with PrP^d may contribute to clinical disease progression and may be associated indirectly with increasing PrP^d accumulation. However, prion disease-infected tissues possess other lesions, such as tubulovesicular bodies, that do not have clear association with PrP^d and are of unknown relationship to disease pathogenesis.

Key words Transmissible spongiform encephalopathy, Atypical scrapie, Prion disease, Immunogold electron microscopy, Prion protein, Cell membranes, Tubulovesicular bodies, Cerebral amyloid angiopathy, Mesaxon, Lymphoid tissues, Adrenal, Brain, Follicular dendritic cell

Pawel P. Liberski (ed.), *Prion Diseases*, Neuromethods, vol. 129,
DOI 10.1007/978-1-4939-7211-1_6, © Springer Science+Business Media LLC 2017

1 Introduction

The principal purpose of electron microscopy is to characterize the nature of subcellular lesions seen by light microscopy. Uniquely among prion research methodologies, immunogold electron microscopy can be used to determine how the abnormal prion protein damages cells, and is particularly valuable for the study of naturally occurring disease. In addition to transmissible spongiform encephalopathy (TSE)-specific subcellular changes, immunogold electron microscopy can also be used to demonstrate other molecules that interact with abnormal prion protein.

Abnormal prion protein accumulations are found in nearly all prion diseases and can be detected by immunohistochemical methods. Immunohistochemically detected, disease-associated prion protein (PrP^d) consists of protease-resistant and protease-sensitive fractions and extracellular whole-length and intracellularly truncated PrP^d variants [1, 2]. In brain, cellular PrP^d accumulations take a wide range of morphological forms, and are associated with neurons, glia, and endothelia [3, 4]. Electron microscopy can be used to resolve these morphological forms to three subcellular locations: the greatest majority of PrP^d is localized to plasma membranes, other smaller fractions are found in endolysosomes, and extracellular PrP^d mainly occurs as larger molecular aggregates often forming amyloid fibrils. In most naturally occurring diseases of animals, extracellular PrP^d is usually relatively inconspicuous but may be prominent in some human diseases.

2 Materials

2.1 Light Microscopy Immunolabeling of Resin-Embedded Tissue

Sodium ethoxide: sodium hydroxide pellets (7 g), ethanol (100 mL).

6% Hydrogen peroxide in distilled water.

Formic acid (neat).

Phosphate-buffered saline (PBS) concentrate: Sodium chloride (80 g), potassium chloride (2 g), disodium hydrogen phosphate (11.5 g), potassium dihydrogen phosphate (2 g), distilled water (1 L).

PBS with Tween (PBST): PBS concentrate (100 mL), distilled water (900 mL), Tween (2 mL).

Blocking buffer: 10% Normal serum in PBST.

1A8 PrP antibody [5] at a 1:2000 dilution in PBST.

Avidin–biotin complex (ABC) kit (Vector Laboratories Ltd, Peterborough, UK), made as per kit instructions.

3′3-Diaminobenzidine tetrachloride (DAB).

Copper sulfate: Copper sulfate (4 g), sodium chloride (7.2 g), distilled water (1 L).

Meyer's hematoxylin.

0.1% Acid alcohol: Hydrochloric acid (1 mL) and ethanol (1 L).

0.05% Lithium carbonate: Lithium carbonate (0.5 g) and distilled water (1 L).

Xylene.

DPX mounting reagent.

2.2 Immunogold Electron Microscopy Procedure

Sodium periodate (saturated in distilled water).

5 nm gold probe (secondary antibody): BBI gold IgG (BBI solutions, Cardiff, UK), 1:50 dilution in blocking buffer.

Boiled distilled water.

2% Hydrogen peroxide (in boiled distilled water).

Primary antibodies: PrP-1A8 at a 1:500 dilution [5], 1C5 at a 1:20 dilution [6], 523.7 at a 1:250 dilution (J. Langeveld, ID–Lelystad, The Netherlands), and immunoglobulin G (IgG) at a 1:75 dilution (Zymed, Fisher Scientific UK Ltd, Loughborough).

Nanoprobes Goldenhance (Universal Biologicals, Cambridge, UK): Prepared according to kit instructions.

Formic acid (filtered, neat).

Phosphate-buffered saline (PBS): Made from concentrate as described above, and filtered.

Normal blocking serum: 5% Appropriate normal serum in PBS, microfiltered and heat inactivated.

British biocell buffer (BBI): PBS (100 mL), appropriate normal serum (2 mL), Tween (0.1 mL), bovine serum albumin (BSA) (1 mL), sodium azide (0.1 mL) pH 8.2.

2.5% Glutaraldehyde (in PBS).

2.3 Electron Microscopy Counterstain Procedure

10% Uranyl acetate in boiled and filtered distilled water.

Reynolds lead citrate: Lead nitrate (1.33 g), trisodium citrate (1.76 g), boiled distilled (44 mL), sodium hydroxide (6 mL).

Boiled distilled water.

Sodium hydroxide pellets.

3 Methods

3.1 Light Microscopy Immunolabeling of Resin-Embedded Tissue

Resin Removal

This part of the procedure is carried out using a stainless steel labeling tray.

1. Sodium ethoxide is applied to slides for 10 min followed by three washes in absolute ethanol to clear residual ethoxide.
2. Rinse in running water for 10 min.

Pretreatment

1. Inhibit endogenous peroxidase by treating with hydrogen peroxide for 10 min.
2. Rinse in running water for 10 min.
3. Enhance antigenicity with formic acid for 5 min.
4. Rinse in running water for 10 min.

Immunolabeling

The following part of the procedure is carried out using a Sequenza rack system (Shandon, Fisher Scientific UK Ltd, Loughborough, UK).

1. Apply 100 μL of PBST to slides for 2 h.
2. Blocking step: Incubate in blocking buffer for 60 min.
3. Primary antibody incubation: Incubate sections with appropriate dilution of specific primary antibody in blocking buffer overnight at 27 °C.
4. Wash slides in PBST for 20 min.
5. ABC amplification: Apply ABC reagents as per kit instructions.
6. Wash slides in two changes of PBST for 10 min each.
7. Remove slides from Sequenza rack and place in standard slide rack.
8. Place slide rack into a trough of fresh DAB visualization product for 10 min (2 mL DAB, 200 mL PBST, and 200 μL hydrogen peroxide).
9. Rinse slides in four quick changes of PBST.

Counterstain

1. Immerse slides in copper sulfate for 10 min.
2. Rinse in tap water.
3. Immerse in filtered hematoxylin for 30 min.
4. Rinse in tap water.
5. Differentiate using acid alcohol for 2 s.
6. Rinse in tap water.
7. Color enhance using lithium carbonate for 10 s.
8. Rinse in tap water.

Dehydrate in two changes of absolute ethanol, clear in xylene, and mount in DPX.

3.2 Immunogold Electron Microscopy Procedure

65 nm sections are cut using a Leica EM UC6 ultramicrotome (Leica Microsystems (UK) Ltd, Milton Keynes, UK) placed onto gold grids (TAAB Laboratories Equipment Ltd, Reading, UK) and allowed to dry in an airtight box. All steps of the immunolabeling

procedure are performed at room temperature (unless otherwise specified) on reagent droplets within a humid chamber.

Resin removal

1. The grid is floated (section side down) on a droplet of sodium periodate for 60 min.
2. The grid is then passed through multiple drops of water for 15 min.

Pretreatment

1. Place grid on a droplet of hydrogen peroxide solution for 5 min.
2. The grid is rinsed in multiple drops of water for 15 min.
3. The grid is placed on a drop of formic acid for 5 min.
4. The grid is then rinsed in multiple drops of water for a further 15 min

Immunogold labeling

1. Blocking step: Place grid on a drop of blocking serum for 30 min.
2. Primary antibody incubation: Incubate grid with appropriate dilution of specific primary antibody in BBI buffer overnight at 27 °C.
3. Rinse grid on BBI buffer: pass through multiple drops for 60 min.
4. Immunogold incubation: Place grid onto a drop of BBI gold IgG for 2 h.
5. Pass grid through multiple drops of water for 30 min.
6. Postfixation step: Apply glutaraldehyde for 10 min.
7. Wash grid in multiple drops of PBS for 10 min and then in water for a further 10 min.
8. Gold amplification: Place the grid on a drop of Goldenhance solution for 5 min.
9. The grid is then washed in multiple drops of water for a further 30 min.

Counterstain with uranyl acetate and lead citrate.

3.3 Electron Microscopy Counterstain Procedure

All steps of the staining procedure are performed at room temperature. Uranyl acetate staining is carried out in a humid chamber while lead citrate is carried out in a chamber devoid of moisture. This is achieved by placing sodium hydroxide pellets within the chamber.

1. Float grid section side down on the drops of uranyl acetate for 15 min.
2. Wash in water for 8 s. Blot grid on the edge of filter paper.

3. Place grid onto a lead citrate drop for 10 min
4. Rinse grid in water for 30 s.
5. Touch dry grid on filter paper and dry carefully using hairdryer (approximately 5–10 s).

4 Immunogold Electron Microscopy of the Nervous System

4.1 Nervous System: Membrane and Intracellular Forms of PrP^d

Most morphological types of PrP^d accumulation observed upon light microscopical examination of prion disease-infected brains, such as diffuse punctuate (often called synaptic), peri-neuronal, peri-glial, larger particulate, or coalescing foci, are each colocalized to plasma membranes of neurites, neuronal perikarya, and glial cells. Initial accumulations on membranes do not visibly alter the structure of the membrane or the cytological appearance of the parent cell or process. In agreement with cell biology studies [7], the early localization of PrP^d on plasma membranes in the absence of morphological change suggests that the misfolding of the normal cellular form of prion protein (PrP^c) takes place at this site and that the conversion of PrP^c into misfolded forms is favored when misfolded and normal molecules are attached to the same membrane [8]. The molecular form of the template is uncertain but small oligomers appear to be more likely candidates than either misfolded monomers or large aggregates, and the conversion most likely takes place in raft domains of the membrane [9]. Although the increasing abundance of misfolded forms of PrP may be associated with increasing infectious titer, the correlation between PrP^d and titer is poor, and membrane-stabilized misfolding of PrP^c into misfolded and aggregated PrP^d can occur without induction of further new infectious prions [10, 11].

Sustained or intense membrane accumulations of PrP^d are associated with specific morphological changes. Dendrites, axon terminals, and neuronal cell bodies may show bizarre plasma membrane invaginations. Invaginations may be spiral or branched; they often appear coated and are associated with an increase in the number of clathrin-coated pits and vesicles (Fig. 1a, b) (see [12] for review). These membrane changes colocalize with both PrP^d and ubiquitin. Immunogold and morphometric analysis suggests that PrP^d and ubiquitin are topologically localized to different sides of the membrane, respectively, the extracellular and intracellular faces [13]. As PrP^d is attached to the exterior of the membrane by its glycophosphoinositol anchor (GPI) it cannot interact directly with ubiquitin suggesting that ubiquitination and clathrin coating is elicited via an intermediary transmembrane molecule(s) as also suggested by cell culture studies [14]. Thus PrP^d molecules or oligomeric aggregates form protein complexes on membranes.

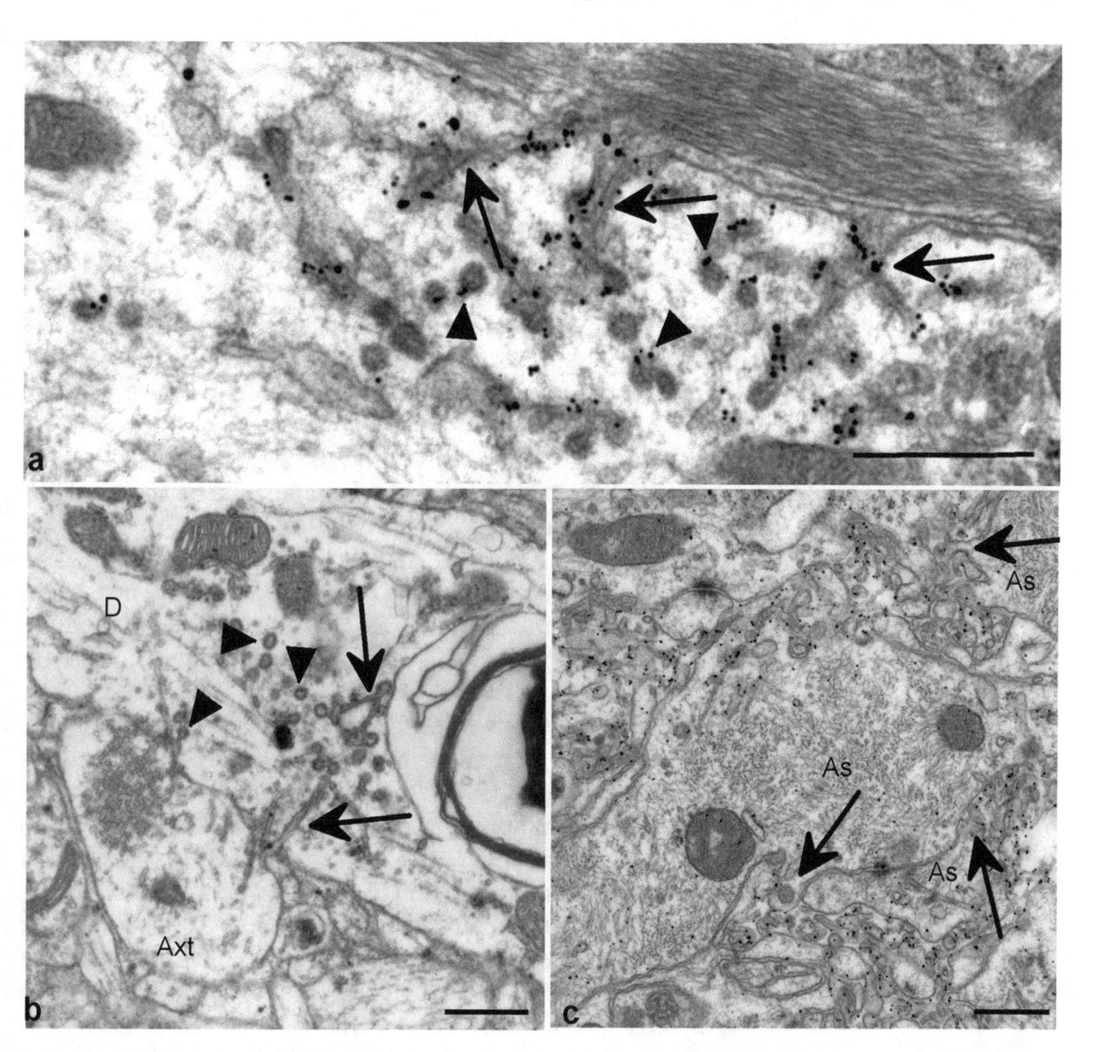

Fig. 1 Membrane invaginations of neurites. (**a**) Natural sheep scrapie. Dendrite showing invaginations of the plasma membrane. The invaginations (*arrows*) are branched with a submembrane coat. There are increased numbers of coated vesicles/pits (*arrowheads*). Immunogold labeling for PrP^d is associated with the invaginations and coated pits. Immunogold labeling—PrP. (**b**) Feline spongiform encephalopathy. Dendrite (D) and axon terminal (Axt) showing spiral membrane invaginations (*arrows*) and increased coated pits (*arrowheads*). Uranyl acetate/lead citrate counterstain. (**c**) Murine scrapie model created by a synthetic PrP source, PrP^{SSLOW} (see Jeffrey et al. [16]). Reactive astrocytic processes (As) are surrounded by immunogold labeling for PrP^d. PrP^d is located on membranes of ruffled and microfolded plasma membranes originating from these astrocytic processes (*arrows*). Immunogold labeling—PrP. (**a**) Bar = 500 nm; (**b**) Bar = 500 nm; (**c**) Bar = 500 nm

These immunogold data are consistent with the proposed role of the normal cellular prion protein molecule (PrP^c) as a scaffolding molecule involved in the assembly of multicomponent signaling modules at the cell surface [15].

Ubiquitinated PrP^d membrane complexes are ultimately internalized to late endosomes, which form abnormal irregular fused vesicular cisterns [13] and to lysosomes where they are truncated [2, 17] and accumulate prior to complete digestion. Scrapie infection is associated with increased numbers of neuronal endolysosomes and abnormal multivesicular bodies [13].

While PrP^d may accumulate on the plasma membranes of glial cells, these membrane PrP^d accumulations are not associated with invaginations. Prominent PrP^d accumulation may occur on astrocyte plasma membranes where it is associated with intense polyp-like microfolding or ruffling of the affected membrane (Fig. 1c), [13]. Microglial cells recognize and respond rapidly to the appearance of PrP^d on the plasma membanes of neurites by inserting fine strands of cytoplasm between PrP^d labeled neurites. PrP^d may be transferred to these cytoplasmic fingers, but in contrast to astrocytes these microglial cells do not show membrane ruffling. Ultimately microglial membrane and cytoplasmic PrP^d is degraded by lysosomes. This microglial response appears to be a general cellular surveillance function of microglia [18] as the same features are seen when GPI-anchored Aβ accumulates on neurite plasma membranes [19]. Oligodendroglia do not appear to be involved in PrP^d amplification in classical TSEs or prion diseases, but under some experimental conditions, rodent oligodendroglia can convert PrP^c into noninfectious disease-associated conformers of PrP^d [10].

Different proportions of different membrane and intracellular lesions are found in animals expressing different PRNP genotypes [20] and infected with different strains of scrapie. However, the distinctive membrane and endolysosomal changes described above are common to all animal TSEs examined so far (see [12] for review) and also to experimental rodent scrapie-like disease induced by intracerebral or intraperitoneal inoculation with refolded recombinant or synthetic PrP sources [16, 21]. Similar morphological defects of membranes have been reported in sporadic CJD [22, 23] though the colocalization of these lesions with PrP^d by immunogold electron microscopy has not yet been tested. The same membrane lesions are also found in a transgenic mouse in which susceptibility to infection is confined to PrP^c-expressing glia cells [24] showing that GPI-anchored PrP^d released from one cell readily reinserts into adjacent uninfected cell membranes by a process known as GPI anchor transfer [25]. Thus the presence of neuronal membrane-specific TSE lesions in a mouse expressing PrP^c on glial cells alone shows that endogenous PrP^c expression is not necessary for these lesions to form [24]. In contrast, membrane invaginations are absent from a mouse model in which PrP is engineered to lack the GPI anchor suggesting that anchoring to the membrane is necessary to allow PrP^d to interact with other membrane proteins [26]. The stereotypical nature of the membrane changes in all classical TSEs raises the possibility that they are a nonspecific response to the presence of aggregated protein at the cell membrane. However, when the Alzheimer's disease-associated peptide Aβ is attached to membranes with a GPI anchor, these specific membrane changes are absent although Aβ plaque formation is enhanced [19]. Thus PrP^d interacts in a specific and apparently unique way with membranes to cause neuritic invaginations and glial membrane ruffling.

4.2 Nervous System: Extracellular Amyloid and the Role of the GPI Anchor

PrP^d released from membranes into the extracellular space forms aggregates and short fibrils that are visible by electron microscopy. Where abundant fibrils are present, they organize into bundles and eventually form amyloid plaques that can be seen at light microscopy. Amyloid plaques are infrequent or absent in most naturally occurring animal TSEs [27], with the exception of a rare atypical bovine TSE known as bovine amyloidotic spongiform encephalopathy [28]. Similarly, amyloid plaques are restricted to particular categories of human prion diseases including variant Creutzfeldt-Jakob disease (vCJD), kuru, and in 10–15% of sporadic Creutzfeldt-Jakob disease (sCJD) individuals belonging to a subgroup sometimes identified as MV 2K [29, 30]. Only in some inherited PRNP mutations, particularly some forms of Gerstmann-Straüssler-Scheinker disease (GSS) [31], and in bovine amyloidotic spongiform encephalopathy are amyloid plaques the preponderant or sole feature. Amyloid plaques in TSEs generally take one of the two basic forms: kuru-type plaques, which are the most commonly encountered, or multicentric plaques which are confined to GSS. Kuru-type plaques are generally composed of a central core of extracellular amyloid fibrils surrounded by a stellate arrangement of radiating bundles of amyloid fibrils (Fig. 2a) [32–34]. Mature plaques are surrounded by swollen astrocytes and/or microglial cell processes and dystrophic neurites containing excess

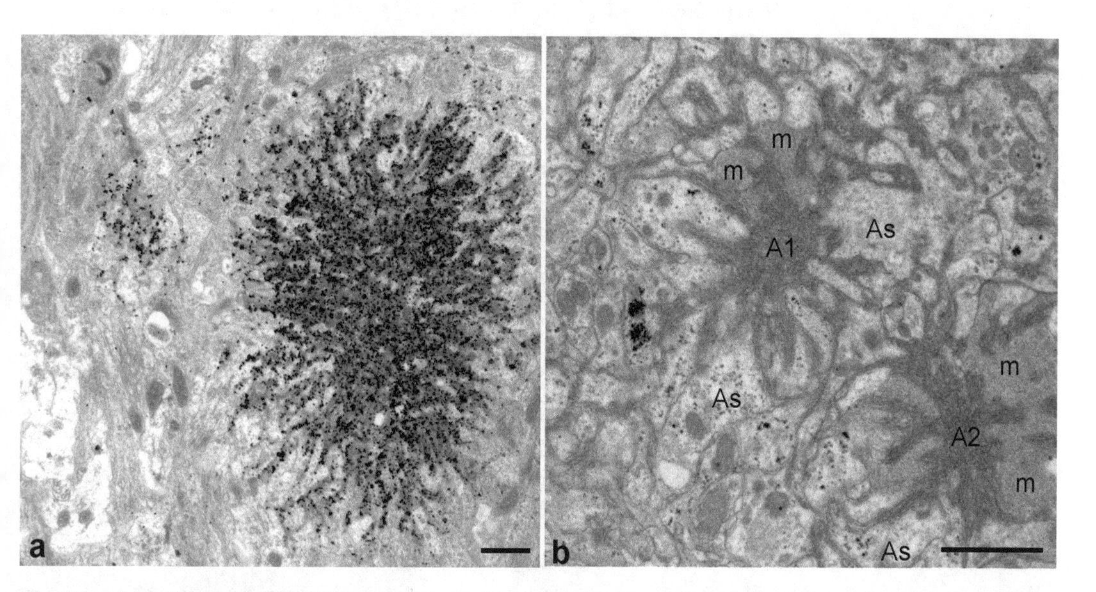

Fig. 2 Amyloid plaques. (**a**) Tg101LL mouse inoculated with 101 L amyloid fibrils. Kuru-type plaques consisting of a central core of extracellular amyloid fibrils surrounded by a stellate arrangement of radiating bundles of amyloid fibrils. Immunogold labeling—PrP. (**b**) Tg101LL mouse inoculated with GSS. Multicentric plaques consisting of two discrete cores (A1, A2), surrounded by microglial (m) or astrocytic (as) processes. Uranyl acetate/lead citrate counterstain. (**a**) Bar = 1 μm; (**b**) Bar = 1 μm

densely packed organelles, mainly autophagic vacuoles and lysosomes. Florid plaques of vCJD also consist of a single core but are less compact than kuru-type plaques. They are composed of thicker amyloid fibrils and are surrounded by frequent dystrophic neurites [34, 35]. GSS plaques [35], including plaques found in GSS-challenged mice carrying GSS transgenes [10], are generally larger and more irregular than kuru-type plaques with multicentric dense cores that often contain interweaving sheets of amyloid [10, 34].

The formation of plaques has been studied in rodent models of scrapie and GSS. In the 87 V murine scrapie strain, early plaques form initially from the localized release of PrP^d from the plasma membrane of individual dendrites which subsequently mature into compact kuru-type plaques [32]. Multicentric GSS plaques also derive from oligomeric PrP^d amyloid precursors converted from membrane PrP^c. In contrast to kuru-type plaques, multicentric plaques (Fig. 2b) grow from multiple locally distributed seeds dispersed within the extracellular space by a "parent" plaque, or from seeds dispersed from the original experimental inoculum [10].

Extracellular PrP^d amyloid may also take the form of cerebral amyloid angiopathy (CAA) (Fig. 3a) with a markedly different pathogenesis from kuru and multicentric plaques. CAA is the major or exclusive pathology of some rare familial forms of prion disease [31] and in a transgenic mouse line in which PrP^c lacks its GPI anchoring to the cell membrane (the anchorless PrP mouse) [26]. CAA is also a feature of intracerebral challenge with synthetic forms of prion disease [16] and occurs as a rare incidental feature of some natural chronic wasting disease [36] and scrapie cases [37].

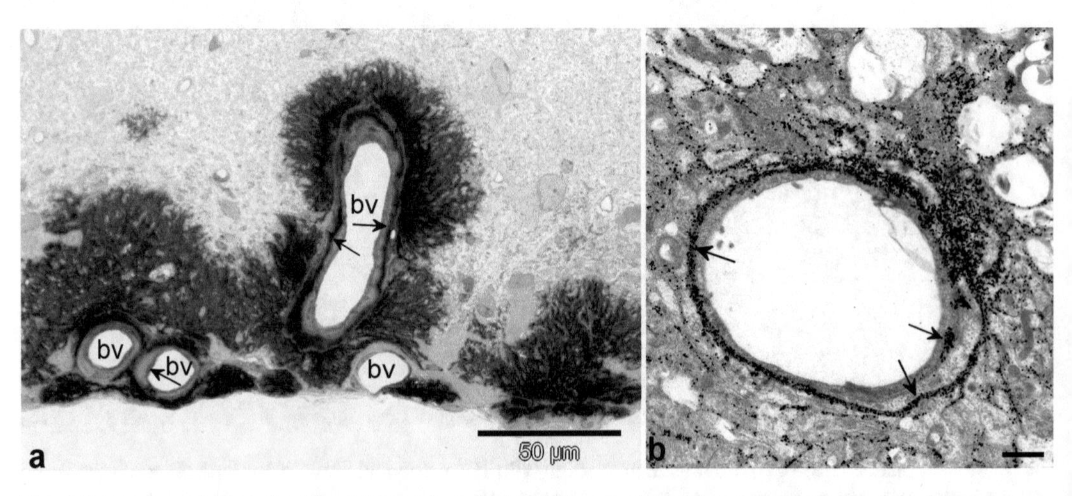

Fig. 3 Cerebral amyloid angiopathy. (**a**) Scrapie-infected GPI anchorless mouse. Light microscopy showing accumulation of PrP^d around blood vessels (bv) and within blood vessel walls (*arrows*). Light microscopy immunolabeling of resin-embedded tissue—PrP. (**b**) Scrapie-infected GPI anchorless mouse. PrP^d accumulation in association with basement membranes (*arrows*). Immunogold labeling—PrP. (**a**) bar = 50 μm; (**b**) Bar = 1 μm

Prion disease-associated CAA is a protein elimination failure arteriopathy [38]. Unlike in other tissues where interstitial fluids are removed by lymphatic ducts, extracellular fluids and solutes of the brain are removed via perivascular channels and through basement membranes of capillaries and arterioles. The term protein elimination failure has been coined for the process of entrapment of proteins in these perivascular pathways. CAA is a common feature of Alzheimer's disease and is caused by the aggregation of Aβ peptides in basement membranes of endothelial cells and other cells of blood vessel walls. CAA may also occur in prion diseases [38].

The PrP anchorless mouse does not retain PrP^c at the surfaces of cells but releases them into the extracellular space and interstitial fluid [39]. When infected with scrapie, anchorless mice become sick but do not accumulate PrP^d in any of the conventional immunohistochemical types usually recognized in naturally occurring TSEs. However, widespread vascular amyloid plaques are present throughout the brain [39]. Immunogold electron microscopy of scrapie-infected anchorless mice shows that there is an initial accumulation of nonfibrillar PrP^d in basement membranes of capillaries (Fig. 3b) and in the abluminal basement membranes of arterioles. This is followed by a progressive involvement of basement membranes located towards the lumen of arterioles and the subsequent assembly of fibrils within basement membranes which precedes replacement of the vessel walls by amyloid resulting in CAA [26]. These aggregates of PrP^d delay drainage of interstitial fluids [40]. PrP^d entrapment in basement membranes is accompanied by altered distribution of laminin, and probably other components of basement membranes, suggesting that the stabilization of PrP^d in basement membranes is due to an interaction with other extracellular matrix molecules [38].

The pathogenesis of CAA in rare human familial prion diseases is probably similar to that of the scrapie-infected PrP anchorless mouse. CAA found in human familial prion disease is associated with several specific genetic inherited mutations of the PRNP gene. Each of these familial disorders is associated with a premature stop mutation or a long truncation mutation that results in carboxyl C-terminal truncation of PrP^c [31]. Endogenous PrP^c and the subsequently formed PrP^d in each of these familial mutations therefore lack a GPI anchor to tether PrP^d to cell membranes and thus will permit its distribution into the interstitial fluid.

Intracerebral inoculation of wild-type or some transgenic mouse lines with refolded and aggregated recombinant forms of prion protein and other aggregate-containing inocula may induce the formation of amyloid plaques or occasionally CAA [10, 11, 41]. Similarly, synthetic prion protein molecules that are generated by bacteria and which lack a GPI anchor and additional sugar side chains may produce amyloid plaques at primary passage. However, secondary passage or recombinant PrP is often necessary for more widespread pathology [42].

Amyloid plaques and CAA induced by intracerebral passage of recombinant PrP mainly affect the corpus callosum and stratum lacunosum moleculare of the hippocampus [43]. When low volumes of Indian ink are inoculated into the neocortex, the ink distributes around blood vessels of the corpus callosum and subsurface of the hippocampus and hippocampal fissure [44]. Similarly, when brain homogenates containing Aβ are inoculated into mutant mice overexpressing Alzheimer's precursor protein, plaques also accumulate at these sites [44]. Studies which have investigated the fate of intracerebral inoculums show that Indian ink particles and fluorospheres expand the perivascular spaces rather than entering the narrow basement membranes of the perivascular drainage pathways and can be forced along these perivascular spaces in a volume- and pressure-dependent manner [45, 46]. Intracerebral synthetic PrP challenge experiments are generally conducted with high volumes of inoculum and it is likely that larger aggregates within these inoculums are distributed along expanded perivascular spaces in a similar way to Indian ink particles. Only some smaller soluble components in the inoculum may enter directly into the interstitial drainage pathways in vascular basement membranes and result in prion CAA [45]. Thus, the distribution of amyloid plaques and of CAA following intracerebral challenge with aggregated forms of prion protein corresponds to distributions of particulates and of solutes within the inoculum, respectively. When considering the pathogenesis of different experimental prion disease models, the molecular structure of the prion protein, the sizes of aggregates or seeds, and the site, volumes, and rates of delivery of inoculums need to be considered when comparing disease phenotypes of different sources or strains. A different distribution of PrP^{d} at the clinical end stage of disease may therefore occur following challenge with the same prion source when used in different experimental conditions. Thus, different patterns of PrP^{d} or PrP^{res} accumulation should only be interpreted to suggest that the original source inoculums contained different prion strains when all other experimental conditions were the same.

4.3 Nervous System: Atypical Forms of Prion Disease and Alternate PrP Trafficking

Increased awareness and surveillance for prion disease following the recognition of cattle BSE and human vCJD has led to the recognition of a number of nonclassical TSEs or prion diseases including protease-sensitive prionopathy in man, bovine amyloidotic spongiform encephalopathy (also known as L-type BSE), H-type BSE in cattle, and atypical scrapie of sheep and goats. Not surprisingly, given their rarity, there is little data on the subcellular pathology of these diseases but some information has been obtained for atypical sheep scrapie and atypical scrapie transmitted to Tg338 mice.

In contrast to classical sheep scrapie sources, atypical scrapie is a sporadic disease, with long incubation periods. It occurs in geographical regions from which classical scrapie may be absent and typically affects PRNP genotypes considered resistant to classical scrapie [47, 48]. Abnormal prion protein biochemically extracted from atypical scrapie has multiple weakly protease-resistant PrP^{res} species, with a distinctive low-molecular-weight band around 12 kDa [47, 49]. Brain immunohistochemistry of PrP^{d} shows predominantly a diffuse finely punctuate accumulation in cerebellum and cerebrum with distinctive white matter PrP^{d} accumulation [50].

In atypical scrapie of sheep and TG338 mice there are distinctive lesions of myelinated fibers comprising proliferation of the inner oligodendroglial mesaxon which is associated with PrP^{d} immunogold labeling (Fig. 4a). PrP^{d} immunolabeling is mainly found in association with electron-dense mesaxonal cytoplasm often containing electron-dense granules, small vesicles, coated pits, or coated vesicles (Fig. 4b); however, it is also occasionally found on axonal plasma membranes. Some oligodendroglial associated PrP^{d} is specifically localized to the cytoplasmic loops at some paranodes [51].

These atypical scrapie lesions are unique with respect to previous descriptions of classical and naturally occurring forms of prion disease. However, similar oligodendroglial inner mesaxon proliferation occurs in Tg(PG14) transgenic mice which expresses a nine-octapeptide insertional mutation homologous to that of a familial prion disease of humans [52, 53]. This Tg (PG14) mouse transports a weakly protease-resistant form of prion protein in axons

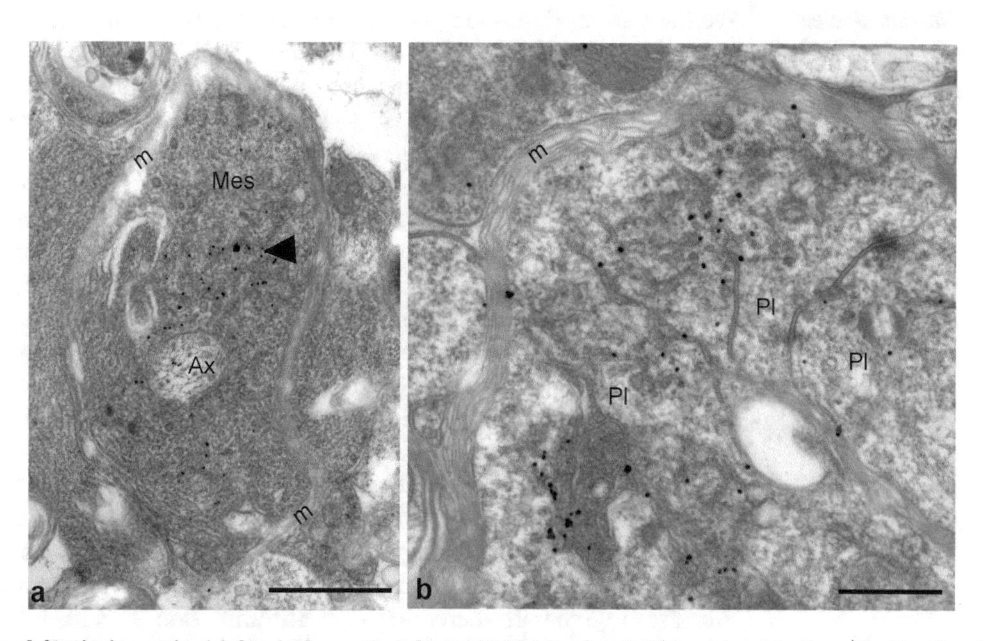

Fig. 4 Atypical scrapie. (**a**) Atypical scrapie-infected Tg338 mouse. PrP^{d} is present in hypertrophic oligodendroglial mesaxonal cytoplasm (mes) surrounding a small axon (ax). Most PrP^{d} labeling is associated with intracytoplasmic membranes (*arrowhead*) or coated pits. The pallor of myelin sheaths (m) is an artifact of the immunogold labeling method. Immunogold labeling—PrP. (**b**) Atypical scrapie-infected Tg338 mouse. PrP^{d} is associated with oligodendroglial paranodal loops (Pl) of cytoplasm. Myelin is again discolored (m). Immunogold labeling—PrP. (**a**) Bar = 1 μm; (**b**) bar = 500 nm

[54]. Oligodendroglial mesaxonal proliferation is detected in some toxic leukoencephalopathies and is thought to be a response to the presence of intra-axonal aggregates or toxins [55]. By extrapolation, PrP^{d}-associated oligodendroglial inner mesaxonal proliferation of atypical scrapie is also potentially a response to the presence of weakly protease-resistant axonal PrP^{d} aggregates [51].

Thus, the cellular localizations of PrP^{d} in classical and atypical scrapie differs. In atypical scrapie, PrP^{d} is prominent on the oligodendroglial mesaxon and on axons of myelinated processes while most classical scrapie is associated with dendritic and perikaryonal membranes. These data show that the cellular targeting and cellular trafficking of atypical PrP^{d} differ from those of classical scrapie. With respect to neurons, classical scrapie traffics aggregates predominantly into dendrites and neuronal somas while a significant fraction of atypical scrapie PrP^{d} aggregates appear to be trafficked to axons. Wallerian-type degeneration is not uncommon in classical prion disease but the extent of this degeneration and other white matter lesions is particularly conspicuous in atypical scrapie-affected sheep and mice [51] and may be related to the putative altered trafficking of atypical PrP^{d}. The mechanisms underpinning altered PrP^{d} trafficking are uncertain but may be related to the protease sensitivity of aggregates.

4.4 Nervous System: Lesions Lacking Associations with PrP^{d}

We have so far described positive correlations between subcellular lesions and PrP^{d} accumulations. However, it is by no means clear that the lesions described above are the proximate cause of clinical neurological defects. There are a number of reasons to suppose that neurological defects do not have a simple correlation with PrP^{d} or PrP^{res} accumulation, not least there being a poor correlation between the magnitude of PrP^{d} or PrP^{res} and incubation period or infectivity titer [56, 57]. It is therefore important to record negative correlations, that is, those lesions that lack colocalization with PrP^{d}. Subcellular features of scrapie that do not colocalize with PrP^{d} include vacuolation, the cardinal diagnostic feature of TSEs, and also neuronal apoptosis, neuronal and synaptic autophagy, axon terminal degeneration (occurs only in mice), and so-called tubulovesicular bodies.

Although there is no precise colocalization of PrP^{d} with vacuolation, neuronal apoptosis, autophagy, and axon terminal degeneration, each of these changes occurs in neuroanatomic areas in which PrP^{d} accumulation takes place and each of these features increases in line with increasing PrP^{d} accumulation. It is likely that these neurodegenerative features make variable contribution to clinical disease according to the severity of each change found in different TSE infections. In contrast, tubulovesicular bodies occur very early in incubation period when PrP^{d} is still minimally detected [58]. However, tubulovesicular bodies can only be observed by electron microscopy and their molecular cause and clinical and pathological significance remains uncertain.

5 Immunogold Electron Microscopy of Visceral Tissues

5.1 Lymphoid System

The naturally occurring and contagious forms of TSEs including scrapie of sheep and goats and chronic wasting disease of deer, vCJD, and the experimental rodent models of these diseases all show accumulation of PrP^d and infectivity in tissues of the lymphoreticular system (LRS). Prior to neuroinvasion TSE agents often accumulate to moderately high titer levels within the LRS. Light microscopy studies show that PrP^d in LRS tissues is associated with follicular dendritic cells (FDCs) in the light zones of germinal centers and with tingible body macrophages in the light zone and dark zones of secondary follicles.

FDCs characteristically trap and retain immune complexes on their surfaces which they present to B-lymphocytes to facilitate clonal selection of B cells and induction of plasma cells. Immune complexes are retained at the surface of FDCs by Fc and complement receptors. Following scrapie infection, FDCs accumulate PrP^d on the cell surface of dendrites. This corresponds with abnormal dendritic hypertrophy, increased retention of electron-dense material at FDC plasma membrane (Fig. 5a), and an increase in numbers of mature B lymphocytes within the secondary follicles [59, 60]. The increased electron-dense material trapped by hypertrophic FDC dendrites contains both excess IgM and IgG immune complexes (Fig. 5b) and is putatively caused by an interaction between PrP^d and Fc or complement receptors [61].

Normal germinal centers grow in response to antigen stimulation and ultimately regress. FDC progresses though immature, mature, and regressing phases during this germinal center cycle [62]. Scrapie infection appears to perturb this maturation cycle. In scrapie-affected mesenteric lymph nodes, some FDCs were found where areas of normal and abnormal immune complex retention occurred side by side [61]. The latter colocalized with plasma membrane accumulations of PrP^d, suggesting that PrP^d accumulation at the plasma membrane of FDC dendrites is the initial stage of the abnormal FDC maturation cycle and leads to abnormal cell membrane ubiquitin and excess immunoglobulin accumulation. Regressing FDCs, in contrast, appeared to lose their membrane-attached PrP^d. Together, these data suggest that TSE infection adversely affects the maturation and regression cycle of FDCs, and that PrP^d accumulation is causally linked to the abnormal pathology observed.

Tingible body macrophages, whose normal function is to remove effete B cells, show abundant PrP^d in association with the cell membrane, noncoated pits and vesicles, and also discrete, large,

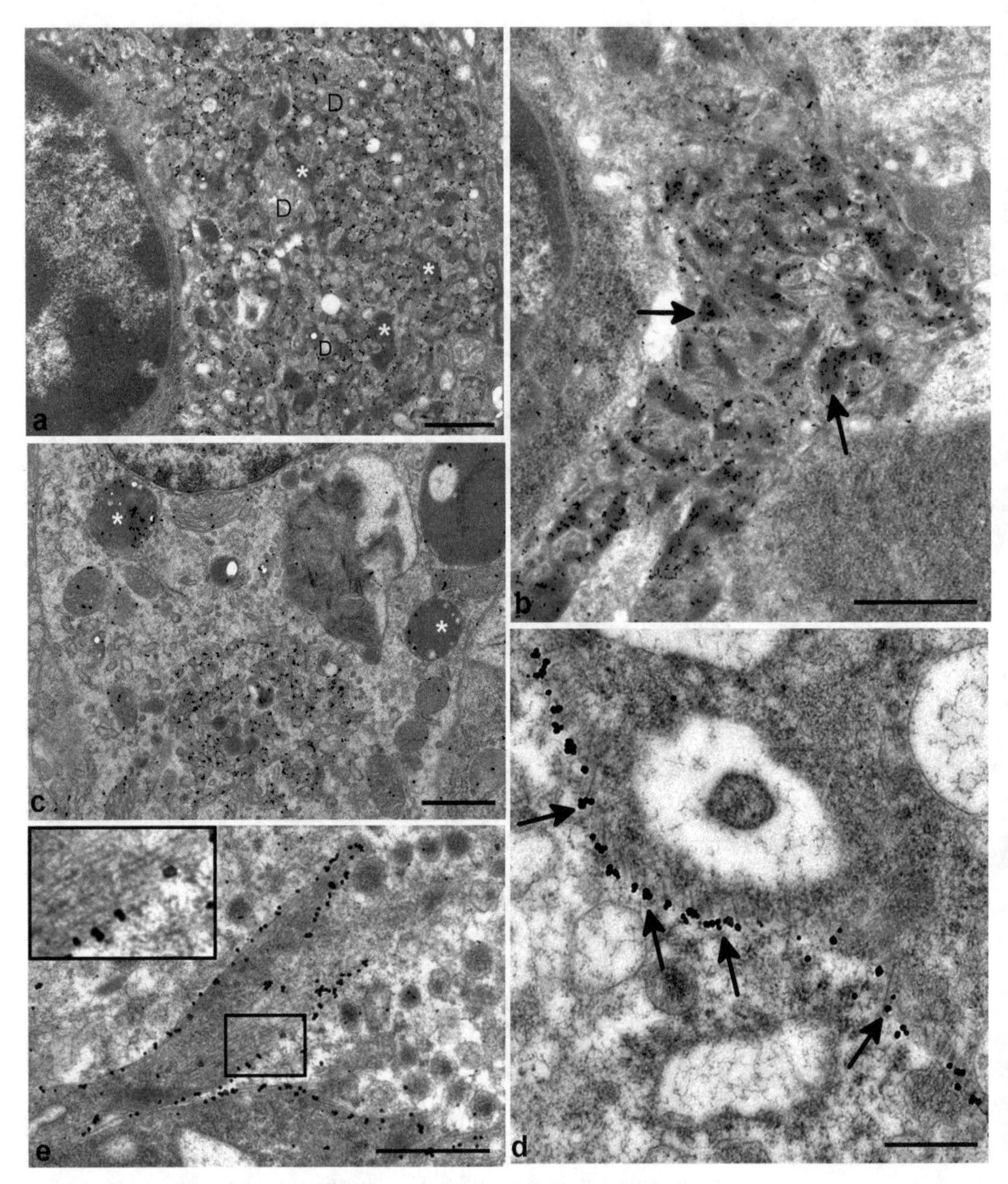

Fig. 5 Responses to PrP^d accumulation in viscera. (**a**) Murine scrapie. Hypertrophic FDC labyrinthine complex within the spleen showing PrP^d labeling primarily in association with the plasma membrane of dendrites (D) and not the adjacent electron-dense deposit (*asterisk*). Immunogold labeling—PrP. (**b**) Murine scrapie. Immunogold labeling for immunoglobulins is primarily associated with the extensive electron-dense deposit between processes of the matured FDC within the spleen (*arrows*). Immunogold labeling—IgG. (**c**) Natural sheep scrapie. PrP^d is associated with the membrane of fused ER networks within the cytoplasm of a TBM. Lysosomes are variably labeled (*asterisks*). Immunogold labeling—PrP. (**d**) Natural sheep scrapie. Chromaffin cells showing PrP^d labeling of membrane palisades. Palisades form short parallel segments of straight membranes. PrP^d labeling is limited to the polar ends of these palisades (*arrows*). Immunogold labeling—PrP. (**e**) Natural sheep scrapie. Segments of abnormal chromaffin cell membranes (shown *inset*) with developed palisades and abundant PrP^d accumulation invaginate into the cytoplasm of chromaffin cells containing multiple cytoplasmic granules. Immunogold labeling—PrP. (**a**) Bar = 1 μm; (**b**) Bar = 1 μm; (**c**) Bar = 1 μm; (**d**) Bar = 500 nm; (**e**) Bar = 1 μm

and fused endoplasmic reticulum networks, which colocalized with ubiquitin (Fig. 5c) [63]. As for microglia in the brain, lysosomal PrP^d is first truncated to a C-terminal core corresponding to a 19 or 21 kDa fragment [64] and is then further degraded [65]. Unlike neurons which endocytose membrane PrP^d by a ubiquitin and clathrin-mediated mechanism, tingible body macrophages internalize PrP^d by a nonclathrin-mediated, probably caveolin mechanism [63].

5.2 Adrenal and Kidney

There are a number of sites in addition to lymphoid tissues that can accumulate PrP^d. These include the retina, kidney, adrenal medulla, muscle, and placenta [27]. Significant amounts of infectivity can be detected in blood which may contribute to this infection of visceral tissues, and also to neuroinvasion via the circumventricular organs [66]. The cellular association of PrP^d in blood has not been established but disease transmission can be achieved in blood fractions lacking cells [67]. The subcellular localization of PrP^d has been established in kidney and adrenal of scrapie-affected sheep.

In the adrenal medulla, nonfibrillar forms of PrP^d accumulate mainly in association with plasma membranes of chromaffin cells, occasional nerve endings, and endolysosomes of macrophages [68]. As with CNS and LRS tissues, membrane PrP^d is associated with morphological changes. PrP^d colocalized with segments of chromaffin cell membrane which demonstrated abnormal electron density, contorting palisades, and parallel or irregular membrane segments (Fig. 5d), and which appeared to invaginate into the cytoplasm of the chromaffin cell (Fig. 5e). Internalization of PrP^d from the chromaffin cell plasma membrane occurred in association with granule recycling following hormone exocytosis. These changes further show that the PrP^d membrane toxicity is tissue and cell specific, further supporting the idea that the normal protein may act as a multifunctional scaffolding molecule.

PrP^d and PrP^{res} are deposited in the renal papillae of approximately half of scrapie-affected sheep, most commonly in sheep showing abundant PrP^d in the LRS [69]. Using electron microscopy, PrP^d was shown to accumulate in the interstitium of the renal papillae, in association with the cell membrane and lysosomes of fibroblasts cells or extracellularly, in close contact with collagen and basal membranes [69]. These sites strongly argue for a hematogenous origin of renal PrP^d, though whether this occurs following glomerular filtration or extravasation from vasa recta capillaries remains unresolved. Abundant amyloid is also found in a number of visceral tissues of the scrapie-infected PrP GPI anchorless mice. The amyloid in such mice is closely associated with collagen and basal membranes in tissues such as the intestine, myocardium, and brown fat [70]. This amyloid is also presumptively derived from hematogenous PrP^d.

6 Summary

Immunogold electron microscopy has shown that the primary toxic effects of PrP^d occur at the plasma membrane, and that the nature of the toxic effect appears constant for all classical TSEs that give rise to biochemical PrP^{res} fragments of 19 or 21 kDa. Thus, the main subcellular toxic effects of PrP^d are not affected by host species, genotype, or infecting strain. Little electron microscopy data is currently available for the novel atypical prion diseases recognized in recent years. However, a proportion of PrP^d in naturally occurring and experimental atypical scrapie is directed along axons and exerts a toxic effect on oligodendroglial mesaxons. Thus, atypical scrapie PrP^d appears to have altered trafficking relative to classical PrP^d sources.

Although the nature of membrane changes is constant across all classical TSEs, membrane toxicities are tissue and cell type dependent (Fig. 6). Similarly, the molecular interactions of PrP^d also appear to differ according to the cell type (Fig. 7). This lack of consistency of membrane changes and molecular interactions

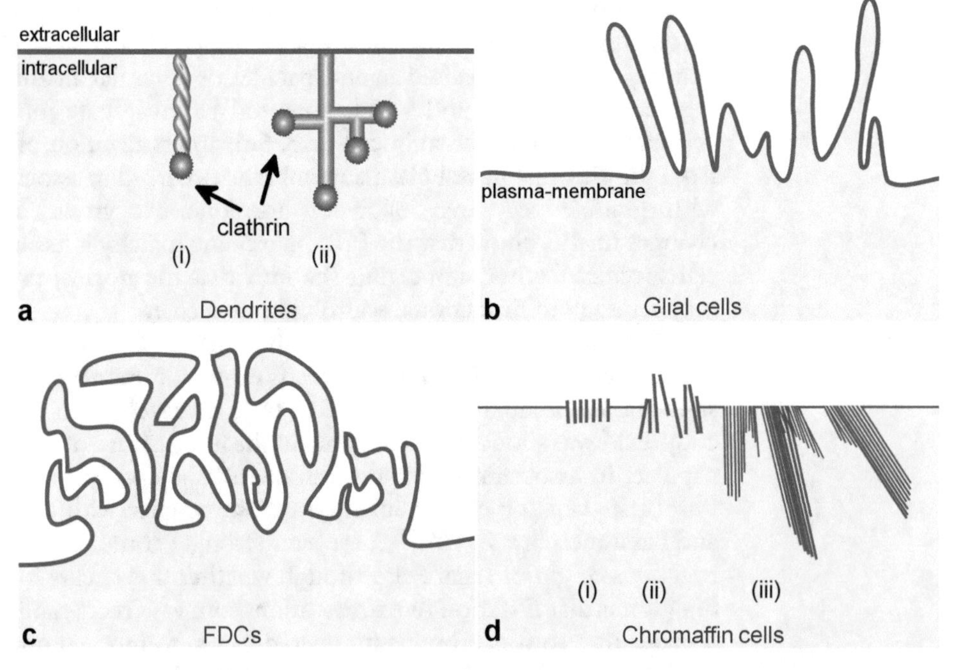

Fig. 6 Diagrammatic representations of PrP^d-associated membrane changes. (**a**) On dendrites, membranes are invaginated. They may be spiral (*i*) or branched (*ii*) and often have clathrin coats at their extremities. (**b**) On glial cells, membranes may show ruffling or microfolds. (**c**) FDC processes show marked convoluted enlargement. (**d**) On Chromaffin cells, membranes form short parallel linear (*i*) or irregular segments (*ii*) perpendicular to adjacent normal membranes, and deep invaginations consisting of many parallel linear segments of membrane (*iii*)

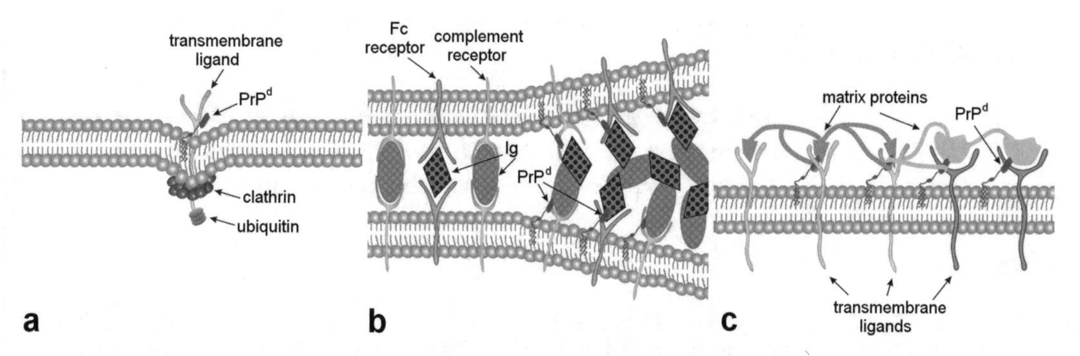

Fig. 7 Diagrammatic representations of PrP^d interactions. (**a**) PrP^d internalization: PrP^d on the exterior leaflet interacts with transmembrane ligands to elicit ubiquitination and clathrin-mediated invagination. (**b**) PrP^d and FDC processes: PrP^d interacts with membrane Fc and complement receptors to trap excess immunoglobulins (Ig). (**c**) PrP^d as a scaffolding molecule: PrP^d interacts with numerous undefined ligands and associated matrix proteins which create a stable multimolecular complex. These complexes are responsible for the diversity of morphological changes of membranes associated with different cell types

among different tissues suggests that PrP^c may be scaffolding protein and the precise toxic effects on membranes are manifest by multimolecular complexes rather than by PrP^d alone. Cell membrane PrP^d may be internalized and digested within endolysosomes or released from the cell membrane into the extracellular space. Released forms of PrP^d may be reinserted by their GPI anchor into adjacent cell membranes or be internalized and digested within lysosomes of the relevant tissue phagocytes. Some extracellular PrP^d may aggregate into amyloid fibrils and plaques. Under particular experimental conditions and in some familial forms of prion disease, absence of GPI anchors or sugar side chains facilitates diffusion of PrP^d in the interstitial fluid where it may eventually be trapped within the basement membranes of blood vessels causing CAA.

It seems unlikely that membrane toxicity alone is sufficient to account for clinical disease effects, not least because of the paucity of PrP^d or PrP^{res} in some diseases and genotype combinations of prion disease. Other lesions such as axon terminal loss, autophagy, and activation of microglia that may have clinical impact do not colocalize with PrP^d but are likely to be indirectly linked to PrP^d accumulation. However some ultrastructural changes such as the accumulation of tubulovesicular bodies have no direct link to PrP^d and may have as yet an unknown role in disease pathogenesis. Important areas of future research include increased understanding of the proximate causes of clinical disease and the relationship between toxic forms of PrP^d and infectivity.

Acknowledgements

This work was financially supported by Defra. We are grateful to the many collaborators who have made this work possible and to Callum Donnelly for expert technical assistance.

References

1. González L, Terry L, Jeffrey M (2005) Expression of prion protein in the gut of mice infected orally with the 301V murine strain of the bovine spongiform encephalopathy agent. J Comp Pathol 132(4):273–282
2. Jeffrey M, Martin S, González L (2003) Cell-associated variants of disease-specific prion protein immunolabelling are found in different sources of sheep transmissible spongiform encephalopathy. J Gen Virol 84(4): 1033–1046
3. González L, Martin S, Houston FE, Hunter N, Reid HW, Bellworthy SJ, Jeffrey M (2005) Phenotype of disease-associated PrP accumulation in the brain of bovine spongiform encephalopathy experimentally infected sheep. J Gen Virol 86(3):827–838
4. Kovacs GG, Head MW, Hegyi I, Bunn TJ, Flicker H, Hainfellner JA, McCardle L, Laszlo L, Jarius C, Ironside JW, Budka H (2002) Immunohistochemistry for the prion protein: comparison of different monoclonal antibodies in human prion disease subtypes. Brain Pathol 12(1):1–11
5. Farquhar CF, Sommerville RA, Ritchie LA (1989) Post mortem immunodiagnosis of scrapie and bovine spongiform encephalopathy. J Virol Methods 24:215–222
6. Choi JK, Park SJ, Jun YC, Oh JM, Jeong BH, Lee HP, Park SN, Carp RI, Kim YS (2006) Generation of monoclonal antibody recognized by the GXXXG motif (glycine zipper) of prion protein. Hybridoma (Larchmt) 25(5):271–277
7. Caughey B, Raymond G (1991) The scrapie-associated form of PrP is made from a cell surface precursor that is both protease- and phospholipase-sensitive. J Biol Chem 266(27):18217–18223
8. Baron GS, Wehrly K, Dorward DW, Chesebro B, Caughey B (2002) Conversion of raft associated prion protein to the protease-resistant state requires insertion of PrP-res (PrPSc) into contiguous membranes. EMBO J 21(5):1031–1040
9. Caughey B, Baron GS, Chesebro B, Jeffrey M (2009) Getting a grip on prions: oligomers, amyloids, and pathological membrane interactions. Annu Rev Biochem 78(1):177–204
10. Jeffrey M, McGovern G, Chambers EV, King D, González L, Manson JC, Ghetti B, Piccardo P, Barron RM (2012) Mechanism of PrP-amyloid formation in mice without transmissible spongiform encephalopathy. Brain Pathol 22(1):58–66
11. Piccardo P, King D, Telling G, Manson JC, Barron RM (2013) Dissociation of prion protein amyloid seeding from transmission of a spongiform encephalopathy. J Virol 87(22):12349–12356
12. Jeffrey M, McGovern G, Sisó S, González L (2011) Cellular and sub-cellular pathology of animal prion diseases: relationship between morphological changes, accumulation of abnormal prion protein and clinical disease. Acta Neuropathol 121(1):113–134
13. Jeffrey M, McGovern G, Goodsir CM, González L (2009) Strain-associated variations in abnormal PrP trafficking of sheep scrapie. Brain Pathol 19(1):1–11
14. Harris DA (1997) Cell biological studies of the prion protein. In: Harris DA (ed) Prions: molecular and cellular biology. Horizon Scientific Press, Wymondham, pp 53–65
15. Linden R, Cordeiro Y, Lima LM (2012) Allosteric function and dysfunction of the prion protein. Cell Mol Life Sci 69(7):1105–1124
16. Jeffrey M, McGovern G, Makarava N, Gonzalez L, Kim JS, Rohwer RG, Baskakov IV (2013) Pathology of SSLOW, a transmissible and fatal synthetic prion protein disorder and comparison with naturally occurring classical Transmissible Spongiform Encephalopathies. Neuropathol Appl Neurobiol 40:296–310
17. Beringue V, Demoy M, Lasmezas CI, Gouritin B, Weingarten C, Deslys JP, Andreux JP, Couvreur P, Dormont D (2000) Role of spleen macrophages in the clearance of scrapie agent early in pathogenesis. J Pathol 190(4):495–502

18. Nimmerjahn A, Kirchhoff F, Helmchen F (2005) Resting microglial cells are highly dynamic surveillants of brain parenchyma in vivo. Science 308(5726):1314–1318
19. Jeffrey M, McGovern G, Barron R, Baumann F (2015) Membrane pathology and microglial activation of mice expressing membrane anchored or membrane released forms of Abeta and mutated human APP. Neuropathol Appl Neurobiol 41(4):458–470
20. Ersdal C, Simmons MM, Goodsir C, Martin S, Jeffrey M (2003) Sub-cellular pathology of scrapie: coated pits are increased in PrP codon 136 alanine homozygous scrapie-affected sheep. Acta Neuropathol 106(1):17–28
21. Wang X, McGovern G, Zhang Y, Wang F, Zha L, Jeffrey M, Ma J (2015) Intraperitoneal infection of wild-type mice with synthetically generated mammalian prion. PLoS Pathog 11(7):e1004958
22. Bastian FO (1979) Spiroplasma-like inclusions in Creutzfeldt-Jakob disease. Arch Pathol Lab Med 103(13):665–669
23. Liberski PP, Yanagihara R, Wells GAH, Gajdusek DC (1993) Neuronal degeneration in unconventional slow virus disorders. In: Liberski PP (ed) Light and electron microscopic neuropathology of slow virus disorders. CRC Press, Boca Raton, pp 349–371
24. Jeffrey M, Goodsir CM, Race RE, Chesebro B (2004) Scrapie-specific neuronal lesions are independent of neuronal PrP expression. Ann Neurol 55(6):781–792
25. Ilangumaran S, Robinson PJ, Hoessli DC (1996) Transfer of exogenous glycosylphosphatidylinositol (GPI)-linked molecules to plasma membranes. Trends Cell Biol 6(5):163–167
26. Chesebro B, Race B, Meade-White K, LaCasse R, Striebel J, Klingeborn M, Race R, McGovern G, Dorward D, Jeffrey M (2010) Fatal transmissible amyloid encephalopathy: a new type of prion disease associated with lack of prion protein membrane anchoring. PLoS Pathog 6(3):e1000800
27. Jeffrey M, González L (2004) Pathology and pathogenesis of bovine spongiform encephalopathy and scrapie. In: Harris D (ed) Mad cow disease and related spongiform encephalopathies. Springer, Berlin, pp 65–97
28. Casalone C, Zanusso G, Acutis P, Ferrari S, Capucci L, Tagliavini F, Monaco S, Caramelli M (2004) Identification of a second bovine amyloidotic spongiform encephalopathy: molecular similarities with sporadic Creutzfeldt-Jakob disease. Proc Natl Acad Sci U S A 101(9):3065–3070
29. Parchi P, Strammiello R, Notari S, Giese A, Langeveld JP, Ladogana A, Zerr I, Roncaroli F, Cras P, Ghetti B, Pocchiari M, Kretzschmar H, Capellari S (2009) Incidence and spectrum of sporadic Creutzfeldt-Jakob disease variants with mixed phenotype and co-occurrence of PrP(Sc) types: an updated classification. Acta Neuropathol 118(5):659–671
30. Head M, Ironside JI, Ghetti B, Jeffrey M, Piccardo P, Will RG (2015) Prion diseases. In: Greenfield's neuropathology, vol 2, 9th edn. CRC Press, Boca Raton, pp 1016–1086
31. Revesz T, Holton JL, Lashley T, Plant G, Frangione B, Rostagno A, Ghiso J (2009) Genetics and molecular pathogenesis of sporadic and hereditary cerebral amyloid angiopathies. Acta Neuropathol 118(1):115–130
32. Jeffrey M, Goodsir CM, Bruce ME, McBride PA, Farquhar C (1994) Morphogenesis of amyloid plaques in 87V murine scrapie. Neuropathol Appl Neurobiol 20(6):535–542
33. Liberski PP, Guiroy DC, Williams ES, Yanagihara R, Brown P, Gajducek DC (1992) The amyloid plaque. In: Liberski PP (ed) Light and electron microscopic neuropathology of slow virus disorders. CRC press, Boca Raton, pp 295–347
34. Sikorska B, Liberski PP, Sobow T, Budka H, Ironside JW (2009) Ultrastructural study of florid plaques in variant Creutzfeldt-Jakob disease: a comparison with amlyoid plaqes in kuru, Creutzfeldt-Jakob disease and Gerstamnn-Straussler-Scheinker disease. Neuropathol Appl Neurobiol 35(1):46–59
35. Boellaard JW, Schlote W, Heldt N (1989) The development of amyloid plaques in human transmissible encephalopathy and the role of astrocytes in their formation, Paris 2–6 December 1986. In: Court LA, Dormont D, Brown P, Kingsbury DT (eds) Unconventional virus diseases of the central nervous system. Commissariat à l'Energie Atomique, Fontenay-aux-Roses, Paris, pp 162–171
36. Guiroy D, Williams E, Liberski P, Wakayama I, Gajdusek D (1993) Ultrastructural neuropathology of chronic wasting disease in captive mule deer. Acta Neuropathol 85(4):437–444
37. Jeffrey M, Goodsir CM, Holliman A, Higgins RJ, Bruce ME, McBride PA, Fraser JR (1998) Determination of the frequency and distribution of vascular and parenchymal amyloid with polyclonal and N-terminal-specific PrP antibodies in scrapie-affected sheep and mice. Vet Rec 142(20):534–537
38. Carare RO, Hawkes CA, Jeffrey M, Kalaria RN, Weller RO (2013) Review: cerebral amyloid angiopathy, prion angiopathy, CADASIL

and the spectrum of protein elimination failure angiopathies (PEFA) in neurodegenerative disease with a focus on therapy. Neuropathol Appl Neurobiol 39(6):593–611

39. Chesebro B, Trifilo M, Race R, Meade-White K, Teng C, LaCasse R, Raymond L, Favara C, Baron G, Priola S, Caughey B, Masliah E, Oldstone M (2005) Anchorless prion protein results in infectious amyloid disease without clinical scrapie. Science 308(5727):1435–1439
40. Rangel A, Race B, Striebel J, Chesebro B (2013) Non-amyloid and amyloid prion protein deposits in prion-infected mice differ in blockage of interstitial brain fluid. Neuropathol Appl Neurobiol 39(3):217–230
41. Makarava N, Kovacs GG, Savtchenko R, Alexeeva I, Ostapchenko VG, Budka H, Rohwer RG, Baskakov IV (2012) A new mechanism for transmissible prion diseases. J Neurosci 32(21):7345–7355
42. Makarava N, Kovacs GG, Bocharova O, Savtchenko R, Alexeeva I, Budka H, Rohwer RG, Baskakov IV (2010) Recombinant prion protein induces a new transmissible prion disease in wild-type animals. Acta Neuropathol 119(2):177–187
43. Jeffrey M (2013) Review: membrane-associated misfolded protein propagation in natural transmissible spongiform encephalopathies (TSEs), synthetic prion diseases and Alzheimer's disease. Neuropathol Appl Neurobiol 39(3):196–216
44. Walker LC, Callahan MJ, Bian F, Durham RA, Roher AE, Lipinski WJ (2002) Exogenous induction of cerebral beta-amyloidosis in beta APP-transgenic mice. Peptides 23(7): 1241–1247
45. Carare RO, Bernardes-Silva M, Newman TA, Page AM, Nicoll JAR, Perry VH, Weller RO (2012) Solutes, but not cells, drain from the brain parenchyma along basement membranes of capillaries and arteries: significnace for cerebral amyloid angiopathy and neuroimmunology. Neuropathol Appl Neurobiol 34:131–144
46. Zhang ET, Richards HK, Kida S, Weller RO (1992) Directional and compartmentalised drainage of interstitial fluid and cerebrospinal fluid from the rat brain. Acta Neuropathol 83(3):233–239
47. Benestad SL, Arsac JN, Goldmann W, Noremark M (2008) Atypical/Nor98 scrapie: properties of the agent, genetics, and epidemiology. Vet Res 39(4):19
48. Goldmann W (2008) PrP genetics in ruminant transmissible spongiform encephalopathies. Vet Res 39(4):30
49. Le Dur A, Beringue V, Andreoletti O, Reine F, Lai TL, Baron T, Bratberg B, Vilotte JL, Sarradin P, Benestad SL, Laude H (2005) A newly identified type of scrapie agent can naturally infect sheep with resistant PrP genotypes. Proc Natl Acad Sci U S A 102(44): 16031–16036
50. Simmons MM, Konold T, Thurston L, Bellworthy SJ, Chaplin MJ, Moore SJ (2010) The natural atypical scrapie phenotype is preserved on experimental transmission and subpassage in PRNP homologous sheep. BMC Vet Res 6(1):14
51. Jeffrey M, González L, Simmons MM, Hunter N, Martin S, McGovern G (2017) Altered trafficking of abnormal prion protein in atypical scrapie: prion protein accumulation in oligodendroglial inner mesaxons. Neuropathol Appl Neurobiol 43(3):215–226
52. Chiesa R, Piccardo P, Ghetti B, Harris DA (1998) Neurological illness in transgenic mice expressing a prion protein with an insertional mutation. Neuron 21(6):1339–1351
53. Jeffrey M, Goodsir C, McGovern G, Barmada SJ, Medrano AZ, Harris DA (2009) Prion protein with an insertional mutation accumulates on axonal and dendritic plasmalemma and is associated with distinctive ultrastructural changes. Am J Pathol 175(3):1208–1217
54. Medrano AZ, Barmada SJ, Biasini E, Harris DA (2008) GFP-tagged mutant prion protein forms intra-axonal aggregates in transgenic mice. Neurobiol Dis 31(1):20–32
55. Hemm RD, Carlton WW, Welser JR (1971) Ultrastructural changes of cuprizone encephalopathy in mice. Toxicol Appl Pharmacol 18(4):869–882
56. Barron RM, Campbell SL, King D, Bellon A, Chapman KE, Williamson RA, Manson JC (2007) High titers of transmissible spongiform encephalopathy infectivity associated with extremely low levels of PrPSc in vivo. J Biol Chem 282(49):35878–35886
57. Jeffrey M, González L (2007) Classical sheep transmissible spongiform encephalopathies: pathogenesis, pathological phenotypes and clinical disease. Neuropathol Appl Neurobiol 33(4):373–394
58. Jeffrey M, Fraser JR (2000) Tubulovesicular particles occur early in the incubation period of murine scrapie. Acta Neuropathol 99(5):525–528
59. Jeffrey M, McGovern G, Goodsir CM, Brown KL, Bruce ME (2000) Sites of prion protein accumulation in scrapie-infected mouse spleen

revealed by immuno-electron microscopy. J Pathol 191(3):323–332
60. McGovern G, Brown KL, Bruce ME, Jeffrey M (2004) Murine scrapie infection causes an abnormal germinal centre reaction in the spleen. J Comp Pathol 130(2-3):181–194
61. McGovern G, Mabbott N, Jeffrey M (2009) Scrapie affects the maturation cycle and immune complex trapping by follicular dendritic cells in mice. PLoS One 4(12):e8186
62. Rademakers LHPM (1992) Dark and light zones of germinal centres of the human tonsil: an ultrastructural study with emphasis on heterogeneity of follicular dendritic cells. Cell Tissue Res 269(2):359–368
63. McGovern G, Jeffrey M (2007) Scrapie-specific pathology of sheep lymphoid tissues. PLoS One 2(12):e1304
64. Jeffrey M, Martin S, González L, Ryder SJ, Bellworthy SJ, Jackman R (2001) Differential diagnosis of infections with the bovine spongiform encephalopathy (BSE) and scrapie agents in sheep. J Comp Pathol 125(4):271–284
65. Sassa Y, Inoshima Y, Ishiguro N (2010) Bovine macrophage degradation of scrapie and BSE PrPSc. Vet Immunol Immunopathol 133(1):33–39
66. Sisó S, González L, Jeffrey M (2010) Neuroinvasion in prion diseases: the roles of ascending neural infection and blood dissemination. Interdiscip Perspect Infect Dis 2010: 747892
67. Houston F, McCutcheon S, Goldmann W, Chong A, Foster J, Sisó S, González L, Jeffrey M, Hunter N (2008) Prion diseases are efficiently transmitted by blood transfusion in sheep. Blood 112(12):4739–4745
68. McGovern G, Jeffrey M (2013) Membrane toxicity of abnormal prion protein in adrenal chromaffin cells of scrapie infected sheep. PLoS One 8(3):e58620
69. Sisó S, Jeffrey M, Steele P, McGovern G, Martin S, Finlayson J, Chianini F, González L (2008) Occurrence and cellular localization of PrP(d) in kidneys of scrapie-affected sheep in the absence of inflammation. J Pathol 215(2):126–134
70. Race B, Jeffrey M, McGovern G, Dorward D, Prion CB (2017) Ultrastructure and pathology of prion protein amyloid accumulation and cellular damage in extraneural tissues of scrapie infected transgenic mice expressing anchorless prion protein. Epub ahead of print

Chapter 7

Electron Microscopy of Prion Diseases

Paweł P. Liberski

Abstract

Electron microscopy provided detailed description of submicroscopic changes in prion diseases or transmissible spongiform encephalopathies (TSEs). For Creutzfeldt-Jakob disease (CJD and its variant, vCJD) and fatal familial insomnia (FFI) only vacuolation (spongiform change) and the presence of tubulovesicular structures are consistent findings. Other changes—i.e., the presence of "myelinated" vacuoles, branching cisterns, neuroaxonal dystrophy, and autophagic vacuoles—were present in different proportions in either CJD or FFI, but they are nonspecific ultrastructural findings that can also occur in other neurodegenerative conditions.

The hallmark of Gerstmann-Sträussler-Scheinker disease (GSS) and vCJD is the amyloid plaque, but plaques of GSS and kuru are different than those of vCJD. Whereas the former are typical unicentric "kuru-type" or multicentric plaques, the latter are unicentric "florid" plaques. Also, kuru plaques are non-neuritic, whereas GSS florid plaques are usually neuritic; however, a proportion of plaques from GSS were also found to have non-neuritic characteristics. Thus, the presence or absence of dystrophic neurites is not a discriminatory factor for GSS and vCJD. Furthermore, plaques from GSS with different mutations were also slightly different. In GSS with mutation P102L, 232T and A117V plaques were "stellate" while in one case with 144 base pair insertion and in GSS-A117V, "round" plaques were also observed, and typical "primitive" "neuritic" plaques, i.e., composed of dystrophic neurites with little or no amyloid, were found only in a P102L case from the original Austrian family. In two cases of sporadic CJD, the kuru "stellate" plaque predominated.

Key words Prion diseases, Electron microscopy, Creutzfeldt-Jakob disease, Gerstmann-Sträussler-Scheinker disease

1 Introduction

Electron microscopy is presently rarely used as a diagnostic aid, except for the suggestion that scrapie-associated fibrils or "prion rods" [1–3] should be searched for in atypical or doubtful cases. This chapter is devoted to electron microscopy, which can still be useful in special situations, such as auxiliary diagnostic aid in patients under consideration for potentially hazardous experimental therapy, like pentosan polysulfate [4]. This chapter is based on the two largest series of sporadic CJD brain biopsy specimens so far published, together with brain biopsy and autopsy specimens from vCJD, GSS, and FFI.

Pawel P. Liberski (ed.), *Prion Diseases*, Neuromethods, vol. 129,
DOI 10.1007/978-1-4939-7211-1_7,

2 Materials and Methods

This chapter is based on two series of sporadic form of Creutzfeldt-Jakob disease (sCJD) and iatrogenic form (iCJD). The first set comprised four cases of CJD (sCJD) collected at the *Laboratoire Raymond Escourolle, Hôpital de la Salpêtrière*, Paris, France (Table 1). This series has been completed between 1982 and 1986 as a part of epidemiology study of CJD in France [5]. The second series (Table 2) consisted of brain (right prefrontal cortex) biopsies

Table 1
Clinical data in the first series of brain biopsy (prefrontal cortex) specimens of CJD

Code	Age	Duration (months)	Cerebellar signs	Dementia	Blindness	Typical EEG
3684	50	N/A	Present	Present	Absent	Absent
3032	63	N/A	N/A	Present	N/A	N/A
3439	64	6	Present	Present	Absent	Present
4100	65	N/A	Present	Present	Absent	Present

The status of 129 codon is not available for this series

Table 2
Clinical data and polymorphism of the codon 129 in the second brain biopsy (prefrontal cortex) series of CJD

Code	Type	Age/sex	Duration (months)	Cerebellar signs	Dementia	Blindness	Typical EEG	Codon 129
AJ53286	Iatrogenic (dura mater)	52/W	5	Present	Present	Absent	Absent	Met/Met
AM60023	Sporadic	32/W	24	Present	Present	Absent	Absent	Met/Met
AH62014	Iatrogenic (Gh)	25/M	12	Present	Present	Absent	Present	?
AK55312	Sporadic	66/M	5	Present	Present	Absent	Absent	Val/Val
AN59129	Sporadic	57/W	?	Present	Present	Absent	Present	?
AK54615	Sporadic	54/M	6	Present	Present	Absent	Present	Val/Val
AK54548	Sporadic	46/W	6	Present	Present	Absent	Absent	Val/Val
AL57150	Sporadic	67/M	8	Present	Present	Absent	Present	Met/Val
AN60459	Sporadic	65/M	5	Present	Present	Present	Present	Met/Met
AK55013	Sporadic	72/M	14	Present	Present	Absent	Absent	Met/Val
AL55013	Sporadic	72/M	2	Present	Present	Absent	Absent	Met/Met

obtained between 1995 and 1999 by open surgery from 11 specimens referred to us for diagnostic purposes: nine cases of sporadic CJD, one case of familial CJD (V203I codon mutation), and one case of iCJD [6, 7]. Clinical and molecular (the codon 129 status of the prion protein gene) data of these cases are summarized in Tables 1 and 2, but no data on the PrP isotype (PrP^{Sc} type 1 or 2) were available. In addition we collected three brain biopsies from The Neurological Institute, University of Vienna [8].

For vCJD, one biopsy and five autopsy cases were collected by Professor James W. Ironside, National CJD Surveillance Unit, Edinburgh, Scotland (Table 3). The biopsy specimen of the first French case was in part published [9, 10].

For GSS (Table 4), we collected one autopsy specimens from the Institute of Neurology, Medical University of Vienna, Austria [11, 12], one autopsy and one biopsy specimens from a large GSS family [13–16], and autopsy specimens of a case of GSS 232T from Poland [17, 18]. Additionally, we reverse-processed and examined paraffin blocks of a case of GSS A117V from a Hungarian family.

Table 3
Clinical data on vCJD cases—one biopsy (AJ53977) and autopsy cases

Code	Type	Age/sex	Duration (months)	Codon 129
96/02	VCJD	41/F	18	MM
96/03	VCJD	29/F	23	MM
96/07	VCJD	30/F	18	MM
96/45	VCJD	31/M	9	MM
96/10	VCJD	35/F	14	MM
AJ53977	VCJD	27/M	23	MM

Table 4
Clinical data of GSS cases

Type	Age/ sex	Duration (months)	Cerebellar signs	Dementia	Blindness	Typical EEG	Codon 129	References
GSS M232 T	50	6	Present	Present	No	No	Met/Val	[35, 36]
GSS A117V	55	24	Present	Present	No	No	Unknown	
GSS P102L	47	60	Present	Present	N/A	Yes	Met/Met	[31–34]
GSS P102L	59	60	Present	Present	N/A	?	Met/Met	[31–34]
GSS P102L	41	34	Present	Present	N/A	Yes	Met/Met	[29, 30]

Table 5
Clinical data of FFI case

Code	Type	Age/sex	Duration (months)	Cerebellar signs	Dementia	Blindness	Typical EEG	Codon 129
AN61237	(D178N)	46/W	7	Present	Present	Absent	Absent	Met/Met

To compare plaques of GSS to those of sCJD, we studied two cases of sCJD—one MV at the codon 129 and one of unknown status of this polymorphism. Furthermore, we include one autopsy case with 144 base pair insertion (bpi) in the *PRNP* with unusual PrP immunodeposits in the cerebellum [19].

For FFI, one biopsy case from a new French family [20, 21] was studied (Table 5).

3 Methods

3.1 Fixation

1. Human material should be fixed in 3.5% glutaraldehyde in 0.13 M cacodylate buffer for 24 h in 4 °C.
2. Rinse the sample in the same buffer for 24 h at 4 °C.
3. Postfix in 1% osmium tetroxide for 2 h.
4. Rinse for 24 h in 0.13 M cacodylate buffer for 10 min at 4 °C.
5. Dehydrate through graded ethanol solution:
   (a) 50%—10 min.
   (b) 70%—20 min.
   (c) 90%—20 min.
   (d) 95%—20 min.
   (e) 99.8%—3 times 15 min.
   (f) Propylene oxide—15 min.

3.2 Embedding

1. Mixture of propylene oxide and Epon (1.5 mL of solution A plus 3.5 of solution B plus 5 mL of propylene oxide)—for 24 h at room temperature.
2. Solution A: Epon 812—12.4 mL plus DDSA—20 mL.
3. Solution B: Epon 812—20 mL plus MNA 17.8 mL.
   Solution A 5 mL.
   Solution B 5 mL.
   DMP 5 drops.
   It is useful to spin down polymerizing solution to get rid of bubbles.

Polymerization—48 h at 60 °C.
Semithin section—1 μm.

3.3 Ultrathin Sections

Grid Contrast with 20% Uranyl Acetate and Lead Citrate

(a) Uranyl acetate solution—20 g of uranyl acetate in 100 mL of methyl alcohol.

(b) Lead citrate solution—1.33 g of lead nitrate plus 30 mL of distilled water plus 1.76 natrium citrate plus 8 mL 1 N NaOH; add distilled water until 50 mL.

3.4 Cacodylate Buffer

11.12 g Sodium cacodylate plus 25 mL of distilled water plus 2 mL 1 N HCL; add distilled water to 40 mL.

Perfusion for Animals (Those Protocols were Personally used by the Author)

1. Relatively larger rodents like hamsters may be readily fixed by intracardiac perfusion using 100 mL of 2.5% glutaraldehyde and 1% paraformaldehyde prepared in a phosphate (or cacodylate buffer, pH 7.4), followed by 50 mL of 5% glutaraldehyde and 4% paraformaldehyde. The brains are removed and kept overnight at 4 °C.

Preparation of Fixative

1. For 100 mL of 1% paraformaldehyde and 2.5% paraformaldehyde: dissolve 1 g of paraformaldehyde and 5 mL of 25% glutaraldehyde, 40 mL of 0.2 M Na cacodylate, and six drops of 5% $CaCl_2$; add distilled water (dwater) up to 100 mL.
2. For 100 mL of 4% paraformaldehyde and 5% paraformaldehyde: dissolve 2 g of paraformaldehyde and 10 mL of 25% glutaraldehyde, 20 mL of 0.2 M Na cacodylate, and six drops of 5% $CaCl_2$; add dwater up to 50 mL.
3. Animals may be also perfused with just 180 mL of 1% paraformaldehyde and 1.5% glutaraldehyde prepared in phosphate buffer. I personally used bot methods and they produced excellent and practically identical results.

Note: Dissolving of paraformaldehyde is best done in a microwave oven overnight.

Preparation for Perfusion

1. Cut off the sternum.
2. Clamp everything below diaphragm with a clamp.
3. Inject heparin into the heart.
4. Cut the right auricle (it is darker than the rest of the heart).

5. Do a small incision in the left ventricle.
6. Insert a tubine through the left ventricle into the aorta (you should see it).
7. Connect a syringe with a fixative to the tubine, and press the syringe plunger by hand or by a Hamilton pump; the best is if you see the clonic seizure when the fixative reaches the brain. This event is readily seen in hamsters, not so readily in mice.

3.5 Marion Simmons

Dept. Pathology, Animal Health and Veterinary Laboratories Agency (AHVLA), Surrey, UK.

Perfusion of Cow Brain (NB this is not a Whole-Body Perfusion) (Extracted from [22])

1. Sedate animal with 0.2 mg/kg xylazine intramuscularly.
2. After 10 min give 100,000 IU heparin intravenously.
3. Induce deep anesthesia with iv pentobarbitone (25–30 mg/kg) and then quickly administer 4 mg adrenaline iv to maintain good cerebral blood pressure.
4. Expose the carotid arteries and jugular veins bilaterally.
5. Cannulate all four vessels and tie cannulae in place.
6. Administer all perfusates via the carotid cannulae, using a peristaltic pump, keeping the pressure at approximately 100 mm Hg.
7. Flush brain with 15 L of 0.85% saline.
8. Primary fixation with 15 L of mixed aldehyde fixative (we used 1% paraformaldehyde, 1.25% glutaraldehyde in 0.25 M Sorensen's phosphate buffer).
9. Secondary fixation with 10 L of stronger fixative (we used 4% paraformaldehyde, 5% glutaraldehyde).
10. Use the jugular cannulae to collect circulating fixative for safe disposal.
11. All personnel should wear appropriate respiratory protection given the large volumes of aldehyde fixatives used.

4 Results

The following ultrastructural findings will be discussed (Tables 6, 7, 8, and 9):

1. Spongiform vacuoles—(Fig. 1) these are membrane bound and contain secondary vacuoles (i.e., membrane-bound compartments or vesicles within vacuoles) and curled membraned fragments.
2. Tubulovesicular structures (TVS) (Fig. 2)—these are vesicular structures of approximately 27 nm in diameter within neuronal

Table 6
The first series of biopsies (prefrontal region) with prion diseases

Code	Vacuoles	Myelinated vacuoles	TVS	Branching cisterns	NAD	Autophagic vacuoles
3684	+++	+	+	+	+	
3032	+++	+	+	+	++	
3439	+	+	+	+	+	
4100	+++	+	+++	+++	+++	++

Table 7
The second series of biopsies (prefrontal region) with prion diseases

Code	Diagnosis	Vacuoles	Myelinated vacuoles	TVS	Branching cisterns	NAD	Autophagic vacuoles
AJ53286	Iatrogenic (dura mater)						
AM60023	Sporadic	+++	+	+	+	++	
AH62014	Iatrogenic (Gh)	+					
AK55312	Sporadic	+++	+	+++	+++	+++	++
AN59129	Sporadic	++		++	+++	++	++
AK54615	Sporadic	++		++		++	
AK54548	Sporadic	+		+++	+++	+++	+
AL57150	Sporadic						
AN60459	Sporadic	+		+	+	+	
AK55013	Sporadic	+	+	–	+	+	
AN61237	(D178N)						
4190	sCJD	+++		+++		+++	+++
3032	sCJD	+++		+		++	
3439	sCJD	+++	++	++		+	
3684	sCJD			++		+++	+

Table 8
vCJD case—a brain biopsy (prefrontal region)

Code	Diagnosis	Vacuoles	Myelinated vacuoles	TVS	Branching cisterns	NAD	Dark synapses	Autophagic vacuoles
AJ59977 (biopsy)	VCJD	+++	+	+++	+++	+++		+

Table 9
FFI case

Code	Diagnosis	Vacuoles	Myelinated vacuoles	TVS	Branching cisterns	NAD	Dark synapses	Autophagic vacuoles
AN61237	FFI	++		+++	++	+	−	+++

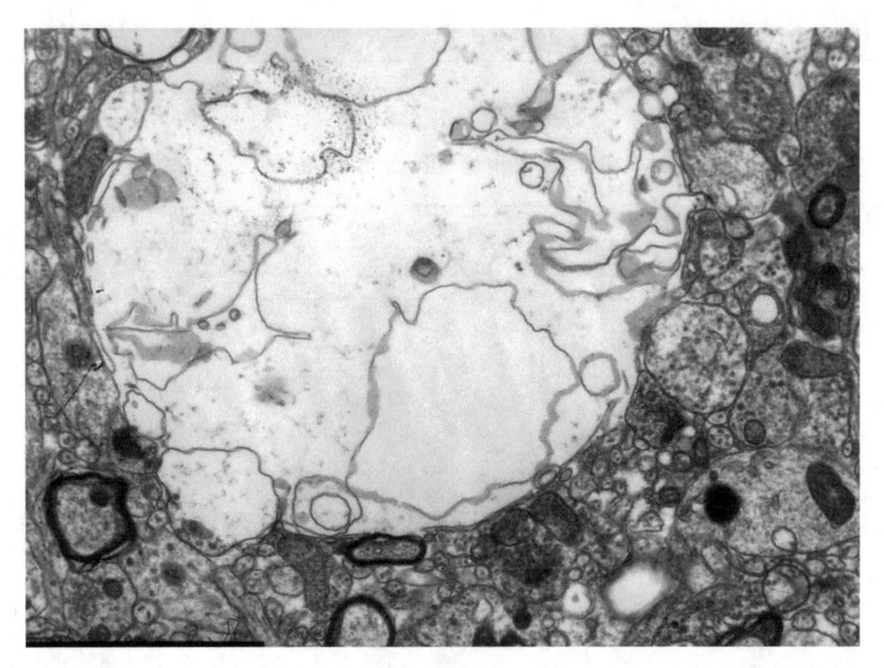

Fig. 1 A low-power view of a spongiform vacuole from a sCJD specimen. Note secondary vesicles and vacuoles. Original magnification, × 3000

processes—i.e., axonal terminal or dendrites [7]. TVS are smaller and of higher electron density than synaptic vesicles. The significance of TVS remains unknown [23].

3. Branching cisterns (Fig. 3)—those structures fill neuronal processes. They were described as specific for human neuroaxonal dystrophies [24] but they were described also in CJD [6].
4. Dystrophic neuritis (Fig. 4)—Dendrites or axonal preterminals and terminals filled with electron-dense bodies, including small autophagic vacuoles [25].

Previously published analyses found that vacuolation (spongiform change—Figs. 1 and 5) and the presence of tubulovesicular structures (TVS) are the only disease-specific structures for CJD (including vCJD) [7]. "Myelinated" vacuoles, branching cisterns (Fig. 3), neuroaxonal dystrophy (Fig. 4), and autophagic vacuoles

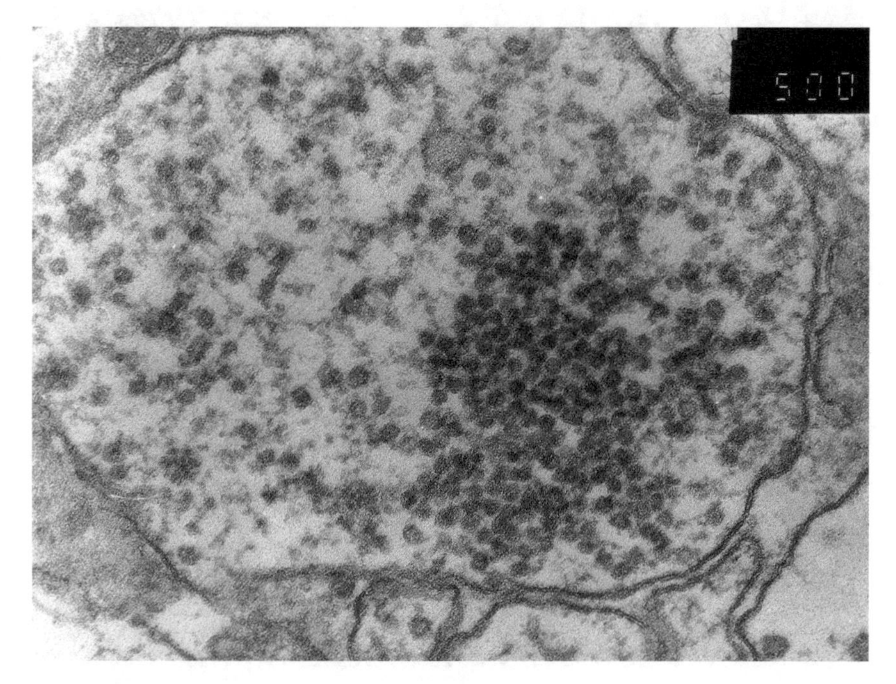

Fig. 2 A neuronal process filled with tubulovesicular structures (TVS). Original magnification, × 50,000

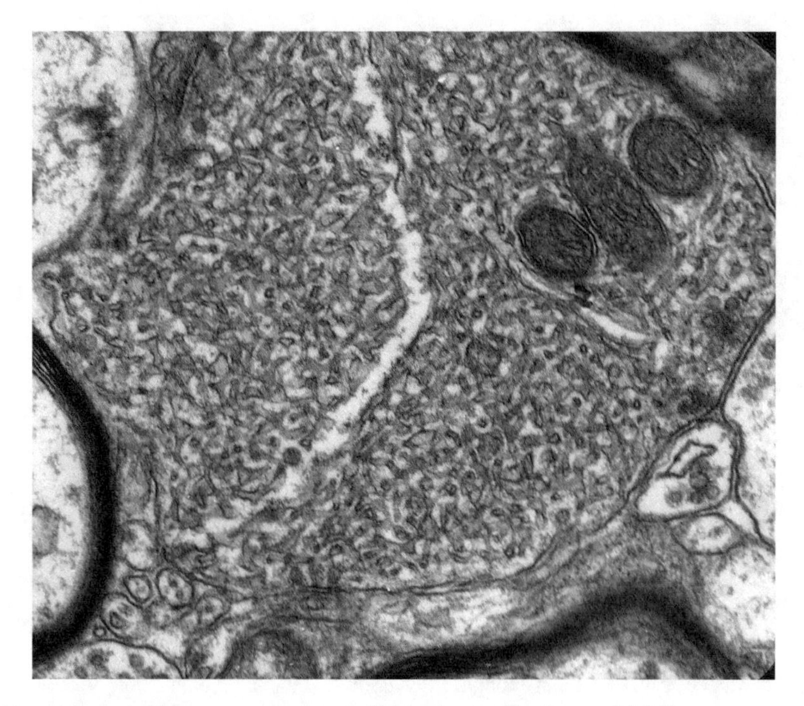

Fig. 3 Branching cisterns within neuronal process. Original magnification, × 30,000

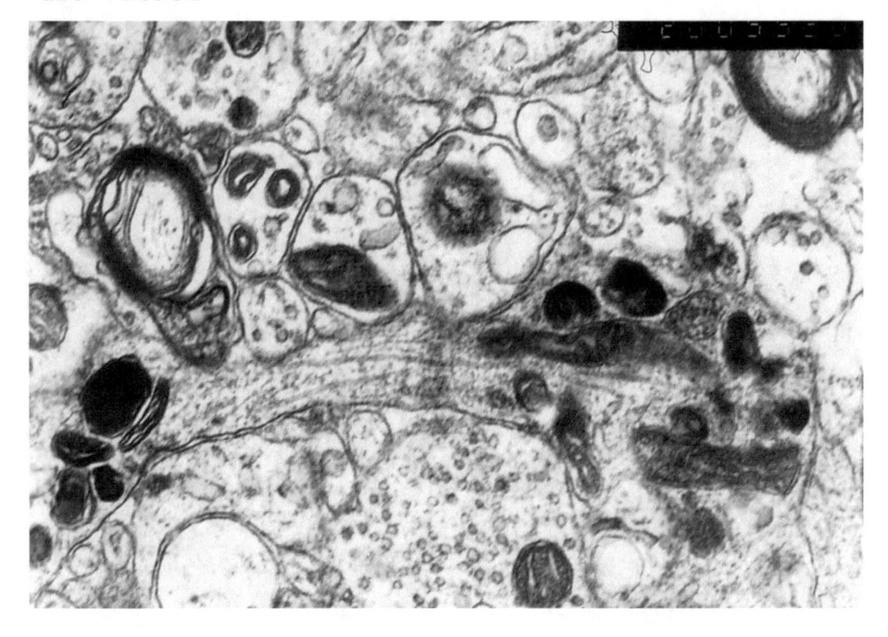

Fig. 4 A dystrophic neurite filled with numerous dense bodies of different appearances. Original magnification, × 20,000

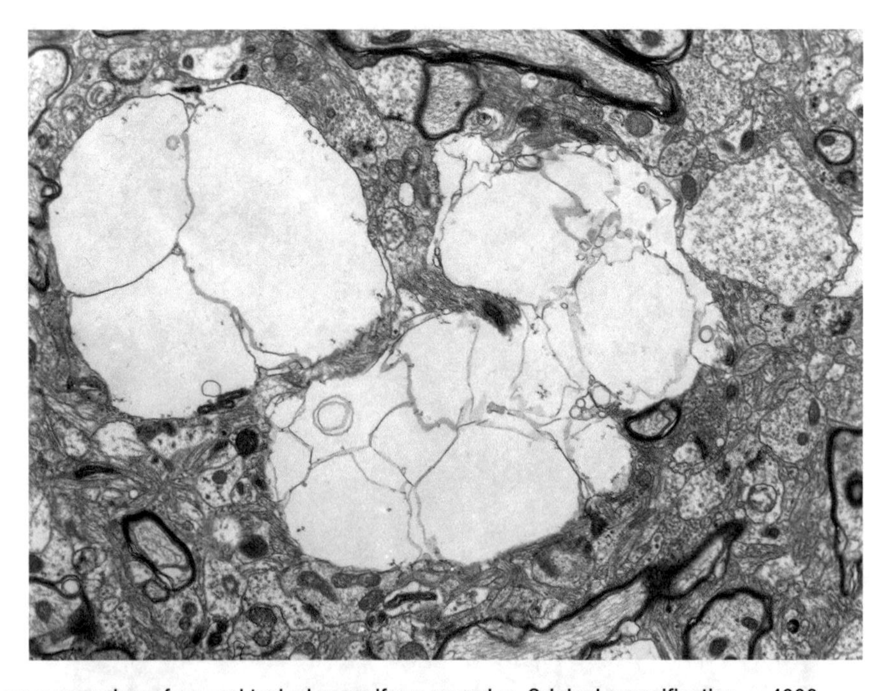

Fig. 5 Low-power view of several typical spongiform vacuoles. Original magnification, × 4000

(see chapter on autophagy) were also found in either CJD or FFI, but they are nonspecific ultrastructural findings that can occur in other neurodegenerative conditions.

4.1 Spongiform Change

They are defined as small round or oval "empty" spaces in the neuropil (Figs. 1 and 5). When confluent, they merge to form "morula-like" structures. By electron microscopy, spongiform change is equivalent to vacuoles. While the majority of these are intracellular and membrane bound, some of vacuoles are unbounded and the originating structures cannot be identified. The membranes surrounding the intracellular vacuoles may be single or multiple.

Vacuoles appear "empty" but amorphous material probably of supposedly proteinaceous nature is often seen. Numerous curled membranes are "pulled off" from the inner leaflets of vacuoles (Figs. 1 and 5) and divide its contents into secondary chambers.

4.2 Amyloid Plaques

Amyloid plaques are encountered in all cases of kuru and, by definition, GSS and in some 10–15% of sCJD. In vCJD, the particular type of plaque—so-called *florid* or *daisy* plaques—exists in 100% of cases. By electron microscopy, several types of amyloid plaques which corresponded to those visualized by PrP immunohistochemistry were delineated. Unicentric "kuru" plaques (Fig. 6) consisted of stellate cores of amyloid fibrils emanating from a dense interwoven center. Amyloid stars are enveloped by astrocytic

Fig. 6 Typical stellate kuru plaque from a case of GSS with P102L mutation [30]. Original magnification, × 12,000

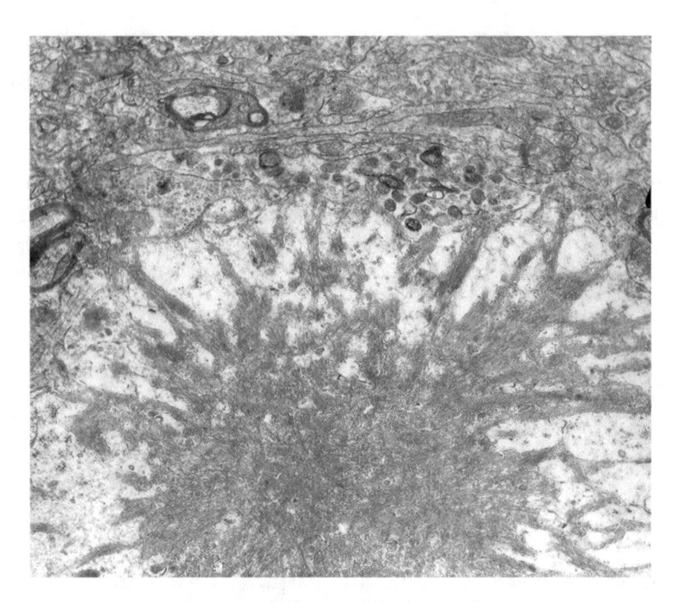

Fig. 7 A stellate plaque from a case of GSS [14]. Note that plaque is surrounded by electron lucen "collar" of astrocytic process. Original magnification, × 3000

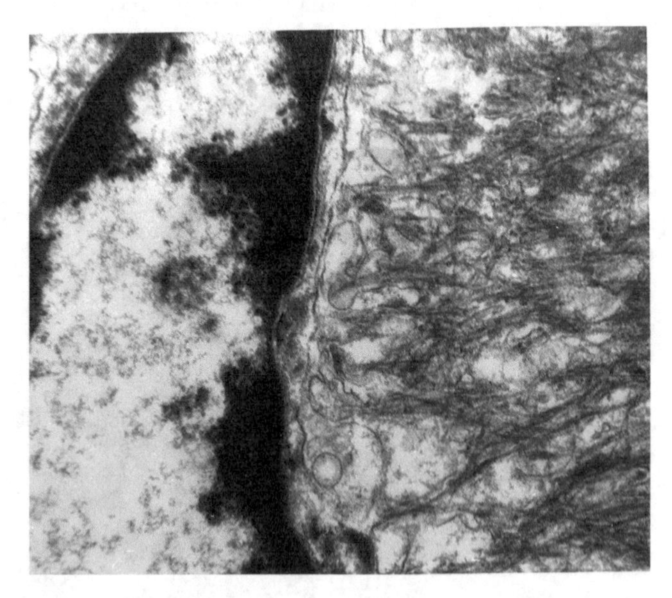

Fig. 8 A microglial cell (*left*) in a close contact with an amyloid plaque from a GSS case

processes (Fig. 7) and invaded by microglial cells. High-power electron micrographs revealed that amyloid bundles (Fig. 8) are observed to be located within pockets of obscure cellular origin. Interestingly, basement membranes lined with electron-dense material were observed at the periphery of virtually all amyloid plaques in GSS. Dystrophic neurites (DN) were seen only rarely.

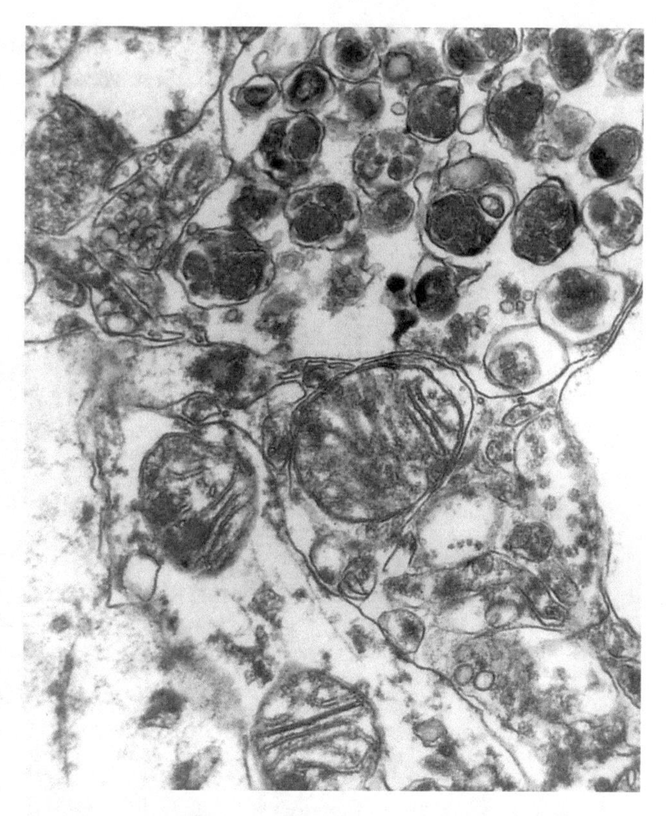

Fig. 9 A dystrophic neurite from a margin of florid plaque from a case of vCJD. Original magnification, × 30,000

Occasionally, clusters composed of several kuru plaques were found. The latter were intermediate forms to multicentric plaques which consisted of several merging stellate cores. Smaller amyloid deposits surrounded larger cores. In contrast to kuru plaques associated with a limited number of dystrophic neurites (DN), numerous such structures were seen at the periphery of florid plaque (Fig. 9). DN are filled with abnormal organelles like electron-dense bodies, multivesicular bodies, and multilamellar bodies and thus were indistinguishable from those seen in scrapie and CJD or Alzheimer's disease, except that they did not contain paired helical filaments (PHF). Instead, piled neurofilaments were often detected in the center of DN. Similar DN were observed but not associated with any plaques.

The last, and by the same token, the rarest type of plaques, were purely neuritic plaques. These consisted of large areas filled with DN of different sizes and shapes (sometimes bizarre) but not amyloid bundles. Analogously to kuru and multicentric plaques, astrocytic processes were observed at the periphery.

In experimental scrapie in hamsters so-called "loose" plaques were reported [26]. They were first reported also by DeArmond and others [27] and they are composed of distended areas of subependymal space in which PrP-composed amyloid fibrils are located (Figs. 10, 11).

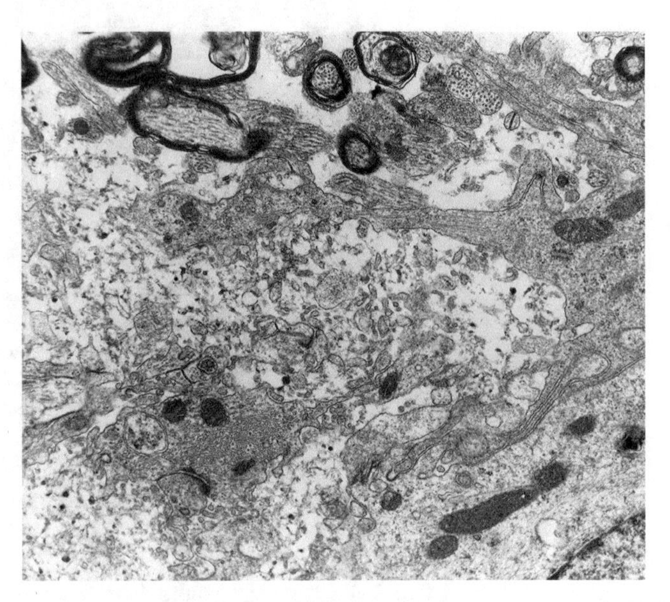

Fig. 10 A distended subependymal space filled with numerous PrP filaments. Original magnification, × 10,000

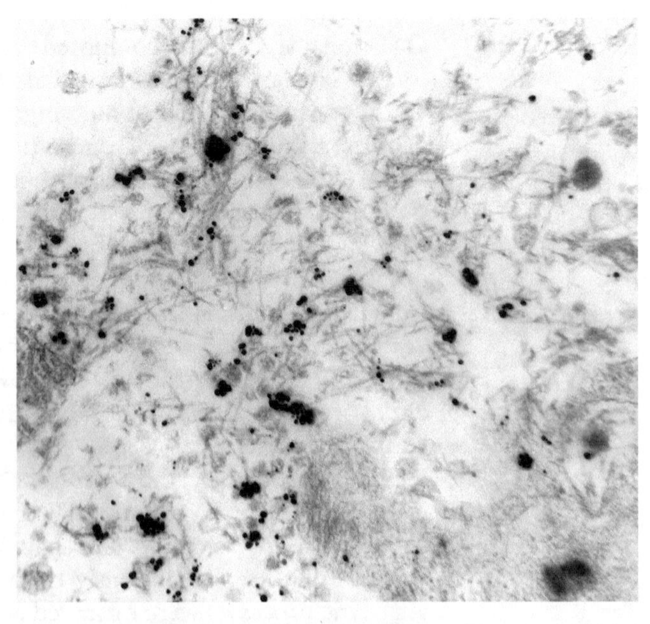

Fig. 11 Immunogold reaction with antibodies against PrP showing that those filaments are composed of PrP. Original magnification, × 30,000

Table 10
GSS and a hereditary case with 144 bp insertion (5-01)

Genetic cases and mutation, if known, and sCJD	Kuru-type (unicentric) plaques	Multicentric plaques	Primitive plaques	Comments
bpi 5-01144 bp insert	+	+	–	Round margin of plaques
P102L, family "H"	+	+	–	
A117V	+	+	–	Round margin of plaques
M232V				
P102L	+	+	–	–
P102L	+	+	–	–
sCJD, MV	+	–	–	Starlike plaques
sCJD with plaques	+	–	–	Starlike plaques

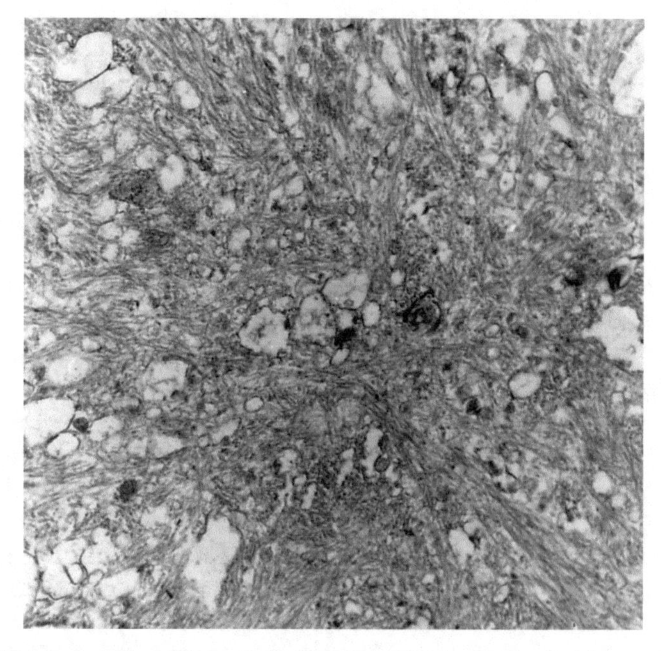

Fig. 12 A typical "florid" plaque of a vCJD case. Original magnification, × 10,000

Plaques of GSS (Table 10) and kuru are basically different from those of vCJD (Fig. 12). Plaques in kuru, CJD, and GSS are unicentric "kuru type" or multicentric plaques, respectively [10, 12], while plaques in vCJD are unicentric "florid" plaques. Kuru plaques are regarded as non-neuritic, in contrast to neuritic florid plaques, albeit a low number of DN may be seen. However, we found that a proportion of plaques from GSS were typically neuritic [12]. Thus, the presence or absence of dystrophic neuritis is not a criterion to discriminate plaques in GSS and vCJD. Furthermore, plaques from GSS caused by different mutations differ. In GSS with mutation P102L, 232T and A117V plaques were "stellate," while in a single and unusual case with 144 bpi (5-01) but also in GSS-A117V, "round" plaques were observed. Furthermore, typical "primitive neuritic" plaques were found only in a case P102L from the original Austrian family.

It is worthy to compare those plaques described so far with plaques modeled in transgenic mice (Tg20) infected with chronic wasting disease (CWD). Electron microscopy performed on perfusion-fixed brains confirmed our preliminary data obtained on material reversed from paraffin blocks to electron microscopy [28]. We observed large amyloid deposits, unicentric kuru plaques, and perivasular unicentric plaques. Unicentric plaques were composed of bundles of amyloid fibrils deeply interwoven in the center and radiating toward periphery. Perivascular plaques were those unicentric kuru plaques that were in close contact with cerebral brain vessels. In the latter situation, amyloid fibrils were in close contact, even merged with, with basal membranes.

As described previously, all those plaques were totally devoid of neuritic elements. However, they were invaded by numerous microglial cells. Those cells looked very active with enlarged cisterns of endoplasmic reticulum. The outer membranes formed pockets where amyloid fibrils seemed to be "engulfed." Some cells were very dense and contained amorphous masses of even higher density. The labyrinth-like network of microglial processes was visible at the periphery and within plaques.

4.3 Production of the Prion Organotypic Slice Culture Assay (POSCA) [29–31]

The detailed procedure is described in [29]; here the major points will be recapitulated:

It is necessary to use the vibratome (i.e., Leica VT1000S) and the injector blades.

Slice culture medium. The concentrated (2×) solution of minimal essential medium (MEM) in water should be made, pH 7.4, filtered. A mixture of 100 mL 2× MEM, 100 mL BME (basal medium Eagle without glutamine; Invitrogen, cat no 21370-028), 100 mL of horse serum, 4 mL of glutamax, 4 mL of penicillin/streptomycin, 5.5 mL of glucose, and 86.5 mL of ddwater adjusted to pH 7.2–7.4 and filtered should be made.

Gey's balanced salt solution (GBSS). For 1 L, dissolve all the reagents in ddwater in the order written below (to prevent insolubility): 137 mM of NaCl; 5 mM of KCl; 0.845 of Na_2HPO_4; 1.5 mL of $CaCl_2{\cdot}H$; and 0.66 mM of KH_2PO_4.

Procedures

1. Put 250 μm of culture medium into the 24-well plates and prepare GBSSK.
2. Prepare 2% agarose in GBSSK.

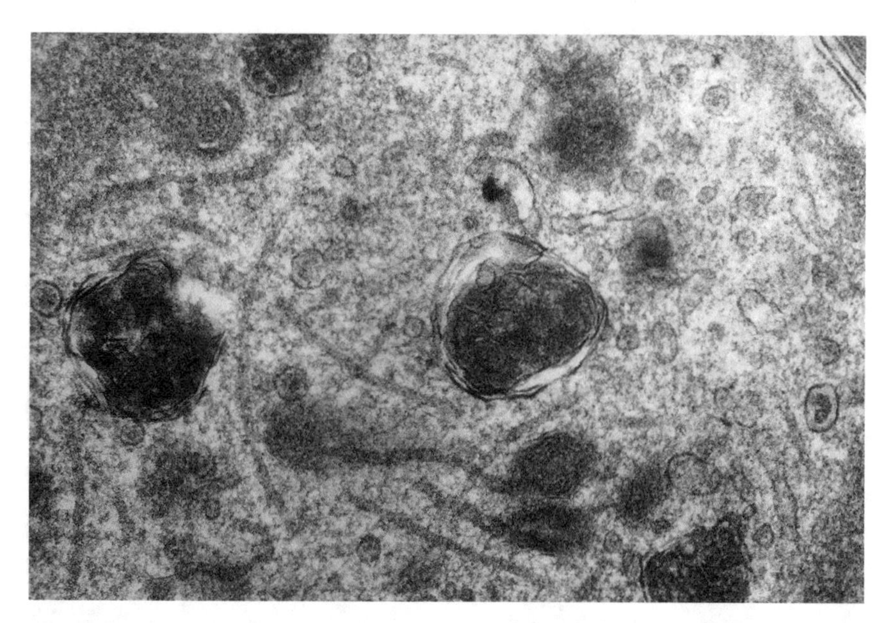

Fig. 13 Typical autophagic vacuoles. RML strain of scrapie in a slice obtained from Tg20 25 days postinoculation

3. Cool the vibratome.
4. Decapitate the pups, open the skull with scissors, remove the cerebellum, and submerge it into liquid agarose.
5. After agarose has solidified, glue the block into a vibratome disk and cut 350 μm thick sections.
6. Release the tissue from the agarose with Dumont no. 5 forceps and transfer it into 24-well plates with membrane inserts for long-term culturing.
   Details are described in [29–31].

4.4 Electron Microscopy

Slices are washed in Na-phosphate buffer, fixed in freshly prepared 2% PFA + 2.5% glutaraldehyde in 0.1 M Na-phosphate buffer 0.1 M pH 7.4, postfixed in osmium tetroxide, embedded in epon, and examined in transmission electron microscope.

4.4.1 Electron Microscopy of Brain Slices

In both wt- and Tg20-derived slices we observed the full spectrum of changes of prion diseases: spongiform vacuoles, different stages of autophgy and apoptosis, dystrophic neuritis, and tubulovesicular structures [29–31] (Figs. 13, and 14).

5 Discussion

In this chapter, we extended our previous observations [4] on the ultrastructural analysis of the largest collection of brain biopsy specimens examined by thin-section electron microscopy. While

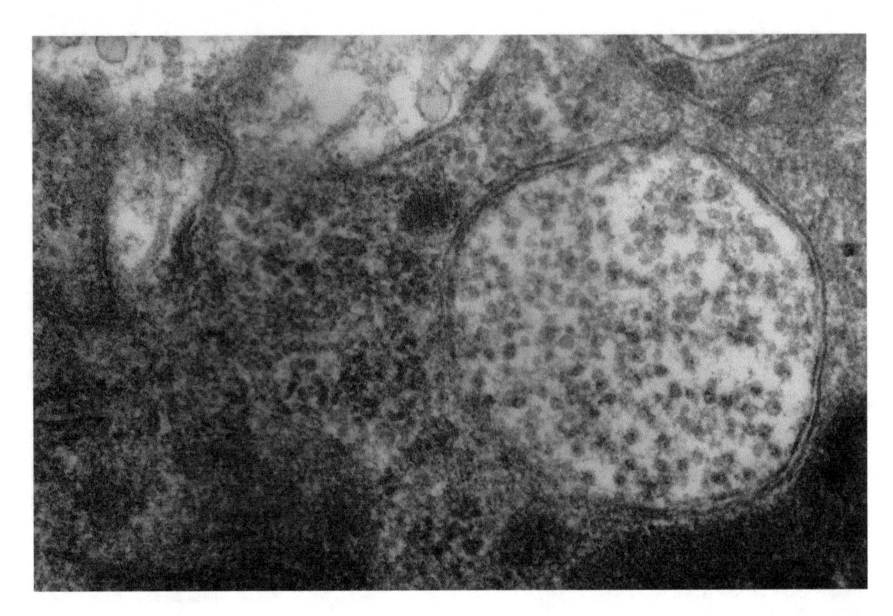

Fig. 14 A process containing tubulovesicular structures. RML strain of scrapie in TG20 mice 39 days postinoculation

ultrastructural analysis is no longer widely used because it is laborious and time consuming, it can be potentially useful when the immunohistochemical analysis of the brain biopsy is either negative or equivocal [32]. Furthermore, the brain biopsy may again become important because potential novel therapies in prion disease, like pentosan sulfate administered intraventricularly, are potentially toxic and thus an accurate pretreatment diagnosis is mandatory.

The vacuole is an ultrastructural correlate of spongiform change as seen by light microscopy [33–35]. While the presence of spongiform change has been recognized from the beginning of modern scrapie research [36], its pathogenesis remains as obscure as it was more than 30 years ago when Elizabeth Beck and her colleagues studied their morphogenesis [37, 38]. It seems that in most experimental models [39–41] vacuoles develop within neuronal elements—viz., dendrites and, rarely, axons. However, in models such as CJD-infected guinea pigs, vacuoles were reported to develop within astrocytes [42].

Vacuoles are always intracellular as they are membrane bound, and can be bordered by a single or double [41] membrane layer. Jeffrey et al. [41] suggested that double membranes are more specific for TSEs; however, the majority of the vacuoles that we examined in human cases were delimited by a single layer of membrane. Both situations likely exist in TSE-affected brains.

The morphogenesis of vacuoles is unknown. However, during hundreds of electron microscopic sessions, we could see only fully developed vacuoles, which suggests that vacuoles developed abruptly with short transitional stages. This can be compared with the apoptotic process which, although widespread, is observed only rarely because of the speed of apoptosis and removal of apoptotic nuclei. Interestingly, the third type of programmed cell death (in addition to apoptosis and autophagy—[43, 44]) is characterized by large-scale formation of electron-lucent vacuoles not different from those vacuoles under discussion. As autophagy is a common but nonspecific finding in TSEs, spongiform change may be associated with the third type of neuronal cell death and experiments are in progress to elucidate this possibility.

Tubulovesicular structures are intriguing entities first reported by David-Ferreira et al. [45] almost 40 years ago and subsequently reported in scrapie [46, 47], CJD [7, 8], GSS [12, 48], and BSE [40]. Immunogold techniques suggest that they do not contain PrP [49]. TVS have never been isolated. However, in 1979 Siakotos et al. [50] observed spherical particles within the range of TVS size in scrapie-infected brains and spleen. The number of TVS-containing neuronal processes increased through the incubation period [51]. Even more interestingly, in transgenic mice in which PrP transgene is targeted to astrocytes through the GFAP promoter [52], PrP appears to be shed by astrocytes but neurons degenerate and accumulate TVS. In CJD-infected cell cultures, TVS-like particles were also readily seen [53]. TVS are thus structures unique to TSEs and their enigmatic nature requires further investigation [23].

References

1. Merz PA, Somerville RA, Wisniewski HM, Iqbal J (1981) Abnormal fibrils from scrapie-infected brain. Acta Neuropathol (Berl) 60:63–74
2. Merz PA, Rohwer RG, Kascak R et al (1984) Infection specific particle from the unconventional slow virus diseases. Science 225(4660): 437–440
3. McKinley MP, Bolton DC, Prusiner SB (1983) A protease resistant protein is a structural component of the scrapie prion. Cell 35(1):57–62
4. Dealler S, Rainov NG (2003) Pentosan polysulfate as a prophylactic and therapeutic agent against prion disease. IDrugs 6(5):470–478. Erratum in: IDrugs (2004) 7:88
5. Brown P, Cathala F, Raubertas RF et al (1987) The epidemiology of Creutzfeldt-Jakob disease: conclusion of a 15-year investigation in France and review of the world literature. Neurology 37(6):895–895
6. Liberski PP, Streichenberger N, Giraud P et al (2005) Ultrastructural pathology of prion diseases revisited: brain biopsy studies. Neuropathol Appl Neurobiol 31(1):88–96
7. Liberski PP, Sikorska B, Hauw JJ et al (2008) Tubulovesicular structures are a consistent (and unexplained) finding in the brains of humans with prion diseases. Virus Res 132(1-2):226–228
8. Liberski PP, Budka H, Sluga E et al (1992) Tubulovesicular structures in Creutzfeldt-Jakob disease. Acta Neuropathol (Berl) 84:238–243
9. Streichenberger N, Jordan D, Verejan I et al (2000) The first case of new variant Creutzfeldt-Jakob disease in France: clinical data and neuropathological findings. Acta Neuropathol (Berl) 99(6):704–708
10. Sikorska B, Liberski PP, Sobow T et al (2009) Ultrastructural study of florid plaques in variant Creutzfeldt-Jakob disease: a comparison with amyloid plaques in kuru, sporadic Creutzfeldt-Jakob disease and Gerstmann-Sträussler-Scheinker disease. Neuropathol Appl Neurobiol 35(1):46–59

11. Hainfellner J, Brantner-Inhaler S, Cervenakova L et al (1995) The original Gerstmann-Sträussler-Scheinker family of Austria: divergent clinicopathological phenotypes but constant PrP genotype. Brain Pathol 5(3): 201–211
12. Liberski PP, Budka H (1995) Ultrastructural pathology of Gerstmann-Sträussler-Scheinker disease. Ultrastruct Pathol 19(1):23–36
13. Boellaard JW, Schlote W (1980) Subakute spongiforme Encephalopathie mit multiformer Plaquebildung. Eigenartige familiar-hereditare Kranknheit des Zentralnervensystems [spinocerebellare Atrophie mit Demenz, Plaques and plaqueahnlichen im Klein- and Grosshirn (Gerstmann, Sträussler, Scheinker)]. Acta Neuropathol (Berl) 49:205–212
14. Schlote W, Boellaard JW, Schumm F, Stohr M (1980) Gerstmann-Sträussler-Scheinker's disease. Electron microscopic observations on a brain biopsy. Acta Neuropathol (Berl) 52(3):203–211
15. Schumm F, Boellaard JW, Schlote W, Stohr M (1981) Morbus Gerstmann-Sträussler-Scheinker. Familie SCh.—Ein Bericht uber drei Kranke. Arch Psychiatr Nervenkr 230(3):179–196
16. Brown P, Goldfarb LG, Brown WT et al (1991) Clinical and molecular genetic study of a large German kindred with Gerstmann-Sträussler-Scheinker syndrome. Neurology 41(3):375–375
17. Liberski PP, Barcikowska M, Cervenakova L et al (1998) A case of sporadic Creutzfeldt-Jakob disease with a Gerstmann-Sträussler-Scheinker phenotype but no alterations in the PRNP gene. Acta Neuropathol 96(4): 425–430
18. Liberski PP, Bratosiewicz J, Barcikowska M et al (2000) A case of sporadic Creutzfeldt-Jakob disease with a Gerstmann-Sträussler-Scheinker phenotype but no alterations in the PRNP gene. Acta Neuropathol 100(2): 233–234
19. Gelpi E, Kovacs GG, Ströbel T et al (2005) Prion disease with a 144 base pair insertion: unusual cerebellar prion protein immunoreactivity. Acta Neuropathol 110(5):513–519
20. Kopp N, Richard M, Giraud P et al (2001) Un cas d'insomnie fatale. Rev Neurol (Paris) 157:332–333
21. Kopp N, Richard M, Liberski PP, Laplanche J, Giraud P, Biacabe A, Streichenberger N, Mallie F, Gross E, Boulliat J, Perret-Liaudet A (2000) A case of fatal familial insomnia: neuropathology and biochemistry. Abstract no C34-05. In: Abstracts of the XVth International Congress of Neuropathology, Birmingham, September 3–6th, 2000. Brain Pathol 10:671
22. Simmons MM, Blamire IW, Austin AR (1996) Simple method for the perfusion-fixation of adult bovine brain. Res Vet Sci 60(3): 247–250
23. Liberski PP, Brown P (2007) Disease-sepcific particles without prion protein in prion diseases—phenomenon or epiphenomenon? Neuropathol Appl Neurobiol 33(4):395–397
24. Jellinger K (1973) Neuroaxonal dystrophy. Its natural history and related disorders. In: Zimmerman HM (ed) Progress in neuropathology, vol 2. Grune & Stratton, New York, pp 129–180
25. Liberski PP, Yanagihara R, Gibbs CJ Jr, Gajdusek DC (1989) Scrapie as a model for neuroaxonal dystrophy: ultrastructural studies. Exp Neurol 106(2):133–141
26. Sikorska B, Liberski PP, Brown P (2008) Subependymal plaques in scrapie-affected hamster brains—why are they so different from compact kuru plaques? Folia Neuropathol 46(1):32–42
27. DeArmond SJ, McKinley MP, Barry RA et al (1985) Identification of prion amyloid filaments in scrapie-infected brain. Cell 41(1):221–235
28. Sigurdson CJ, Manco G, Schwarz P et al (2006) Strain fidelity of chronic wasting disease upon murine adaptation. J Virol 80(24):12303–12311
29. Falsig J, Aguzzi A (2008) The prion organotypic slice culture assay—POSCA. Nat Protoc 3(4):555–562
30. Falsig J, Sonati T, Herrmann US et al (2012) Prion pathogenesis is faithfully reproduced in cerebellar organotypic slice cultures. PLoS Pathog 8(11):e1002985
31. Sonati T, Reimann RR, Falsig J et al (2013) The toxicity of antiprion antibodies is mediated by the flexible tail of the prion protein. Nature 501(7465):102–106
32. Kovacs GG, Kalev O, Budka H (2004) Contribution of neuropathology to the understanding of human prion disease. Folia Neuropathol 42(Suppl A):69–76
33. Liberski PP, Yanagihara R, Gajdusek DC (1993) The spongiform vacuole—the hallmark of slow virus diseases. In: Liberski PP (ed) Light and electron microscopic neuropathology of slow virus disorders. CRC Press, Boca Raton, pp 155–180
34. Armstrong RA, Cairns NJ, Ironside JW, Lantos PL (2002) The spatial patterns of prion protein deposits in cases of variant Creutzfeldt-Jakob disease. Acta Neuropathol (Berl) 104:665–669

35. Armstrong RA, Cairns NJ, Lantos PL (2001) Quantification of the vacuolation (spongiform change) and prion protein deposition in 11 patients with sporadic Creutzfeldt-Jakob disease. Acta Neuropathol (Berl) 102(6):591–596
36. Zlotnik I (1957) Significance of vacuolated neurones in the medulla of sheep affected with scrapie. Nature 180(4582):393–394
37. Beck E, Bak IJ, Christ JF et al (1975) Experimental kuru in the spider monkey. Histopathological and ultrastructural studies of the brain during early stages of incubation. Brain 98(4):595–612
38. Beck E, Daniel PM, Davey AJ et al (1982) The pathogenesis of transmissible spongiform encephalopathy: an ultrastructural study. Brain 105(4):755–786
39. Liberski PP, Yanagihara R, Asher DM et al (1990) Reevaluation of the ultrastructural pathology of experimental Creutzfeldt-Jakob disease. Brain 113(1):121–137
40. Liberski PP, Yanagihara R, Wells GAH et al (1992) Ultrastructural pathology of axons and myelin in experimental scrapie in hamsters and bovine spongiform encephalopathy in cattle and a comparison with the panencephalopathic type of Creutzfeldt-Jakob disease. J Comp Pathol 106(4):383–398
41. Jeffrey M, Scott JR, Williams A, Fraser H (1992) Ultrastructural features of spongiform encephalopathy transmitted to mice from three species of bovidae. Acta Neuropathol (Berl) 84(5):559–569
42. Kim JH, Manuelidis EE (1986) Serial ultrastructural study of experimental Creutzfeldt-Jacob disease in guinea pigs. Acta Neuropathol 69(1-2):81–90
43. Liberski PP, Brown DR, Sikorska B (2008) Cell death and autophagy in prion diseases (transmissible spongiform encephalopathies). Folia Neuropathol 46(1):1–25
44. Liberski PP, Sikorska B, Bratosiewicz-Wasik J et al (2004) Neuronal cell death in transmissible spongiform encephalopathies (prion diseases) revisited: from apoptosis to autophagy. Int J Biochem Cell Biol 36(12):2473–2490
45. David-Ferreira JF, David-Ferreira KL, Gibbs CJ Jr, Morris JA (1968) Scrapie in mice: ultrastructural observations in the cerebral cortex. Proc Soc Exp Biol Med 127(1):313–320
46. Gibson PH, Doughty LA (1989) An electron microscopic study of inclusion bodies in synaptic terminals of scrapie-infected animals. Acta Neuropathol (Berl) 77(4):420–425
47. Liberski PP, Asher DM, Yanagihara R et al (1989) Serial ultrastructural studies of scrapie in hamsters. J Comp Pathol 101(4):429–442
48. Liberski PP, Budka H (1994) Tubulovesicular structures in Gerstmann-Sträussler-Scheinker disease. Acta Neuropathol (Berl) 88(5):491–492
49. Liberski PP, Jeffrey M, Goodsir C (1997) Tubulovesicular structures are not labeled using antibodies to prion protein (PrP) with the immunogold electron microscopy techniques. Acta Neuropathol (Berl) 93(3):260–264
50. Siakotos AN, Raveed D, Longa G (1979) The discovery of a particle unique to brain and spleen subcellular fractions from scrapie-infected mice. J Gen Virol 43(2):417–422
51. Liberski PP, Yanagihara R, Gibbs CJ Jr, Gajdusek DC (1990) Appearance of tubulovesicular structures in experimental Creutzfeldt-Jakob disease and scrapie precedes the onset of clinical disease. Acta Neuropathol (Berl) 79(4):349–354
52. Jeffrey M, Goodsir CM, Race RE, Chesebro R (2004) Scrapie-specific neuronal lesions are independent of neuronal PrP expression. Ann Neurol 55(6):781–792
53. Manuelidis L, Yu ZX, Barquero N, Mullins B (2007) Cells infected with scrapie and Creutzfeldt-Jakob disease agents produce intracellular 25-nm virus-like particles. Proc Natl Acad Sci U S A 104(6):1965–1970. Erratum in: Proc Natl Acad Sci USA (2007) 104:6090

Chapter 8

Cell Death and Autophagy in Prion Diseases

Pawel P. Liberski

Abstract

Neuronal autophagy, like apoptosis, is one of the mechanisms of programmed cell death. In this chapter, we summarize current information about autophagy in naturally occurring and experimentally induced scrapie, Creutzfeldt-Jakob disease and Gerstmann-Sträussler-Scheinker syndrome, against the broad background of neural degenerations in prion diseases. Typically a sequence of events is observed: from a part of the neuronal cytoplasm sequestrated by concentric arrays of double membranes (phagophores) through the enclosure of the cytoplasm and membrane proliferation to a final transformation of the large area of the cytoplasm into a collection of autophagic vacuoles of different sizes. These autophagic vacuoles form not only in neuronal perikarya but also in neurites and synapses. On the basis of ultrastructural studies, we suggest that autophagy may play a major role in transmissible spongiform encephalopathies and may even participate in the formation of spongiform change.

Key words Autophagy, Apoptosis, Prion diseases, Neurons, Ultrastructure

1 Introduction

1.1 Neuronal Cell Death in Prion Diseases

The 2016 Nobel Prize to Yoshinori Ohsumi highlighted this fascinating field [1]. The premature, primary death of nerve cells underlies the clinical picture of prion diseases. Unfortunately, despite great effort of the researchers, the cellular pathways leading to this neuronal loss are not entirely clear. What's more, the question whether there is a direct relation between the deposits of PrP^{Sc} and the loss of neurons still remains conjectural. As in other neurodegenerative diseases, in prion diseases, apoptosis has become the most popular concept of cell death. However, there is no direct and convincing evidence of apoptosis of nerve cells in most of the neurodegenerative diseases. In addition, the term "apoptosis" is used in a wider sense than it was originally coined and it has become synonymous with the non-necrotic cell death or even with the programmed cell death. The data on the role of apoptosis in prion diseases are conflicting. Among recently recognized other types of programmed cell death only autophagy has been reported in prion diseases but its role in prion diseases pathology is not established.

Pawel P. Liberski (ed.), *Prion Diseases*, Neuromethods, vol. 129,
DOI 10.1007/978-1-4939-7211-1_8, © Springer Science+Business Media LLC 2017

There is currently no consensus on the classification of different types of programmed cell death. One of the oldest but also still considered most accurate classifications is based on morphology. According to this classification introduced by Schweichel i Merker [2], three types of programmed cell death (PCD) are discriminated [3–6]:

1. Apoptosis
2. Autophagy
3. Cytoplasmic cell death

Although it must be stressed that this categorization was based on the ultrastructural features of embryonic cells during morphogenesis, such a classification is widely accepted for developing and mature organisms.

1. The term "apoptosis" in its original meaning refers to a morphological phenomenon [7] characterized by chromatin condensation, cell shrinkage, pyknosis, plasma membrane blebbing, and fragmentation of the nucleus (karyorrhexis). There is little or no ultrastructural modification of other subcellular organelles. The integrity of plasma membrane is maintained until the late stages of the process [8]. In the end stage, the cell breaks into small membrane-bound fragments, called apoptotic bodies, which are phagocytosed by macrophages or, in the case of neurons, by microglial cells without inciting any inflammatory response [7]. Apoptosis is regulated by highly conservative network of molecules consisting of Bcl-2 family and caspases and Apaf-1 [3–6]. The apoptotic cell death process has a very rapid time course and is complete within a few hours.
2. Autophagic cell death is also one of the programmed cell death mechanisms, sometimes called "type II programmed cell death," in contrast to "type I programmed cell death" (apoptosis). Contrary to apoptosis, autophagic cell death is characterized by abundant autophagic vacuoles in the cytoplasm, mitochondrial dilatation, and enlargement of both the Golgi apparatus and endoplasmic reticulum, which precedes nuclear destruction. Intermediate filaments and microfilaments are largely preserved [9–13].
3. A third type of programmed cell death is called "cytoplasmic cell death" and it was subsequently divided into two subtypes, 3A and 3B [14]. Type 3A of neuronal death that occurs during development is characterized by swelling of subcellular organelles, formation of large empty spaces within the cytoplasm, fusion of these spaces to extracellular space, and, finally, disintegration of the cellular structures. There are no features of autophagic or heterophagic activity. In the 3B subtype of

cell death there are similar vacuoles and empty spaces in the cytoplasm but, in addition, there is a retraction of plasma membrane and karyolysis.

2 Aggresomes

It is well known that protein aggregates, like crystals, are generally difficult to unfold or to degrade. Misfolded and aggregated proteins are usually handled in the cell through chaperone-mediated refolding or when this is impossible they are destroyed by proteasomal degradation. Recent findings suggest that there is a third way for a cell to deal with misfolded proteins. This pathway involves the sequestration of aggregated proteins into specialized "holding stations" as they are sometimes called or aggresomes [15]. In this mechanism, proteins form small aggregates that are transported along microtubules (MTs) towards a microtubule organizing center (MTOC) by the process mediated by the minus-end motor protein, dynein. At the organizing center the particles form a spherical structure, usually 1–3 μm in diameter, called aggresome. Aggresomes are not just static garbage deposits; they recruit various chaperones, ubiquitination enzymes, and proteasome components. They are also supposed to trigger autophagy [16].

A report of Kristiansen et al. [17] suggests that neuronal propagation of prions invokes a neurotoxic mechanism involving intracellular formation of PrP^{Sc} aggresomes. The authors showed that only in prion-infected cells mild proteasome impairment resulted in formation of large cytosolic, perinuclear structures, containing PrP^{Sc}, heat-shock protein 70, ubiquitin, proteasome subunits, and vimentin. These structures are consistent with the definition of aggresomes. Those authors also claimed to show aggresomes in vivo in brains of prion-infected mice but it is well known that vimentin is present in neurons only in a trace quantity but it is robust in glial cells. It means that showing aggresomes in vivo needs further studies. Cohen and Taraboulos [18] showed that hampering the activity of the cyclophilin isomerases with the fungal immunosuppressant CsA in different cell lines led to accumulation of a PrP population with prion-like properties that was not ubiquitylated and partially resisted proteasomal degradation. These aggregated molecules formed perinuclear aggresomes. Although a growing body of evidence seems to confirm the presence of aggresomes in prion diseases, it must be mentioned that majority of those studies were performed in vitro and PrP^{Sc} in human or animal diseased brains does not intracellular aggregates reminiscent of aggresomes. Similar aggregates are only observed by electron microscopy but they are extremely rare.

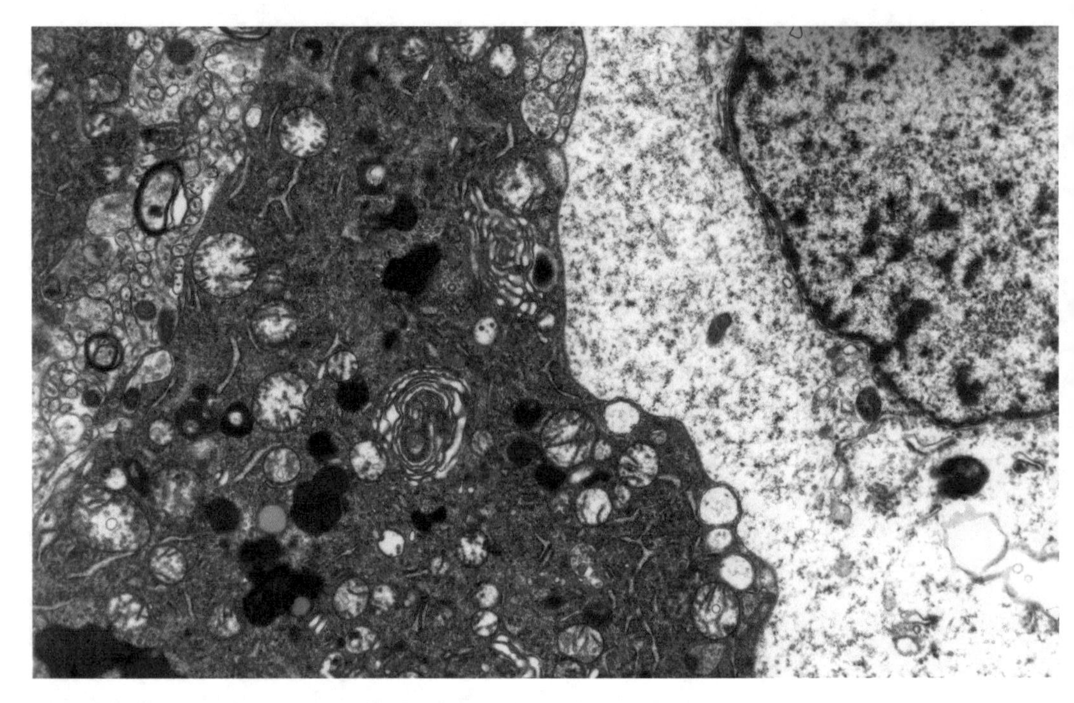

Fig. 1 A dark neuron (*left*) at the *left*; lipofuscin granules and vacuoles are visible. Original magnification, × 3000

2.1 Apoptosis and (Macro)Autophagy

Using TUNEL methodology, apoptotic neurons have been repeatedly identified in both naturally occurring and experimentally induced TSEs [19–24], and some investigators believe that at least a proportion of "dark neurons" that are shrunken (Fig. 1), with homogeneously dark cytoplasm, may represent cells undergoing apoptosis (other workers regard them as fixation artifacts). The data on whether neurons, and perhaps even glia, die of apoptosis in prion diseases are conflicting, however. While Migheli et al. [25] found no evidence of apoptosis in scrapie-infected mouse brains, Giese et al. [26, 27], Lucassen et al. [28], Fraser et al. [21], Williams et al. [29], Kretzschmar et al. [30–32], and Jamieson et al. [33] readily found apoptosis in various scrapie-infected rodent models. The characteristic DNA fragmentation ladder has been seen in both BSE [34] and natural scrapie [35], and in situ end labeling (ISEL) revealed apoptotic cells in human and experimental CJD [22, 36, 37]. Recently, Yuan et al. demonstrated that amplification of PrP by PMCA (protein misfolding cyclic amplification) methods induced apoptosis through activation of caspase-3 and increased the Bcl/Bac ration [38]. Collectively, these data strongly suggest that neurons in prion diseases die because of apoptosis.

It is evident that PrP^{Sc} accumulation in TSE-affected brain largely precedes development of other changes—i.e., spongiform change and astrogliosis for which PrP serves as a signal for proliferation [39–44].

As an evolutionarily ancient cellular response to intra- and extracellular noxious stimuli, autophagy may precede or coexist with apoptosis, and the process may be induced by apoptotic stimuli [45, 46]. Furthermore, the level of autophagy may define the sensitivity of a given neuronal population to apoptotic stimuli, which may underlie the phenomenon of "selective neuronal vulnerability." Thus, autophagy and apoptosis are often interwoven [47].

Cellular autophagy [13] is a physiological degradation process employed, like apoptosis, in embryonic growth and development, cellular remodeling, and biogenesis of some subcellular organelles—viz., multilamellar bodies [48–50]. Autophagosomes coalesce with lysosomes to form degraded autophagic vacuoles, and as in apoptosis, only excessive or misdirected autophagy causes a pathological process. Autophagy is highly enhanced in other brain amyloidoses or conformational disorders [51], Alzheimer disease [52], Parkinson's disease [53], and Huntington's disease, in which the signal for autophagy is huntingtin [54]. Here, we extend these observations using different models of scrapie and CJD.

2.2 Neuronal Autophagy in Prion Diseases

Data on autophagy in prion diseases are very limited, consisting of a few electron-microscopic papers, including reports from our own laboratories [12, 55–59]. The pioneering work was published in Acta Neuropathologica by Boellaard et al. [60]. Our initial experimental approach using the hamster-adapted 263 K or 22C–H strains of scrapie [55, 56, 58] was subsequently extended by studies of human brain biopsies from patients with sporadic CJD, variant CJD, and FFI [57]. Experimentally infected animal models are widely used because of their relatively short incubation periods that, for mice, range from 16 to 18 weeks, and for hamsters from 9 to 10 weeks for the 263 K strain and 24 to 26 weeks for the 22C–H strain.

2.3 Formation of Autophagic Vacuoles in Prion-Infected Brains

Autophagic vacuoles are areas of the cytoplasm sequestered within double or multiple membranes (phagophores, Fig. 2) of unknown origin; one possible source is the endoplasmic reticulum. They contain ribosomes, small secondary vacuoles, and occasional mitochondria. Some vacuoles present a homogenously dense appearance (Figs. 3, 4, and 5).

We observed neuronal autophagic vacuoles in different stages of formation in the same specimens and our interpretation of the "maturity" of their formation may or may not equate to actual developmental stages. Initially, a part of the neuronal cytoplasm was sequestered within double or multiple membranes (phagophores, Fig. 1) and often exhibited increased electron density. The intracytoplasmic membranes multiplied in a labyrinth-like manner. The autophagic vacuoles then expanded and eventually a vast area of the cytoplasm was transformed into a merging mass of autophagic vacuoles. Occasionally, a single large autophagic vacuole was

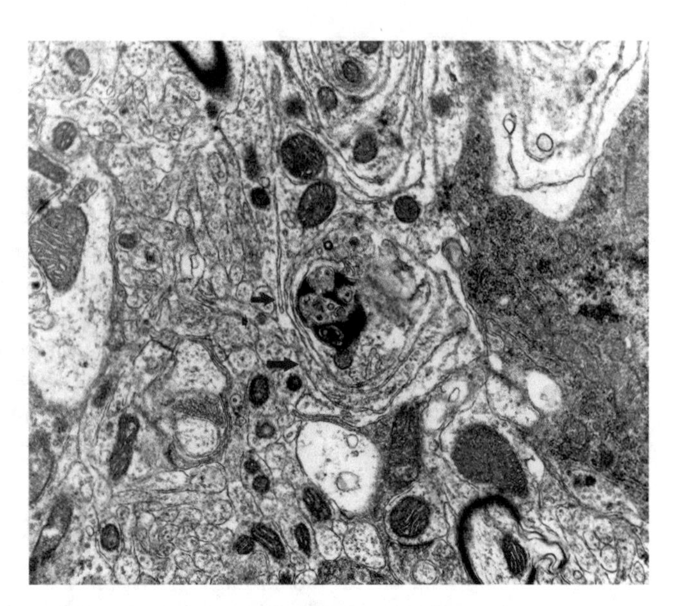

Fig. 2 An autophagic vacuole surrounded by multilayered membranes (phagophores, *arrows*). Original magnification, × 3000

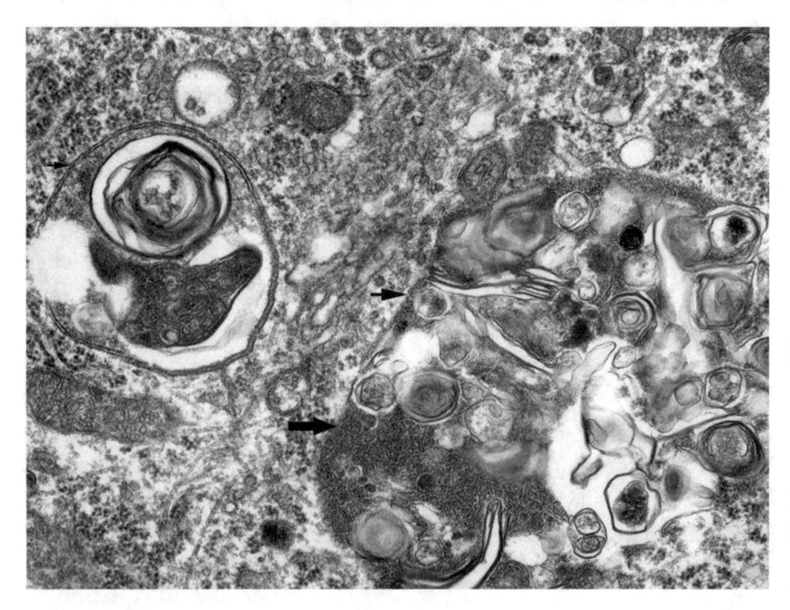

Fig. 3 Two large vacuoles surrounded by membranes (*arrows*), original magnification, × 30,000

visible. Autophagic vacuoles developed not only in neuronal perikarya but also in neuronal processes eventually replacing the whole cross section of affected neurites. In a few specimens we found round electron-dense structures that we identified as aggresomes and numerous multivesicular bodies (Fig. 4). In general, there was

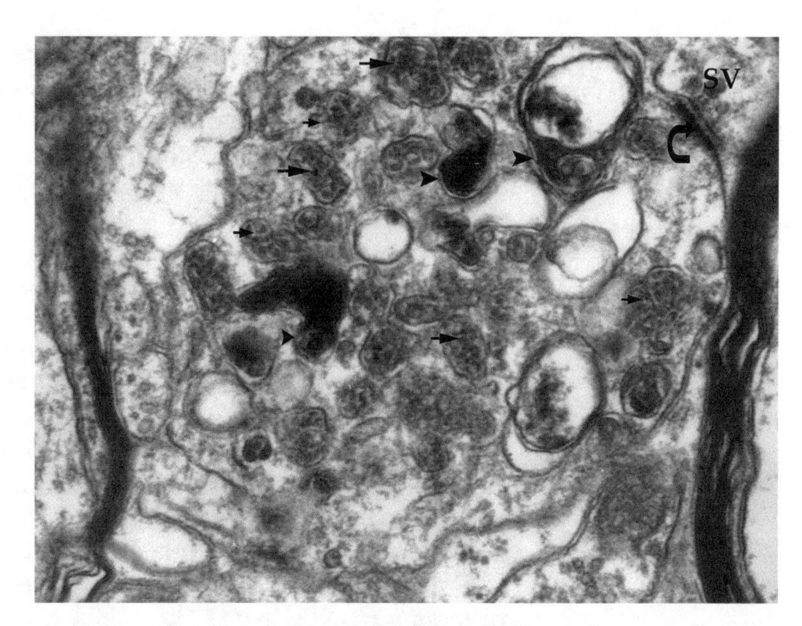

Fig. 4 A neuronal process containing dense bodies (*arrowheads*) and multivesicular bodies (*arrows*). A postsynaptic terminal is labeled with bent arrow; *SV* synaptic terminal. Original magnification, × 20,000

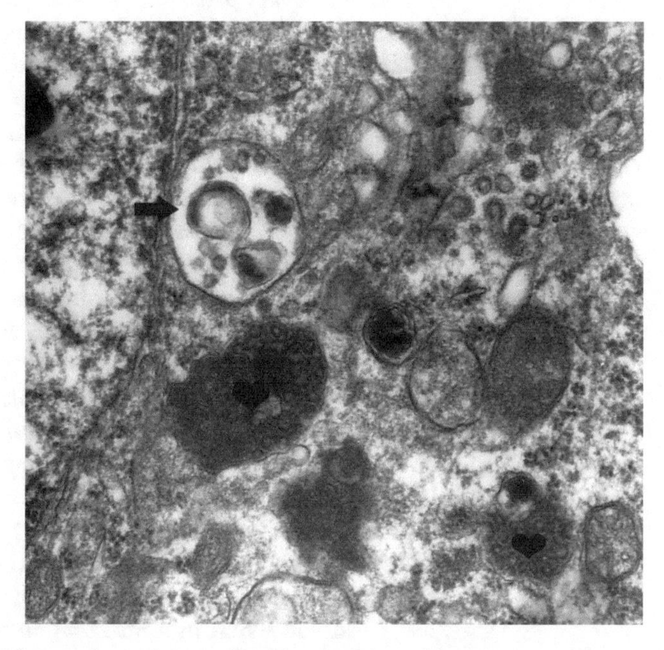

Fig. 5 A part of cytoplasm undergoing autophagy. Vacuoles containing vesicules, multivesicular body (*arrow*), and densely packed multivesicular bodies (hearts) are visible. Original magnification, × 20,000

little qualitative difference between these two models, although hamsters inoculated with the 263 K strain showed a more robust pathology. It seems that conversion of PrP^{c} into PrP^{Sc} triggers autophagy which leads to neuritic degeneration [61].

Additionally autophagy was observed in Tg20 mice expressing PrP gene [62].

2.4 Conclusions and Hypotheses

One of the major problems of prion disease pathogenesis is the cause of neuronal degeneration with eventual neuronal loss [20, 56]. Using a highly sophisticated mathematical model, Stumpf and Krakauer [63] tried to reason whether PrP^{Sc} causes neurons to die because of neurotoxic effect of PrP^{Sc} (gain of function), or loss of function of PrP^{c}. They assumed that if cells die of apoptosis because of neurotoxic gain of function of PrP^{Sc}, the cells should die rapidly, and the amount of PrP^{Sc} should be low. Indeed, in both CJD and FFI, there are more apoptotic cells and a lower amount of PrP^{Sc} [22] than in GSS, where the amount of amyloid is vast and number of apoptotic cells is low [37].

As already mentioned, three types of programmed cell death (PCD) are known: apoptosis, autophagy, and swelling of intracellular organelles. Apoptosis was discussed already, and in this part, we concentrate on autophagy in regard to prion diseases.

Whereas apoptosis in prion diseases is relatively well understood, autophagy is not. As already mentioned, autophagic cell death or type II PCS is a process used by a cell to remove the bulk of organelles, e.g., during growth and development, but which becomes pathological if too robust or wrongly placed [64]. According to Bursch et al. [64], the appearance of "*autophagic vacuoles in dying cells by electron microscopy is taken as condition* sine qua non *(i.e., an absolute prerequisite) to denote cell death as autophagic/type II PCD.*" To this end, the mere presence of autophagic vacuoles is relatively well demonstrated in prion diseases.

There are several uncertainties in our thinking on neuronal autophagy in prion diseases. First, autophagy is regarded as a short-term response toward nutrient limitations [65], which is, by definition, not the case in prion diseases. However, it seems that autophagy is activated to prevent apoptosis; when autophagy is blocked, apoptosis ensues. On the other hand, when apoptosis is blocked as in Bax/Bak-deficient mice, autophagy is activated as a cell survival mechanism [66]. However, when all subcellular organelles are self-eaten, the cell eventually dies. An analogous situation was observed when inhibition of autophagosomes and lysosomes was accomplished via targeting of *LAMP2* by RNA interference [67]. A dual role for autophagy was envisaged—i.e., protective role against apoptosis and detrimental role in cell death—and both of these are mediated by the same set of *ATG* genes [65]. Thus, the abundant presence of autophagy in prion diseases suggests that neurons either try to escape apoptosis or die by autophagy via one of the programmed cell death pathways.

A second uncertainty pertains to the relationship between autophagy and abnormal processing of PrP. The significance of autophagy for controlling the outcome of misfolded PrP^{Sc} is unclear. PrP^{c} is entirely translocated into the lumen of endoplasmic reticulum where the N-terminal signal peptide is cleaved away and the C-terminal GPI is added [68]. Then, PrP^{c} is transported on the cell surface. The misfolded PrP^{Sc} is localized to the cell surface and the endosomal-lysosomal compartment; and the conversion of PrP^{c} into PrP^{Sc} presumably takes place somewhere within one or more of these subcellular compartments [69–71]. However, misfolded PrP^{Sc} is directed to the aggresomes where, according to the hypothesis put forward by Cohen and Taraboulos [18], it may form "a seed" for seeded nucleation process of forming more PrP^{TSE}. Then, aggresomes are engulfed by autophagic vacuoles that in turn fuse with lysosomes containing PrP^{Sc}, and the process of nucleation may be either initiated or perpetuated. It was found that aggresomes may indeed induce the formation of autophagosomes [72].

A third uncertainty is how autophagy contributes to overall pathology underlying prion diseases. The hallmark of prion diseases is the vacuole which is an intracellular "empty" space surrounded by a single or a double membrane. The histogenesis of vacuoles is not well understood and most of the ultrastructural studies suffer from the inability to ascertain the subcellular organelles from which vacuoles originate; dilated endoplasmic reticulum or mitochondria have been suggested [73]. Our own unpublished work suggests that vacuoles are formed relatively abruptly with no detectable transitional stages (Gibson and Liberski, personal communication). It is tempting to speculate that vacuolation in prion diseases is somehow related to type III PCD characterized by the presence of large, membrane-bound intracellular empty spaces without the participation of lysosomes. The other option, that tissue destruction by autophagy results in vacuolation, would imply that autophagic vacuoles should be detected prior to spongiform vacuoles. Indeed, in the terminal stages of sheep scrapie (M. Jeffrey—personal communication) and in human prion diseases (CJD, GSS, and FFI) [57], little or no autophagy is seen but robust vacuolation is typical if not pathognomic. In contrast, when earlier stages of prion diseases are monitored in experimental rodent models, both autophagy and vacuolation have been reported [58, 74, 75]. Jeffrey et al. [74] have suggested that their presence reflects the robust accumulation of misfolded PrP^{Sc}, overloading of the neuronal catabolic machinery, and, eventually, bulk removal of damaged neurons via autophagy.

If this scenario is correct, the pathology of prion diseases is akin to the reactivation of certain embryonic processes in which

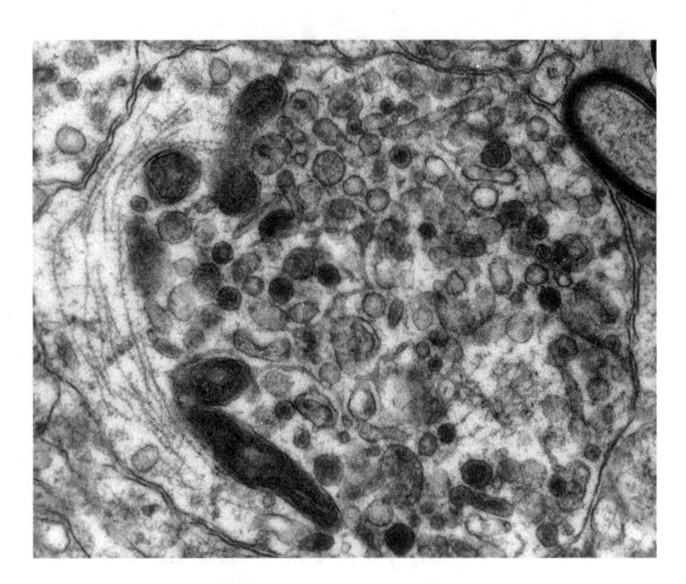

Fig. 6 A dystrophic neurite containing abundant lysosomes and small autophagic vacuoles. Original magnification, × 20,000

bulk removal of cells is present, i.e., remodeling of insect larvae [76]. Nearly 30 years ago, Elizabeth Beck [77] suggested that robust vacuolation like that characteristic of prion diseases is present when rats are inoculated with a suspension of cells from normal immature cerebellum. Neurons of inoculated rats not only demonstrated intracytoplasmic vacuoles but, at the ultrastructural level, showed abundance of coated pits and vesicles, phenomena of widespread appearances in prion diseases.

The other form of neuronal degeneration is neuroaxonal dystrophy (Figs. 6 and 7) [78–81]. The ultrastructural correlate of NAD is the dystrophic neurite—an axon or a dendrite filled with electron-dense inclusions, many of which were recently recognized as small autophagic vacuoles [82]. Both immature autophagic vacuoles immunogold-labeled with abs against Cat-D and mature vacuoles containing cathepsin were observed within those neurites [82]. As dystrophic neurites are abundant in prion diseases [79, 83], it is plausible that (macro)autophagy plays a role in neuronal degeneration in prion diseases. However, taking into account the data that autophagy may prevent apoptosis, it is also possible that abundant presence of autophagic vacuoles within dystrophic neurites actually reflects neurons struggling to survive in the noxious environment of misfolded PrP^{Sc}.

In summary, autophagy certainly does occur in prion diseases, but its pathogenetic role as a cause of cell death is uncertain. In particular, more research will be necessary to determine the connection, if any, between programmed cell death and formation of spongiform change.

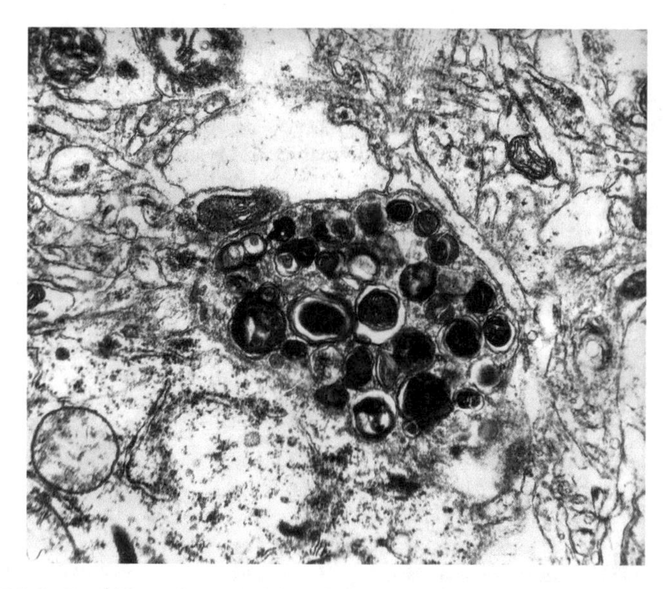

Fig. 7 A dystrophic neurite containing abundant lysosomes and small autophagic vacuoles. Original magnification, × 20,000

Acknowledgments

This chapter is supported in part by Healthy Aging Research Center (HARC, FP7-REGPOT-2012-2013-1). Ms. Ewa Skarzynska, Mr. Ryszard Kurczewski, Ms. Elzbieta Naganska, Ms. Leokadia Romanska, and Mr. Kazimierz Smoktunowicz are kindly acknowledged for skilful technical assistance.

References

1. Ohsumi Y (2014) Historical landmarks of autophagy research. Cell Res 24(1):9–23
2. Schweichel JU, Merker HJ (1973) The morphology of various types of cell death in prenatal tissues. Teratology 7(3):253–266
3. Klionsky DJ, Emr SD (2000) Autophagy as a regulated pathway of cellular degradation. Science 290(5497):1717–1721
4. Reggiori F, Klionsky DJ (2002) Autophagy in the eukaryotic cell. Eukaryot Cell 1(1):11–21
5. Wang C-W, Klionsky DJ (2003) The molecular mechanism of autophagy. Mol Med 9(3-4):65–76
6. Yuan J, Lipinski M, Degterev A (2003) Diversity of the mechanisms of neuronal cell death. Neuron 40(2):401–413
7. Kerr JF, Wyllie AH, Currie AR (1972) Apoptosis: a basic biological phenomenon with wide-ranging implications in tissue kinetics. Br J Cancer 26(4):239–257
8. Kroemer G, Jaattela M (2005) Lysosomes and autophagy in cell death control. Nat Rev Cancer 5(11):886–897
9. Bursch W (2001) The autophagosomal-lysosomal compartment in programmed cell death. Cell Death Differ 8(6):569–581
10. Bursch W (2004) Multiple cell death programs: Charon's lifts to hades. FEMS Yeast Res 5(2):101–110
11. Inbal B, Bialik S, Sabanay I et al (2002) DAP kinase and DRP-1 mediate membrane blebbing and the formation of autophagic vesicles during programmed cell death. J Cell Biol 157(3):455–468
12. Liberski PP, Sikorska B, Bratosiewicz-Wasik J et al (2004) Neuronal cell death in transmissible

spongiform encephalopathies (prion diseases) revisited: from apoptosis to autophagy. Int J Biochem Cell Biol 36(12):2473–2490
13. Klionsky DJ, Abdelmohsen K, Abe A et al (2016) Guidelines for the use and interpretation of assays for monitoring autophagy (3rd edition). Autophagy 12(1):1–222
14. Clarke PG (1990) Developmental cell death: morphological diversity and multiple mechanisms. Anat Embryol (Berl) 181:195–213
15. Sikorska B, Liberski PP, Brown P (2007) Neuronal autophagy and aggresomes constitute a consistent part of neurodegeneration in experimental scrapie. Folia Neuropathol 45(4):170–178
16. Garcia-Mata R, Gao YS, Sztul E (2002) Hassles with taking out the garbage: aggravating aggresomes. Traffic 3(6):388–396
17. Kristiansen M, Messennger MJ, Klöhn PC et al (2005) Disease-related prion protein forms aggresomes in neuronal cells leading to caspase activation and apoptosis. J Biol Chem 280(46):38851–38861
18. Cohen E, Taraboulos A (2003) Scrapie-like prion protein accumulates in aggresomes of cyclosporin A-treated cells. EMBO J 22(3): 404–417
19. Dorandeu A, Wingertsmann L, Chretien F et al (1998) Neuronal apoptosis in fatal familial insomnia. Brain Pathol 8(3):531–537
20. Fraser JR (2002) What is the basis of transmissible spongiform encephalopathy induced neurodegeneration and it be repaired? Neuropathol Appl Neurobiol 28(1):1–11
21. Fraser JR, Halliday WG, Brown D, Belichenko P, Jeffrey M (1996) Mechanisms of scrapie-induced neuronal cell death. In: Court L, Dodet B (eds) Transmissible subacute spongiform encephalopathies: prion disease. IIIrd International Symposium on transmissible subacute spongiform encephalopathies: prion disease, Val-de-Grace Paris, France. Elsevier, Amsterdam, Oxford, Paris, pp 107–112
22. Jesionek-Kupnicka D, Buczynski J, Kordek R, Liberski PP (1999) Neuronal loss and apoptosis in experimental Creutzfeldt-Jakob disease in mice. Folia Neuropathol 37(4):283–286
23. Jesionek-Kupnicka D, Buczynski J, Kordek R et al (1997) Programmed cell death (apoptosis) in Alzheimer's disease and Creutzfeldt-Jakob disease. Folia Neuropathol 35(4): 233–235
24. Jesionek-Kupnicka D, Kordek R, Buczynski J, Liberski PP (2001) Apoptosis in relation to neuronal loss in experimental Creutzfeldt-Jakob disease in mice. Acta Neurobiol Exp 61:13–19
25. Migheli A, Attanasio A, Lee WH et al (1995) Detection of apoptosis in weaver cerebellum by electron microscopic in situ end-labeling of fragmented DNA. Neurosci Lett 199(1):53–56
26. Giese A, Groschup MH, Hess B, Kretzschmar HA (1995) Neuronal cell death in scrapie-infected mice is due to apoptosis. Brain Pathol 5(3):213–221
27. Giese A, Kretzschmar HA (2001) Prion-induced neuronal damage-the mechanisms of neuronal destruction in the subacute spongiform encephalopathies. Curr Top Microbiol Immunol 253:203–217
28. Lucassen PJ, Williams A, Chung WC, Fraser H (1995) Detection of apoptosis in murine scrapie. Neurosci Lett 198(3):185–188
29. Williams A, Lucassen PJ, Ritchie D, Bruce M (1997) PrP deposition, microglial activation, and neuronal apoptosis in murine scrapie. Exp Neurol 144(2):433–438
30. Kretzschmar HA, Giese A, Brown DR et al (1997) Cell death in prion disease. J Neural Transm Suppl 50:191–210
31. Kretzschmar HA, Giese A, Brown DR, Herms J, Schmidt B, Groschup MH (1996) Cell death in prion disease. In: Court L, Dodet B (eds) Transmissible Subacute spongiform Encephalopathies: prion disease. IIIrd International Symposium on transmissible subacute spongiform encephalopathies: prion disease. Val-de-Grace, Paris, France. Elsevier, Amsterdam, Oxford, Paris, pp 97–106
32. Kretzschmar HA, Giese A, Herms J, Brown DR (1998) Neuronal degeneration and cell death in prion disease. In: Morrison DRO (ed) Prions and brain diseases in animals and humans. Plenum Press, New York, pp 253–268
33. Jamieson E, Jeffrey M, Ironside JW, Fraser JR (2001) Apoptosis and dendritic dysfunction precede prion protein accumulation in 87V scrapie. Neuroreport 12(10):2147–2153
34. Theil D, Fatzer R, Meyer R et al (1997) Nuclear DNA fragmentation and immune reactivity in bovine spongiform encephalopathy. J Comp Pathol 121:357–367
35. Fairbairn DW, Carnahan KG, Thwaits RN et al (1994) Detection of apoptosis induced DNA cleavage in scrapie-infected sheep brain. FEMS Microbiol Lett 115(2-3):341–346
36. Ferrer I (1994) Nuclear DNA fragmentation in Creutzfeldt-Jakob disease: does a mere positive in situ nuclear end-labeling indicate apoptosis? Acta Neuropathol (Berl) 97:5–12
37. Gray F, Chretien F, Adle-Biassette H et al (1999) Neuronal apoptosis in Creutzfeldt-Jakob

disease. J Neuropathol Exp Neurol 58(4): 321–328
38. Yuan Z, Yang L, Chen B et al (2015) Protein misfolding cyclic amplification induces the conversion of recombinant prion protein to PrP oligomers causing neuronal apoptosis. J Neurochem 133(5):722–729
39. Brown DR (1998) Prion protein-overexpressing cells show altered response to a neurotoxic prion protein peptide. J Neurosci Res 54(3):331–340
40. Brown DR (1999) Prion protein peptide neurotoxicity can be mediated by astrocytes. J Neurochem 73(3):1105–1113
41. Brown DR, Besinger A, Herms JW, Kretzschmar HA (1998) Microglial expression of the prion protein. Neuroreport 9(7): 1425–1429
42. Brown DR, Mohn CM (1998) Astrocytic glutamate uptake and prion protein expression. Glia 25:282–292
43. Hafiz FB, Brown DR (2000) A model for the mechanism of astrogliosis in prion disease. Mol Cell Neurosci 16(3):221–232
44. Ye X, Scallet AC, Kascsak RJ, Carp RI (1998) Astrocytosis and amyloid deposition in scrapie-infected hamsters. Brain Res 809(2):277–287
45. Bursch W, Ellinger A (2005) Autophagy—a basic mechanism and a potential role for neurodegeneration. Folia Neuropathol 43(4):297–310
46. Xue L, Fletcher GC, Tolkovsky AM (1999) Autophay is activated by apoptotic signalling in sympathetic neurons: an alternative death execution. Mol Cell Neurosci 14(3):180–198
47. Bursch W, Hochegger K, Torok L et al (2000) Autophagic and apoptotic types of programmed cell death exhibit different fates of cytoskeletal filaments. J Cell Sci 113 (Pt 7):1189–1198
48. Filonova LH, Bozhkov PV, Brukhin VB et al (2000) Two waves of programmed cell death occur during formation and development of somatic embryos in the gymnosperm, Norway spruce. J Cell Sci 113 Pt 24:4399–4411
49. Hariri M, Millane G, Guimond M-P et al (2000) Biogenesis of multilamellar bodies via autophagy. Mol Biol Cell 11(1):255–268
50. Sattler T, Mayer A (2000) Cell-free reconstitution of microautophagic vacuole invagination and vesicle formation. J Cell Biol 151(3): 529–538
51. Graeber MB, Moran LB (2002) Mechanisms of cell death in neurodegenerative diseases: fashion, fiction, and facts. Brain Pathol 12(3):385–390
52. Stadelmann C, Deckwerth TL, Srinivasan A et al (1999) Activation of caspase-3 in single neurons and autophagic granules of granulovacuolar degeneration in Alzheimer's disease. Evidence for apoptotic cell death. Am J Pathol 155(5):1459–1466
53. Anglade P, Vyas S, Javoy-Agid F et al (1997) Apoptosis and autophagy in nigral neurons of patients with Parkinson's disease. Histol Histopathol 12(1):25–31
54. Kegel KB, Kim M, Sapp E et al (2002) Huntingtin expression stimulates endosomal-lysosomal activity, endosome tubulation, and autophagy. J Neurosci 20:7268–7278
55. Liberski PP, Asher DM, Yanagihara R et al (1989) Serial ultrastructural studies of scrapie in hamsters. J Comp Pathol 101(4):429–442
56. Liberski PP, Gajdusek DC, Brown P (2002) How do neurons degenerate in prion diseases or transmissible spongiform encephalopathies (TSEs): neuronal autophagy revisited. Acta Neurobiol Exp 62:141–148
57. Liberski PP, Streichenberger N, Giraud P et al (2005) Ultrastructural pathology of prion diseases revisited: brain biopsy studies. Neuropathol Appl Neurobiol 31(1):88–96
58. Liberski PP, Yanagihara R, Gibbs CJ Jr, Gajdusek DC (1992) Neuronal autophagic vacuoles in experimental scrapie and Creutzfeldt-Jakob disease. Acta Neuropathol 83(2):134–139
59. Sikorska B, Liberski PP, Giraud P et al (2004) Autophagy is a part of ultrastructural synaptic pathology in Creutzfeldt-Jakob disease: a brain biopsy study. Int J Biochem Cell Biol 36(12):2563–2573
60. Boellaard JW, Kao M, Schlote W, Diringer H (1991) Neuronal autophagy in experimental scrapie. Acta Neuropathol (Berl) 82(3):225–228
61. Cronier S, Carimalo J, Schaeffer B, Jaumain E, Béringue V, Miquel MC, Laude H, Peyrin JM (2012) Endogenous prion protein conversion is required for prion-induced neuritic alterations and neuronal death. FASEB J 26(9):3854–3861
62. Joshi-Barr S, Bett C, Chiang WC, Trejo M, Goebel HH, Sikorska B, Liberski P, Raeber A, Lin JH, Masliah E, Sigurdson CJ (2014) De novo prion aggregates trigger autophagy in skeletal muscle. J Virol 88(4):2071–2082
63. Stumpf MPH, Krakauer DC (2000) Mapping the parameters of prion-induced neuropathology. Proc Natl Acad Sci U S A 97(19):10573–10577
64. Bursch W, Ellinger A, Gerner C, Schulte-Hermann R (2004) Autophagocytosis and programmed cell death. In: Klionsky D (ed) Autophagy. Landes Bioscience, Georgetown, TX, pp 290–306

65. Eskelinen EL (2005) Doctor Jekyll and Mister Hyde: autophagy can promote both cell survival and cell death. Cell Death Differ 12: 1468–1472
66. Lum JJ, Bauer DE, Kong M et al (2005) Growth factor regulation of autophagy and cell survival in the absence of apoptosis. Cell 120(2):237–248
67. Gonzalez-Polo RA, Boya P, Pauleau AL et al (2005) The apoptosis/autophagy paradox: autophagic vacuolization before apoptotic death. J Cell Sci 118(14):3091–3102
68. Harris DA (2003) Trafficking, turnover and membrane topology of PrP. Br Med Bull 66(1):71–85
69. Borchelt DR, Taraboulos A, Prusiner SB (1992) Evidence for synthesis of scrapie prion protein in the endocytic pathway. J Biol Chem 267(23):16188–16199
70. Caughey B, Raymond GJ (1991) The scrapie-associated form of PrP is made from a cell surface precursor that is both protease- and phospholipase-sensitive. J Biol Chem 266:18217–18223
71. Caughey B, Raymond GJ, Ernst D, Race RE (1991) N-terminal truncation of the scrapie-associated form of PrP by lysosomal protease(s): implications regarding the site of conversion of PrP to the protease-resistant state. J Virol 65(12):6597–6603
72. Kopito RR (2000) Aggresomes, inclusion bodies and protein aggregation. Trends Cell Biol 10(12):524–530
73. Jeffrey M, Goodbrand IA, Goodsir A (1995) Pathology of the transmissible spongiform encephalopathies with special emphasis on ultrastructure. Micron 26(3):277–298
74. Jeffrey M, Scott JR, Williams A, Fraser H (2002) Ultrastructural features of spongiform encephalopathy transmitted to mice from three species of bovidae. Acta Neuropathol (Berl) 84:559–569
75. Liberski PP, Mori S (1997) The Echigo-1: a panencephalopathic strain of Creutzfeldt-Jakob disease: a passage to hamsters and ultrastructural studies. Folia Neuropathol 3583: 250–254
76. Myohara M (2004) Real-time observation of autophagic programmed cell death of drosophila salivary glands in vitro. Dev Genes Evol 214(2):99–104
77. Beck E (1988) Lesions akin to transmissible spongiform encephalopathy in the brains of rats inoculated with immature cerebellum. Acta Neuropathol (Berl) 76(3):295–305
78. Gibson PH, Liberski PP (1987) An electron and light microscopic study of the numbers of dystrophic neurites and vacuoles in the hippocampus of mice infected intracerebrally with scrapie. Acta Neuropathol (Berl) 73(4):379–382
79. Liberski PP (1987) Electron microscopic observations on dystrophic neurites in hamster brains infected with the 263K strain of scrapie. J Comp Pathol 97(1):35–39
80. Liberski PP, Budka H (1999) Neuroaxonal pathology in Creutzfeldt-Jakob disease. Acta Neuropathol (Berl) 97(4):329–334
81. Liberski PP, Budka H, Yanagihara R, Gajdusek DC (1995) Neuroaxonal dystrophy in experimental Creutzfeldt-Jakob disease: electron microscopical and immunohistochemical demonstration of neurofilament accumulations within affected neurites. J Comp Pathol 112(3):243–255
82. Nixon RA, Wegiel J, Kumar A et al (2005) Extensive involvement of autophagy in Alzheimer disease: an immuno-electron microscopy study. J Neuropathol Exp Neurol 64(2):113–122
83. Liberski PP, Kloszewska I, Boellaard I et al (1995) Dystrophic neurites of Alzheimer's disease and Gerstmann-Sträussler-Scheinker's disease dissociate from the formation of paired helical filaments. Alzheimer's Res 1:89–93

Chapter 9

Methods for Isolation of *Spiroplasma* sp. from Prion-Positive Eye Tissues of Sheep Affected with Terminal Scrapie

Frank O. Bastian

Abstract

Spiroplasma, a tiny wall-less bacterium, has been consistently found in tissues affected with transmissible spongiform encephalopathy (TSE), and a closely related spiroplasma laboratory strain isolated from rabbit ticks experimentally induces TSE-like clinical and pathological findings in rodents and ruminants. The consistent presence of a *Spiroplasma* sp. in naturally occurring TSE has been documented by morphological and molecular studies, and the direct isolation of these spiroplasma from TSE-affected tissues into cell-free media confirms our hypothesis that spiroplasma is a candidate causal agent of TSE. In this treatise, we show evidence of isolation of *Spiroplasma* sp. from scrapie-affected eyes and note that the isolate is immunologically distinct from other spiroplasma species. It is noteworthy that experimental inoculation of these scrapie spiroplasma isolates intracranially (IC) into sheep and goats induced the classic vacuolar neuropathology of naturally occurring scrapie in the obices of these ruminants. The growth characteristics of this isolate differ from other spiroplasma species in that the scrapie spiroplasma isolate is extremely fastidious and cannot be propagated beyond a few passages in the special medias designed for spiroplasma growth. Herein we describe the protocols that we have used in our laboratory for isolation of the scrapie-related spiroplasma from scrapie-affected eyes. The tissues are homogenized in special media with high osmolality especially designed for spiroplasma growth; after low speed centrifugation, the supernatant is passed through a 0.45 μm membrane filter and overlaid onto a confluent monolayer of mosquito larvae cells (*Aedes albopictus* clone C6/36). The presence of spiroplasma on the surface of mosquito cells is documented by scanning electron microscopy (SEM) and can be shown as well by dark field microscopy and transmission electron microscopy (TEM). There is minimal cytopathogenic effect (CPE) with some vacuolization of mosquito cells, apoptosis, and multi-nucleation as seen by phase microscopy. The methodology is applicable for use in mammalian cell culture lines, such as bovine corneal endothelial (BCE) cells and mouse neuroblastoma (Neuro-2a) cells, the latter commonly used in prion research. In our hands this methodology has been successful in recovering spiroplasma from the TSE tissue >80% of the time based upon recognition of spirals by dark field microscopy while not from control tissues. These procedures along with future efforts to optimize the specialized growth medias should allow characterization of the scrapie spiroplasma isolates and determine the interrelationship with the prion.

Key words Spiroplasma, TSE, Prion, Scrapie, CJD, CWD, BSE, TME

Pawel P. Liberski (ed.), *Prion Diseases*, Neuromethods, vol. 129,
DOI 10.1007/978-1-4939-7211-1_9, © Springer Science+Business Media LLC 2017

1 Introduction

A major goal of prion disease researchers is to develop an antemortem serological test and/or a vaccine for TSEs, but their efforts over the past 30 years have been thwarted by lack of progress since prions are misfolded host proteins, wherein there is danger of developing autoantibodies. Our laboratory has identified a wall-less bacterium consistently associated with TSE-affected tissues and not in controls that is involved in the pathogenesis of TSEs [1]. This treatise is written to offer our experience and methodology used to isolate this *Spiroplasma* sp. from TSE-affected tissues for the purpose of further characterization of this organism, therein facilitating future use of these foreign antigens to develop a diagnostic test and/or a vaccine for TSE. Preliminary studies have indicated that there are circulating antibodies against spiroplasma-related heat shock proteins in CJD sera [2].

1.1 Association of Spiroplasma with TSE

Prion researchers have ignored the data showing involvement of spiroplasma infection in the TSEs. Spiroplasma were initially found by TEM of a brain biopsy from a patient with Creutzfeldt–Jakob disease (CJD) [3] and then in autopsied brains from several CJD patients [4–6]. Ribosomal DNA specific for spiroplasma has been detected in brain tissues from all forms of TSE including CJD, scrapie in sheep, and chronic wasting disease (CWD) in deer [7]. A negative study is discounted since the authors did not use the primers specific for spiroplasma [8], although our study was referenced [7]. Spiroplasma are well-known phytopathogens that are transmitted via insects [9]. The rabbit tick spiroplasma isolate is unique in that it proliferates at 37 °C and has experimentally produced a persistent brain infection and cataracts in mice [10]. The role of this organism in nature is currently unknown. This spiroplasma was isolated from the mice by passage in embryonated eggs into cell-free culture [11]. These spiroplasma, designated *Spiroplasma mirum*, contain two immunologically related strains, suckling mouse cataract agent (SMCA named after the source) and GT-48, which is a more virulent strain [1, 10]. Both SMCA and GT-48 have been propagated for years both in animal models and in in vitro studies. Much of the work in our laboratory was carried out using those strains as reference.

In our laboratory *Spiroplasma mirum* was inoculated IC into suckling rats for the purpose of examining the neuropathology of experimental spiroplasmosis in the rodent model [12, 13]. The rats infected with *S. mirum* showed clinical signs of neurodegeneration with a clear dose response. Rat brains at autopsy showed vacuolar encephalopathy with minimal inflammatory response not seen in controls. In the study spiroplasma were visualized by TEM primarily as membranous forms that reacted with SMCA-specific polyclonal antibodies generated in rabbits. A single tightly coiled spiroplasma identical to those seen in our initial CJD studies was found in brains of low dose inoculated rats by TEM although spiroplasma were readily recovered from those rat tissues into Sp4 broth cultures. The rat brain neuropil under low dose *S. mirum* infection examined by TEM [12, 14] is essentially identical ultrastructurally to CJD-affected brains [1]. Spiroplasma have been shown to be neurotropic in the rat model [15]. Intracranial inoculation of SMCA into deer resulted in neurological deterioration 4 months after IC inoculation closely resembling clinical signs of CWD [16]. The spiroplasma-infected deer showed spongiform encephalopathy with demonstration of the organism in the obex by immunohistochemistry (IHC) and TEM. There was a striking glial reaction to the spiroplasma infection in the deer model. Experimental spiroplasma infection in deer and sheep showed a particular affinity for the eye where collections of *S. mirum* were seen in the vitreous and corneal endothelia [17].

1.2 Spiroplasma Share Biologic Properties with the TSE Agent

Spiroplasma share morphological characteristics and resistant biologic properties with the transmissible agent of CJD [1]. Spiroplasma can easily pass through virus-sized 100 nm filters [1, 18]. Spiroplasma show wide variability in morphology in broth cultures [12, 14]. Tiny filamentous spiroplasma forms measuring only 40 nanometers (nm) in diameter are seen in broth culture during all stages of the growth cycle [14]; identical forms are seen by TEM of rat brains experimentally inoculated with *S. mirum* [18] and closely resemble structures seen by TEM of scrapie-affected mouse tissues [19]. Spiroplasma digested with detergents show an internal fibrillar network [14] responsible for the locomotor machinery of the cell [20]. The internal spiroplasma fibrils are unique to this bacterium and are essentially identical to filamentous inclusions seen in scrapie-affected brain tissues by negative stain EM [21]. It is noteworthy that these scrapie-associated fibrils (SAF) are consistent and unique markers of TSE infection [1]. Scrapie-specific hyperimmune polyclonal sera against SAF (courtesy of Patricia Merz, New York) reacted with the spiroplasma internal fibrils obtained from protease-digested spiroplasma cell-free broth cultures by Western blots [22]. *Spiroplasma mirum* strains have shown survival after exposure to extremes in heat, with no viability loss until 87.5 °C and surviving near 100 °C [23] (exactly simulating experiments by Hunter with scrapie [24]), while control *Escherichia coli* were dead at 80 °C [23]. *Spiroplasma mirum* showed significant resistance to irradiation [23].

The unusual resistance of *S. mirum* grown in vitro to antibiotics and to physical and chemical treatments is explained by the tendency of this bacterium to produce biofilm on glass, mica, and stainless steel surfaces [25]. Spiroplasma within biofilm on nickel wires are resistant to 50% glutaraldehyde. Scanning electron microscopic examination of *S. mirum* biofilms revealed filamentous and coccoid forms, with abundant curli fibers (bacterial amyloid) [26–28] attaching the organisms to the surface. There currently are no bacteriocidal antibiotics effective for spiroplasma [1]; spiroplasma growth is inhibited by tetracycline, which is a bacteriostatic drug. It is noteworthy that experimental scrapie infection in hamsters can be prevented by adding tetracycline to the inoculum [29]. Spiroplasma within biofilm attached to mica show long friable membranous interconnections [25], suggesting that these organisms are able to communicate directly with one another in complex micro colonies that could occur in a soil environment, therein providing the rationale for possible transmission of CWD via soil ingestion [30], which is a common practice among cervids and other ruminants [31]. These data raise the issue of "what is the relationship between spiroplasma infection and prion in TSEs?". Prion amyloid deposits have not been conclusively documented in brain tissues of ruminants experimentally inoculated with *S. mirum*. On the other hand, *S. mirum* infection of BCE and Neuro-2a tissue cultures induces formation of alpha-synuclein (***Feng J, personal communication, 2012***), suggesting that spiroplasma could be involved in prion amyloid formation. It is noteworthy that curli fibers (bacterial amyloid) found in bacterial biofilms has been shown to misfold host proteins [27].

1.3 Isolation of Spiroplasma from TSE Tissues

The primary subject of this treatise is that *Spiroplasma* sp. have been grown in cell-free media from brain tissues from all forms of TSE [16] and from eyes of sheep with terminal scrapie [17]. Filtered supernatants from homogenized sheep eyes affected with scrapie when incubated in cell-free Sp4 broth show presence of spiral forms by dark field microscopy (Fig. 1) but the spiroplasma isolate cannot be propagated in cell-free media beyond a few passages (*see* **Note 1**). These isolates do not go through the log phase growth in Sp4 broth characteristic of the *S. mirum* laboratory strains [9]. Spiroplasma scrapie isolates inoculated onto mouse neuroblastoma (Neuro-2a) cell cultures induce CPE characterized by vacuolar cytopathology, increased granularity, multi-nucleation, and cell enlargement with numerous apoptotic cells [17, 18]. Bovine corneal endothelial cells infected with scrapie spiroplasma isolates show similar CPE [18]. Identical CPE has been described in scrapie infection of subcultures from scrapie-affected sheep brain [32]. Immune florescent antibody (IFA) studies of BCE cell cultures using hyperimmune sera against *S. mirum* showed presence of spiroplasma both intracellularly and extracellularly in the positive BCE control culture infected with *S. mirum*. However, there is no immunostaining of BCE cultures experimentally infected with the scrapie spiroplasma isolate [17], indicating there is no immune cross-reactivity between this novel *Spiroplasma* sp. and the laboratory *S. mirum* strains. The presence of scrapie spiroplasma isolates in BCE cell cultures was confirmed by TEM [17, 18] (Fig. 2). Recently, we have shown by SEM (Fig. 3) that the scrapie isolates

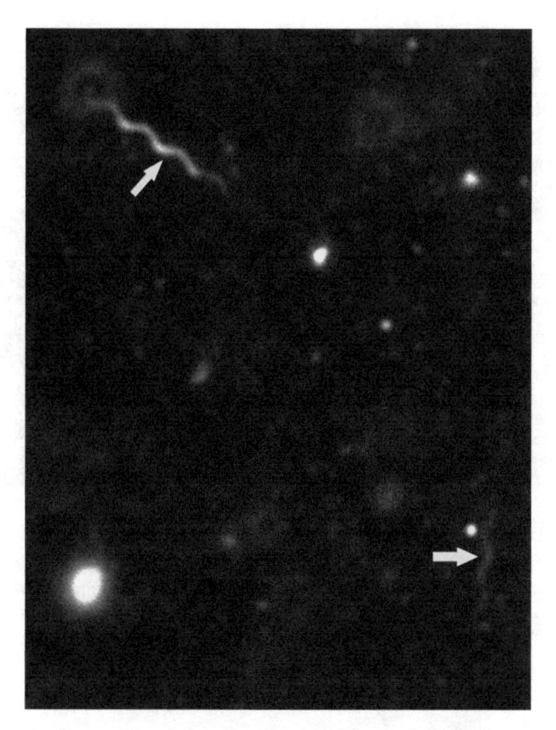

Fig. 1 Spiroplasma are isolated from scrapie-affected eye tissues by direct inoculation into specialized media as shown by dark field microscopy (*arrows*). Original magnification × 1200. Reproduced with permission from Journal of Neuropathology and Experimental Neurology [18]

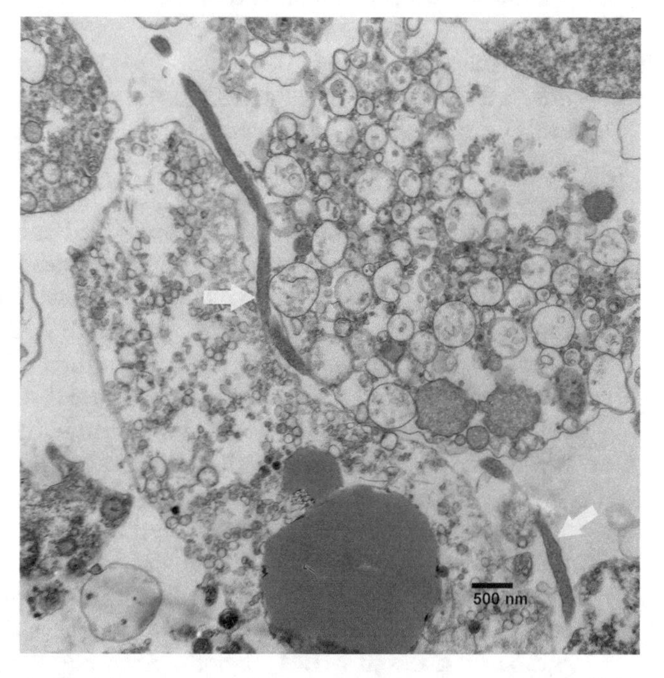

Fig. 2 *Spiroplasma* sp. isolated from scrapie-affected eye tissues via inoculation of BCE cells is shown by TEM as long filament in infected BCE cell culture (*arrows*). Bar = 500 nm. Reproduced with permission from Journal of Neuropathology and Experimental Neurology [18]

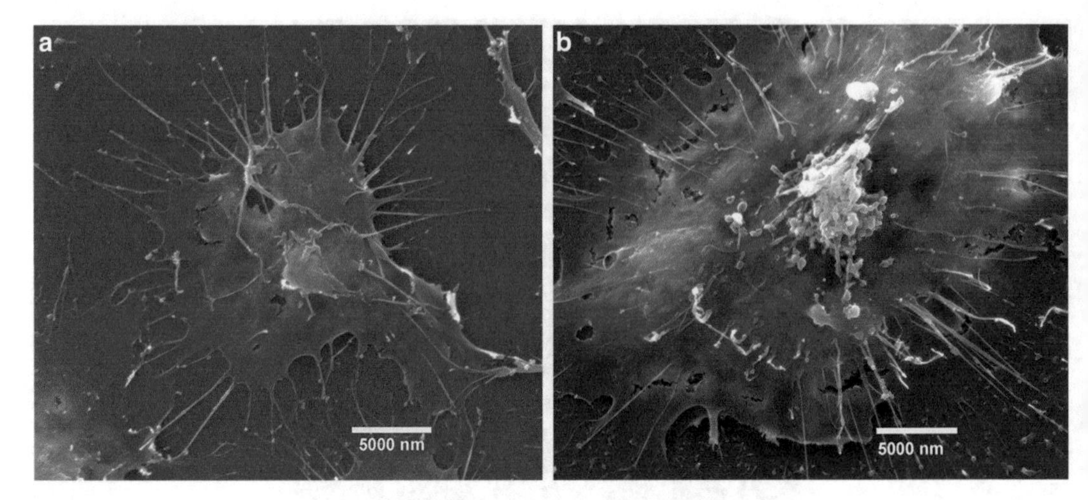

Fig. 3 (**a**) Control mosquito larvae cell culture inoculated with media alone shows normal attachment of cells by SEM. (**b**) Mosquito larvae cell culture inoculated with filtrate of scrapie-affected sheep eye tissues shows attachment of clumps of spiroplasma filaments and coccoid forms on cell surface by SEM. Note fragmentation of cell attachments. Bar = 5000 nm

grow in mosquito larvae cultures, therein providing a workable method for isolation of the scrapie-associated *Spiroplasma* sp., which is the subject of this treatise.

1.4 Outline of Major Procedures in Protocol

The isolation of *Spiroplasma* sp. from prion-positive scrapie-affected sheep eyes (obtained from the scrapie research sheep herd at the University of Idaho) involves initial homogenization of the tissues in specialized media favorable for spiroplasma growth [9]. A genomic study of *S. mirum* (ATCC 29335 strain SMCA, complete genome done in our laboratory) (GenBank CP006720) indicated that the organism requires amino acid supplements for cultivation in growth media. The homogenate is centrifuged to remove large particles, then filtered through 0.450 μm membrane filters to remove any contaminating bacteria. Spiroplasma easily pass through these filters since they are viral like in size (as small as 40 nm in diameter) [14, 18]. The filtrate is inoculated onto the mosquito cell monolayer and incubated at 28 °C for 2–3 weeks, while monitoring the cultures for CPE by phase microscopy. Cells and fluids are collected and are stored by freezing or examined by polymerase chain reaction (PCR) and/or TEM. It is noteworthy that these cultures are maintained in a humidified incubator and fed minimal essential media (MEM). While spiroplasma do not ordinarily grow in MEM, they do especially flourish in the spent MEM media from the cell culture [33].

2 Materials

2.1 Special Media Requirements for Spiroplasma Growth

The majority of spiroplasma were considered not to be culturable until the development of SP4 media with high osmolality as described by Tully in 1976 [34]. SP4 and a variant formulation, M1D, are PPLO-based media developed for isolation of spiroplasma, and have become used for cultivation of most *Mollicutes*. Hardy Diagnostics, the primary commercial supplier of SP4, describes the media on their WEB site as "highly nutritious due to addition of beef heart infusion, peptone supplemented with yeast extract, CMRL 1066 medium, and fetal bovine serum. Yeast extracts provide diphosphopyridine nucleotides and serum provides cholesterol and an additional source of protein." Their formulation differs somewhat from the formula originally reported by Tully with several additives offered including glucose, arginine, or urea. Growth is noted by color change due to presence of phenol red. Most researchers working with *Mollicutes* prepare their own SP4 media.

The ATCC formulation for spiroplasma media SP4 based on Tully's formulation [34]: The media is prepared in two portions, one autoclaved and one added aseptically via a filter.

Autoclavable portion: PPLO broth w/o CV (11.0 g), tryptone (10.0 g), glucose (5.0 g), distilled water (625.0 mL). Adjust pH to 7.5 and autoclave at 121 °C for 15 min.

Portion added aseptically: Mycoplasma growth supplement (CMRL 1066-10× 50.0 mL), yeast extract solution (35.0 mL), TC yeastolate (2% solution) (100.0 mL), fetal bovine serum (heat-inactivated) (170.0 mL), phenol red solution (0.1% solution) (20.0 mL). Filter through a 0.220 μm membrane filter then mix with autoclaved portion to make complete media. The components should be fresh. Store at 4 °C. The shelf life is approximately 3 months.

The M1D media is equally suited for spiroplasma growth and is simpler to make, thus used by most mycoplasmologists. The M1D media is made up of two portions as well:

Autoclavable portion (base Medium): PPLO broth base (3.5 g), tryptone (1.67 g), peptone (1.33 g), glucose (0.167 g), fructose (0.167 g), sucrose (1.67 g), sorbitol (11.67 g), deionized water (133.33 mL). Dissolve and adjust pH to 7.8, then autoclave for 20 min.

Add to base media the following: Schneider's Drosophila medium (266.67 mL), fetal bovine serum (heat inactivated) (83.33 mL), phenol red (0.5% sol) (2 mL), penicillin G (100,000 units/mL) (4.17 mL). Filter sterilize through a 0.22 μm membrane filter. Final pH is 7.6.

2.2 Agar Plates

Add 7.5 g agar to base medium prior to autoclaving. In the case of M1D plates, preheat the Schneider's medium to 37 °C and discard any precipitate. Equilibrate Schneider's medium and fetal bovine serum to 56 °C, then add penicillin and phenol red prior to pouring plates.

2.3 Media for Maintenance of Cell Cultures

The base medium for this cell line is ATCC-formulated Eagle's Minimum Essential Medium, Catalog No. 30–2003. To make the complete growth medium, add fetal bovine serum to a final concentration of 10%.

3 Methods: Isolation Protocol–Inoculation of Mosquito Cell Cultures

3.1 Mosquito Larvae Cells

(*Aedes albopictus* clone C6/36 (ATCC® CRL-1660™)) were chosen since prior experience showed CPE with cellular vacuolization following inoculation with filtered brain homogenate derived from scrapie-infected sheep (unpublished data), while controls inoculated with filtered normal sheep brain homogenate showed no pathology. These cells form a confluent monolayer when grown in an atmosphere: air 95%; carbon dioxide (CO_2) 5%, Temperature 28 °C (*see* **Note**). Subcultures are prepared by scraping or by vigorous pipetting. Remove the old medium, add fresh complete culture medium, dislodge cells from the floor of the flask with sterile spatula, aspirate, and dispense into new flasks. A subcultivation ratio of 1:4 to 1:10 is recommended. Medium Renewal: Twice per week.

3.2 Homogenization of Scrapie Tissues

Sheep eye tissues affected with scrapie were selected since our research using experimental spiroplasmosis (SMCA) in ruminants showed predilection for the eye with accumulation of spiroplasma in the vitreous and corneal endothelial cells [17]. Therefore, cornea, vitreous, and retinal tissues are dissected from frozen materials using a sterile scalpel and forceps. The tissue is put in a sterile Petri dish and minced using two scalpel blades. The minced tissues are added to a 50 mL enclosed tissue grinder (Fisher Scientific Cat # 02-542-11), and 2 mL of specialized SP4 media is added and the tissues ground. Then an additional 8 mL of SP4 specialized media are added in increments, ground repeatedly, then transferred to a closed 50 mL conical tube and centrifuged at 3000 × *g* for 5 min at 20 °C. The supernatant is aspirated using a 10 cm^3 syringe with a 22 gauge needle. The needle is removed and discarded, and replaced by a 0.450 μm membrane filter. The filtrate amounting to 5–8 mL is placed in a fresh 50 mL conical tube. At this point filtrate can be inoculated into specialized media or onto agar plates (*see* **Note 1**).

3.3 Inoculation of Mosquito Larvae Tissue Culture

The media is removed by pipette from 25 mL flasks with confluent monolayer of mosquito embryo cells and 4–5 mL of the filtrate is added. The flasks are incubated in a 28 °C humidified incubator for ½ h, then 10 mL of MEM media is added. The flasks are monitored for CPE [17] using phase microscopy, and the flasks are fed 2× per week. Spent media is saved frozen for future reference, since spiroplasma may be present in this cell-free media [33].

In *summary*, the procedure is as follows:

- Grind scrapie material in specialized media (8–10 mL).
- Centrifuge × 3000 *g* for 5 min, discard pellet.
- Filter supernatant through 0.450 μm membrane filter.
- Add antibiotic—optional (vancomycin 50 μL per flask).
- Inoculate tissue culture flask by pouring off fluid, layer thin layer of inoculum (4 mL) on monolayer and incubate at 28 °C for ½ h, then add additional 10 mL of MEM.
- Incubate at 28 °C in a humidified incubator and change MEM media (15 mL) twice weekly during maintenance of cultures.
- Harvest cells and fluids when recognizable CPE for future studies.

Alternately, mammalian cell cultures can be inoculated (*see* **Note 2**). Note mosquito cells are sensitive to trypsin and removal from the tissue culture flasks require scrapping off the cells with spatulas (*see* **Note 3**). Another option for spiroplasma isolation is passage in embryonated eggs (*see* **Note 4**).

3.4 Documentation of Presence of Spiroplasma in the Cell Culture or Culture Fluids

The presence of spiroplasma is documented in tissue culture fluids either by phase or dark field microscopy. A variety of EM studies can be used to identify and enumerate spiroplasma in the tissue culture cells and/or fluids. These include TEM, SEM, and negative stain EM. Molecular studies such as PCR should be tried but the isolate is a novel *Spiroplasma* sp. so may require purification and further characterization before specific primers can be designed for detection. Ribosomal primers using southern blotting has worked in the past [7] (*see* **Note 5**).

3.5 Preservation of Infected Mosquito Cells

When spiroplasma are identified in the spent fluids by dark field microscopy or on the cell surface by SEM, the infected tissues and fluids should be preserved for future studies (*see* **Note 6**), by freezing at −80 °C or in liquid nitrogen.

4 Notes

Note 1. Isolation directly into special media: The filtrate can be directly inoculated into 1 mL fractions of special media, either SP4 or M1D. In our experience this method has worked >80% of the time, and the spirals indicative of the scrapie spiroplasma isolate can be seen by dark field microscopy (see Fig. 1). However the spiroplasma scrapie isolates do not flourish in these media and do not survive beyond a few passages suggesting that the current specialized media do not contain critical nutrients. It is difficult to determine the concentration of bacterial isolates in those media since spiroplasma show variable forms including the spiral form [12]. A defined media designed by Hackett shows that spiroplasma need complex lipids such as sphingomyelin [35]; that media formulation has been only moderately successful compared to other specialized media [36]. The interest in the isolation in tissue culture is that spiroplasma growth is encouraged by possible stimulatory growth factors associated with the mosquito or mammalian tissue culture cells and/or the spent media [33] as evidenced by growth of the scrapie *Spiroplasma* sp. in these cell cultures. Spiroplasma grow on agar plates made with specialized growth media [34] as tiny colonies beneath the agar surface examined at 21 days post inoculation; these colonies are best seen using a dissecting microscope equipped for dark field. Wrap the plates with parafilm and incubate at 30 °C in a humidified incubator for 21 days.

Note 2. Spiroplasma isolation from scrapie by inoculating certain mammalian cell lines: We chose mouse neuroblastoma cell cultures (Neuro-2a (ATCC® CCL-131™)) because of their wide use in in vitro studies of prions. We also chose bovine corneal endothelial cell line (BCE C/D-1b (ATCC® CRL-2048™)) because of the tendency of spiroplasma laboratory strains experimentally inoculated into deer and sheep to migrate to the eye and its contents [17]. We also found SMCA grew in a mouse astrocyte line (C8-DlA (Astrocyte type I clone) (ATCC® CRL-2541™)) [18], wherein the cells produce abundant glial fibrils. The inoculation method describe above works well with these cell lines, offering an alternative to those more familiar with growing mammalian cells. We have seen CPE in spiroplasma infections, both by laboratory strains and scrapie isolates, in these cell lines along with apoptotic cells. Culture conditions for the mammalian cell lines: humidified incubator, air 95%, CO_2 5%, Temperature 37 °C.

Note 3. Trypsin is used in releasing the mouse Neuro-2a cells from the culture flask, and the cells show morphological changes consistent with autophagy, as we saw by TEM of the infected mosquito cells [23]; autophagy is characteristic of prion diseases [37]. Future work with these cell lines may add significantly to the understanding of the pathogenesis of the spiroplasma associated with TSEs. It is noteworthy that experimental propagation of SMCA in mouse neuroblastoma cell cultures induces alpha-synuclein as identified by immune histochemistry (***Feng J, personal communication***).

Trypsin protocol for mouse Neuro-2a cells: Remove culture medium, rinse cell layer with 0.25% (w/v) trypsin–0.53 mM EDTA, add 2.0–3.0 mL of trypsin-EDTA solution to flask and observe cells under an inverted microscope until cell layer is dispersed (to avoid clumping, do not agitate the cells by hitting or shaking the flask), flasks may be placed at 37 °C to facilitate dispersal, add 6.0–8.0 mL of complete MEM growth medium and aspirate cells by gently pipetting, add aliquots of the cell suspension to new flasks (a sub-cultivation ratio of 1:3 to 1:6 is recommended), incubate cultures at 37 °C with medium renewal twice per week.

Note that the bovine cells are a finite cell line and trypsin is not recommended to detach the cells. Subcultures are prepared for BCE cells similar to the procedure described above for the mosquito larvae cells. The subcultivation ratio is 1:4 to 1:10. Medium Renewal: Twice per week.

Note 4. Passage in embryonated eggs: The isolation of *Spiroplasma* sp. from TSE-affected brain tissues, including scrapie in sheep, CWD in deer, and CJD in humans, was carried out using this methodology [16], which was based upon the original method for isolation of SMCA from mouse brains [11]. The spiroplasma are subcultured from the allantoic fluid from the eggs into either Sp4 and M1D broths specially adapted for spiroplasma growth [34].

Inoculation protocol summary: Briefly, TSE-infected brain tissues are homogenized in Sp-4 broth or M1D broth. The homogenate is initially centrifuged at 3000 × *g* for 5 min and the supernatant filtered through a 0.450 μm membrane filter. The pathogen-free embryonated eggs (commercially available from Charles River laboratories) are incubated for 8 days at 37 °C and are documented to have viable embryos by candling. The eggs are cleansed with alcohol prior to inoculation to prevent contamination and a small hole is drilled over the air sac using a sterile 16 gauge needle. The outline of the air sac is marked when the egg is candled. The TSE or normal brain filtrates are inoculated into the amniotic and allantoic sacs and into the yolk sac with a 22 gauge needle passed through the hole over the air sac. The site of inoculation is sealed with tape, and the eggs are incubated at 37 °C for 9 days. Following incubation, the eggshell over the air sac is removed with a scalpel, and the amniotic and allantoic fluids are aspirated and added directly to 5 mL of fresh Sp-4 or M1D broth in stoppered tubes and incubated at 37 °C in a humidified incubator. An option is to add 50 μL of vancomycin to each tube to prevent bacterial contamination. Blind passages are carried out every 2 weeks by placing 0.5 mL of the original culture into 4.5 mL. of fresh SP-4 media. The temperature of incubation is critical. We incubated the eggs following inoculation at 37 °C for 9 days, and when the fluids were harvested, the control egg contained a fully mature but dead chick, while the infected eggs showed only a mucinous slurry. If these procedures are carried out at 41 °C, there are well-formed live chicks in both control and infected eggs suggesting that *S. mirum*-related strains do not grow at the higher temperature. This may account for the inability to induce TSE infection in birds [1].

Note 5. Identification of spiroplasma in cell-free broth: The cultures are monitored for spiroplasma by dark field microscopy × 1200 magnification for the presence of motile helices. When sufficient numbers of organisms were seen, transmission electron microscopy (TEM) is performed on either a pellet of the culture embedded in epon or as a negative stain preparation. Briefly, a tube of the broth culture containing the spiroplasma isolate is centrifuged at 14,000 × *g* for 20 min, and the pellet is fixed in 2% glutaraldehyde for 1 h, placed in buffer, post-fixed in osmium and embedded in epon for TEM. The negative stained preparations are done by placing a drop of media from the test culture on a formvar-coated copper grid and allowed to dry. The grids are fixed in glutaraldehyde by floating on a drop of fixative, then transferred to a drop of phosphotungstic acid for staining. The grids are dehydrated and examined by TEM. The identification of the spiroplasma has been enhanced by the availability of molecular methods [5]. Briefly, 2 mL of the TSE-infected or normal control Sp-4 broth are centrifuged at 14,000 × *g* for 20 min. DNA is extracted from the pellet using the PureLink® Genomic DNA Mini Kit from Invitrogen (K1820–02). Polymerase chain reaction (PCR) is carried out using specific oligonucleotide primers [7] that specifically identify a 270 bp portion of *S. mirum*-related 16S rDNA. These studies are done using standard PCR combined with Southern blot, which increases the specificity and sensitivity of the test [7]. In our study the amplified PCR product was cloned and submitted to the Tulane Health Science Center for Gene Therapy for DNA sequencing. The DNA sequences were blasted on the Genbank database for identification of the organism.

Note 6. Future studies, improving growth media: There is enhanced growth of the scrapie spiroplasma isolate in tissue culture systems. Direct isolation in specialized media works well for the laboratory strains but is not suitable for extended growth of scrapie isolates. Therefore improving the media to enhance growth of these fastidious spiroplasma is a major goal in our laboratory. Our genome study of the laboratory strain showed that spiroplasma do not make their own amino acids and thus require a very rich supply in vitro. It is imperative that we are successful in isolating the *Spiroplasma* sp. from scrapie-affected tissues in pure form so that the role of spiroplasma in TSEs can be determined and the interrelationship with prion amyloid can be understood.

References

1. Bastian FO (ed) (1991) Creutzfeldt-Jakob disease and other transmissible spongiform encephalopathies. Mosby/Year Book, New York
2. Bastian FO (2011) Assay for transmissible spongiform encephalopathies. US Patent 7888039 B2, 15 February 2011
3. Bastian FO (1979) Spiroplasma-like inclusions in Creutzfeldt-Jakob disease. Arch Pathol Lab Med 103(13):665–669
4. Bastian FO, Hart MN, Cancilla PA (1981) Additional evidence of spiroplasma in Creutzfeldt-Jakob disease. Lancet 1(8221):660
5. Gray A, Francis RJ, Scholtz CL (1980) Spiroplasma and Creutzfeldt-Jakob disease. Lancet 2:660
6. Reyes JM, Hoenig EM (1981) Intracellular spiral inclusions in cerebral cell processes in Creutzfeldt-Jakob disease. J Neuropathol Exp Neurol 40(1):1–8

7. Bastian FO, Dash S, Garry RF (2004) Linking chronic wasting disease to scrapie by comparison of *Spiroplasma mirum* ribosomal DNA sequences. Exp Mol Pathol 77(1):49–56
8. Alexeeva I, Elliott EJ, Rollins S et al (2006) Absence of spiroplasma or other bacterial 16S rRNA genes in brain tissue of hamsters with scrapie. J Clin Microbiol 44(1):91–97
9. Tully JG, Whitcomb RF, Clark HF et al (1977) Pathogenic mycoplasmas: cultivation and vertebrate pathogenicity of a new *Spiroplasma*. Science 195(4281):892–894
10. Bove JM (1997) Spiroplasmas: infectious agents of plants, arthropods and vertebrates. Wien Klin Wochenschr 109(14-15):604–612
11. Clark HF (1974) The suckling mouse cataract agent (SMCA). Prog Med Virol 18(0):307–322
12. Bastian FO, Purnell DM, Tully JG (1984) Neuropathology of spiroplasma infection in the rat brain. Am J Pathol 114(3):496–514
13. Tully JG, Bastian FO, Rose DL (1984) Localization and persistence of spiroplasmas in an experimental brain infection in suckling rats. Ann Microbiol (Paris) 135A:111–117
14. Bastian FO (2005) Spiroplasma as a candidate causal agent of transmissible spongiform encephalopathies. J Neuropathol Exp Neurol 64(10):833–838
15. Bastian FO, Jennings R, Huff C (1987) Neurotropic response of *Spiroplasma mirum* following peripheral inoculation in the rat. Ann Microbiol (Inst Pasteur) 138(6):651–655
16. Bastian FO, Sanders DE, Forbes WA et al (2007) *Spiroplasma* spp. from transmissible spongiform encephalopathy brains or ticks induce spongiform encephalopathy in ruminants. J Med Microbiol 56(9):1235–1242
17. Bastian FO, Boudreaux CM, Hagius SD et al (2011) Spiroplasma found in the eyes of scrapie affected sheep. Vet Ophthalmol 14(1):10–17
18. Bastian FO (2014) The case for involvement of spiroplasma in the pathogenesis of transmissible spongiform encephalopathies. J Neuropathol Exp Neurol 73(2):104–114
19. Jeffrey M, Scott JR, Fraser H (1991) Scrapie inoculation of mice: light and electron microscopy of the superior colliculi. Acta Neuropathol 81(5):562–571
20. Trachtenberg S, Gilad R (2001) A bacterial linear motor: cellular and molecular organization of the contractile cytoskeleton of the helical bacterium *Spiroplasma melliferum* BC3. Mol Microbiol 41(4):827–848
21. Merz PA, Somerville RA, Wisniewski HM et al (1983) Scrapie-associated fibrils in Creutzfeldt-Jakob disease. Nature 306(5942):474–476
22. Bastian FO, Jennings R, Gardner W (1987) Antiserum to scrapie associated fibril protein cross-reacts with *Spiroplasma mirum* fibril proteins. J Clin Microbiol 25(12): 2430–2431
23. Bastian FO (2016) Could Alzheimer's disease be an infectious disease? In: Piñol MS, Molina AJM (eds) The cellular players in Alzheimer's disease: one for all and all for one. Omics Group International eBooks, Foster City, CA; (in press)
24. Hunter GD (1965) Progress toward the isolation and characterization of the scrapie agent. In: Gajdusek DC, Gibbs Jr CJ, Alpers M (eds) Latent and temperate virus infections, NINDB Monograph No 2. Public Health Service Publication No 1378
25. Bastian FO, Elzer PH, Wu X (2012) *Spiroplasma* spp. biofilm formation is instrumental for role in the pathogenesis of plant, insect and animal diseases. Exp Mol Pathol 93(1):116–128
26. Barnhart MM, Chapman MR (2006) Curli biogenesis and function. Annu rev Microbiol 60(1):131–147
27. Wang X, Chapman MR (2008) Curli provide the template for understanding controlled amyloid propagation. Prion 2(2):57–60
28. Lundmark K, Westermark G, Olsen A et al (2005) Protein fibrils in nature can enhance amyloid protein a amyloidosis in mice: cross-seeding as a disease mechanism. PNAS 102(17):6098–6102
29. Guo YJ, Han J, Yao HL et al (2007) Treatment of scrapie pathogen 263K with tetracycline partially abolishes protease-resistant activity *in vitro* and reduces infectivity *in vivo*. Biomed Environ Sci 20(3):198–202
30. Johnson CJ, Phillips KE, Schramm PT et al (2006) Prions adhere to soil minerals and remain infectious. PLoS Pathog 2(4):e32–302. doi:10.1371/journal.ppat.0020032
31. Fries GF, Marrow GS, Snow PA (1982) Soil ingestion by dairy cattle. J Dairy Sci 65(4):611–618
32. Gustafson DP, Kanitz CL (1965) Evidence of the presence of scrapie in cell cultures of brain. In: Gajdusek DC, Gibbs Jr CJ, Alpers M (eds) Latent and temperate virus infections, NINDB Monograph No 2. Public Health Service Publication No 1378
33. Megraud F, Gamon L, McGarrity GI (1983) Characterization of *Spiroplasma mirum* (suck-

ling mouse cataract agent) in a rabbit lens cell culture. Infect Immun 42(3):1168–1175
34. Whitcomb RF (1983) Culture media for spiroplasma. In: Razin S, Tully JG (eds) Methods in mycoplasmology, vol 1. Academic Press, New York
35. Hackett KJ, Ginsberg AS, Rottem S et al (1987) A defined medium for a fastidious spiroplasma. Science 237(4814):525–527
36. Bastian FO, Baliga BS, Pollock HM (1988) Evaluation of [^{3}H]thymidine uptake method for studying growth of spiroplasmas under various conditions. J Clin Microbiol 26(10):2124–2126
37. Dron M, Bailly Y, Beringue V et al (2006) SCRG1, a potential marker of autophagy in transmissible spongiform encephalopathies. Autophagy 2(1):58–60

Chapter 10

Detection and Diagnosis of Prion Diseases Using RT-QuIC: An Update

Byron Caughey, Christina D. Orru, Bradley R. Groveman, Matilde Bongianni, Andrew G. Hughson, Lynne D. Raymond, Matteo Manca, Allison Kraus, Gregory J. Raymond, Michele Fiorini, Maurizio Pocchiari, and Gianluigi Zanusso

Abstract

Until recently, it has been difficult to detect all infectious levels of prions and diagnose prion diseases in living humans and other animals. Real-time quaking-induced conversion (RT-QuIC), an ultrasensitive test based on amplification of the amyloid seeding activity of prions, is now achieving the sensitivity and practicality required for routine diagnostics. Adaptations and refinements of RT-QuIC assays currently allow the detection of most of the known prion diseases of mammals, often with sensitivities that are greater than those of animal bioassays. Many tissues and fluids have been shown to be suitable for RT-QuIC analysis. The most significant and extensively validated application is its use in diagnosing sporadic Creutzfeldt–Jakob disease (sCJD) using cerebrospinal fluid. Recent progress with this test has improved diagnostic sensitivities up to 96% with assays that take less than 24 h. Moreover, as tests of a disease-specific marker, RT-QuIC assays have repeatedly demonstrated specificities of 98–100%. Other diagnostic specimens, such as nasal brushings, have shown even higher sensitivity and specificity, but have not been as extensively evaluated. These tests have already shown considerable potential to provide definitive *antemortem* diagnosis of sCJD, but further testing is required for full validation. Step-by-step RT-QuIC protocols will soon be published elsewhere, so here we provide updates on the range of RT-QuIC tests for prions as well as some additional practical tips on performing and optimizing various RT-QuIC applications.

Key words RT-QuIC, Prion, Diagnosis, Cerebrospinal fluid, Nasal swab, Creutzfeldt–Jakob disease, Seed

1 Introduction

A key aspect of preventing and diagnosing mammalian prion diseases, or transmissible spongiform encephalopathies (TSEs), is the detection of disease-specific prion proteins. Marked improvements in the sensitivity, specificity, and practicality of prion tests have been made in recent years [1, 2]. Among the most practical of ultrasensitive prion disease assays that can often detect even subinfectious quantities

Pawel P. Liberski (ed.), *Prion Diseases*, Neuromethods, vol. 129,
DOI 10.1007/978-1-4939-7211-1_10, © Springer Science+Business Media LLC 2017

of prions or prion-associated seeding activity is real-time quaking-induced conversion (RT-QuIC) [3–7]. Early studies demonstrated the ability of prions to seed the polymerization of noninfectious, proteinase K-sensitive recombinant prion protein ($rPrP^{Sen}$) into protease-resistant amyloid fibrils that could be detected by immunoblotting [8–10] or the amyloid-sensitive dye, thioflavin T (ThT) [11]. RT-QuIC assays have resulted from melding and improving on the practicality of these earlier ultrasensitive assays [3, 4]. Current RT-QuIC assays are performed in 96-well plates with a fluorescence readout, and can directly detect seeds in prion-infected brain samples that have been diluted by as much as 10^6–10^9-fold (Fig. 1). When combined with immune capture steps dilution as extreme as 10^{14}-fold have been detected [5, 6]. In such reactions, the starting amounts of protease-resistant forms of prion protein have been amplified by approximately a trillion fold.

RT-QuIC assays have now been adapted by many labs to the detection of almost all known mammalian prion diseases (e.g., [13] and references therein). A number of common types of prion diseases have been detected by RT-QuIC not only using the most prion-laden tissues such as brain, but also in more diagnostically accessible specimens such as cerebrospinal fluid (CSF) [3, 4, 7, 14–19, 40–44], nasal brushings [16, 20], blood [5, 6, 21], urine [22], recto-anal mucosa-associated lymphoid tissue (RAMALT) [23, 24], and feces [22]. Most notably, extensive studies now support the utility of RT-QuIC in the antemortem diagnosis of human Creutzfeldt–Jakob disease (CJD) using CSF or nasal brushings with greater sensitivity and specificity than had been possible previously [4, 7, 14–18, 20, 25, 42, 43]. Moreover, major improvements in the conditions for key assays have reduced the time required from a matter of days to hours [7]. Along with continuing development of applications and improvements, RT-QuIC assays are being implemented widely as primary prion diagnostic, surveillance, and research tools.

Detailed step-by-step protocols and practical notes on performing RT-QuIC assays will be published elsewhere [40, 41]. Here we will provide recent updates on the protocol that explain new improvements and applications of RT-QuIC to prion diseases of particular interest in medicine, agriculture, and wildlife biology.

2 RT-QuIC Applications

2.1 CJD Diagnosis Using RT-QuIC Analysis of CSF

Diagnostic applications of RT-QuIC to human prion disease were first described by Atarashi and colleagues, who tested for sCJD using CSF specimens [4]. This, as well as subsequent studies by multiple groups [7, 14, 15, 17, 18, 25, 43] have demonstrated diagnostic sensitivities of 77–89% and specificities of 98–100% using

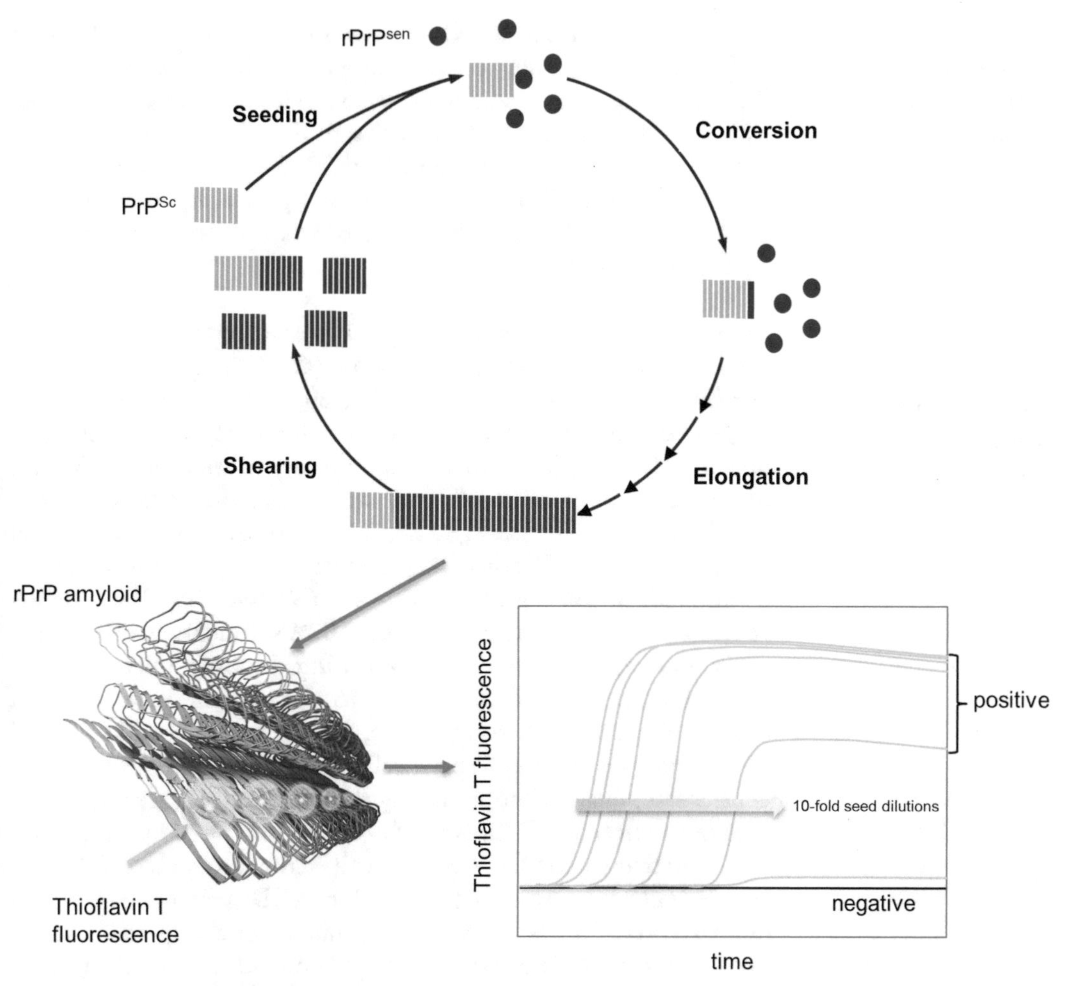

Fig. 1 Diagram of RT-QuIC-seeded polymerization assay. Oligomeric or multimeric seeds of PrP^{Sc} in a test sample are mixed with a vast excess of recombinant PrP^{Sen} ($rPrP^{Sen}$) in 96-well plates. The PrP^{Sc} seeds induce the conversion and incorporation of rPrPsen into much larger amounts of amyloid fibrils that are not identical to, or as infectious as, PrP^{Sc}. The amyloid enhances the fluorescence of the dye thioflavin T (ThT) as the reaction progresses in "real time." With dilution of the seed-containing sample, the lag phase prior to increase in fluorescence tends to increase and the maximum fluorescence achieved decreases. The rPrP amyloid fibril model shown is a hypothetical computational model based on a parallel in-register intermolecular beta sheet architecture [12]

various first generation reaction conditions. This specificity is higher than those obtained for other diagnostically helpful but relatively nonspecific markers of neuronal damage such as the 14-3-3 protein and tau. Helpful analyses of practical considerations in handling and analyzing human CSF specimens by RT-QuIC have been provided by Cramm and colleagues [18]. Cumulatively, these studies have indicated that many labs can readily implement RT-QuIC testing of CSF as an accurate antemortem diagnostic test for sCJD.

2.2 Improved RT-QuIC Conditions for CJD Diagnosis Using CSF

The initially described RT-QuIC assays for diagnosing sCJD using CSF specimens typically required 2.5–5 days. Recently, an improved RT-QuIC assay was described that identifies positive CSF samples within 4–14 h with better analytical sensitivity [7, 45]. Applications of this new assay have obtained 92–95.8% diagnostic sensitivity and nearly 100% specificity [7, 42, 44, 45] indicating that IQ-CSF RT-QuIC assay should enhance prospects for rapid and accurate antemortem CJD diagnosis.

2.3 CJD Diagnosis Using Nasal Samples Collected Using Brushes and Swabs

Another type of accessible diagnostic specimen that has been described recently is olfactory mucosa (OM) samples collected by brushing the upper nasal cavity [16, 20, 44]. This approach was suggested both by the detection of pathologic prion protein in olfactory neuroepithelium by immunological methods [26] and the RT-QuIC detection of prion seeding activity in lavages from scrapie-infected hamsters [3, 27, 28]. RT-QuIC analysis of nasal brushings from humans has allowed the discrimination of Creutzfeldt–Jakob disease (CJD) (n = 69) and non-CJD patients (n = 54) with ≥97% sensitivity and 100% specificity. In the initial study, OM samples were collected using disposable cyto-brushes that might cause mild discomfort to the patient or recovery of blood cells. More recently, softer, less abrasive, flexible flocked swabs that were designed specifically for OM sampling can provide gentler sampling without reducing the prion seeding activity recovered or the overall RT-QuIC diagnostic sensitivity [44]. Moreover, we found that when OM, CSF, or both were analyzed by RT-QuIC, the overall diagnostic sensitivity for sCJD patients (n = 61) improved to 100% (CI, 93–100%) while maintaining 100% specificity in the analysis of an even larger number of non-CJD cases. Thus, RT-QuIC analyses of CSF and/or OM specimens with these newer reaction conditions appear to allow decisive discrimination of sCJD and non-CJD patients.

2.4 Classical and Atypical Forms of BSE

Bovine prion diseases include classical (C) bovine spongiform encephalopathy (BSE), as well as the atypical H-BSE and L-BSE forms. Each of these BSE forms can now be sensitively detected using RT-QuIC assays with different substrates [29, 30]. For example, H-BSE can be detected in as little as 10^{-9} dilutions of brain tissue and neat cerebrospinal fluid from clinically affected cattle. Furthermore, the infectious BSE strains can be discriminated by comparing RT-QuIC responses using different recombinant prion protein substrates and/or immunoblot band profiles of proteinase K-treated RT-QuIC reaction products.

2.5 Classical and Atypical Forms of Scrapie in Sheep

RT-QuIC has been used successfully to detect PrP^{Sc} in sheep infected with scrapie [3, 13]. The best sensitivity for classical scrapie has been obtained using a hamster-sheep chimeric $rPrP^{Sen}$ substrate, which allows detection down to 10^{-8} dilutions of scrapie brain homogenate from clinically affected animals [13]. However, this substrate was not able to detect atypical Nor98 scrapie in this study. In contrast, both classical and Nor98 scrapie were detectable using an $rPrP^{Sen}$ substrate derived from bank voles [13]. By comparing reactivities with these two $rPrP^{Sen}$ substrates, these two sheep scrapie strains can be discriminated by RT-QuIC.

2.6 Scrapie in Goats

Dassanayaake and colleagues have also recently adapted RT-QuIC assays to scrapie in goats [31]. Both full-length sheep $rPrP^{Sen}$ substrates (PrP genotypes $A_{136}R_{154}Q_{171}$ and $V_{136}R_{154}Q_{171}$) discriminated brain tissue from the goats infected with classical scrapie from those of normal goats within 15 h. In this context, RT-QuIC is at least 10^4-fold more sensitive than the ELISA and Western blot assays for the detection of scrapie seeding activity.

2.7 Chronic Wasting Disease (CWD)

Since the initial demonstration of the sensitive detection of CWD prion seeding activity in deer brain tissue [3], a number of further applications of RT-QuIC to CWD-infected cervids have been described. These include the detection of prions in the urine, feces [22], saliva [32], blood [21], CSF [19], lymph nodes [33], nasal brushings, and RAMALT [23, 24], in some cases in preclinical animals. Varying degrees of sensitivity have been obtained from the analyses of these nonneural tissues but the prevailing indication is that RT-QuIC tests of these samples should be useful in diagnostics and surveillance, as well as in fundamental studies of the CWD transmission cycle.

2.8 Bank Vole Prion Protein as a Broad-Spectrum Substrate

Many of the various applications of RT-QuIC tests use different $rPrP^{Sen}$ substrate sequences and reaction conditions. For reasons that are not fully clear, the optimal RT-QuIC substrate sequence does not always precisely match the sequence of the prion being tested. In some cases, this is in part because the exactly homologous $rPrP^{Sen}$ has undesirable behavior, such as a greater tendency to form amyloid fibrils spontaneously, without prion seeding, in the reactions. Thus, optimization for each potential prion in a given sample type has been performed, indicating the need for multiple permutations of RT-QuIC assays to test for different types of prions, even within a given host species. Recently, however, a single RT-QuIC assay, using a single bank vole $rPrP^{Sen}$ substrate, has been shown to sensitively detect 28 different prion diseases from humans, cattle, sheep, cervids, and rodents. These prion types included several that had previously been undetected by RT-QuIC or protein misfolding cyclic amplification (PMCA). As noted above, comparison of the relative abilities of different prions to seed positive RT-QuIC reactions with bank vole $rPrP^{Sen}$, and not other $rPrP^{Sen}$ molecules, has allowed discrimination of (a) classical and atypical L- and H-type bovine spongiform encephalopathy in cattle [29], (b) classical and atypical Nor98 scrapie in sheep [13], and (c) sporadic and variant Creutzfeldt–Jakob disease in humans [13]. Comparison of protease-resistant RT-QuIC conversion products of RT-QuIC reactions using bank vole PrP^{Sen} 23–230 has also aided the discrimination of classical and atypical Nor98 scrapie in sheep [13] and H- and L-type BSE in cattle [29].

2.9 Quantitation

RT-QuIC assays can be used to quantitate the prion seeding activity in a given sample using end-point titrations [3, 5, 7, 16, 27, 34, 46]. Such analyses can provide what we have called the 50% seeding dose (SD_{50}), that is, the amount of sample that causes 50% of replicate RT-QuIC reactions to become positive. This is analogous to the 50% lethal dose (LD_{50}) in animal bioassays. Back calculations then provide the concentration of SD_{50}s per unit of the original specimen. Further statistical considerations in discriminating prion disease-positive and -negative samples when seed concentrations are near the end-point have been described by Gray and colleagues [35]. End-point dilution RT-QuIC assays have been used to identify and characterize a new anti-prion disinfectant, hypochlorous acid [46].

An alternative approach to quantitation that has been described is the comparison of relative lag phases in RT-QuIC reactions. Since the beginning of RT-QuIC reactions, an inverse relationship between sample (prion seed) concentration and the lag phase of the RT-QuIC has been observed [3, 36]. Subsequent studies have formalized the use of this correlation as a means of quantitating relative prion seeding activity levels by comparison of experimental prion-seeded polymerization lag phases to standard curves [35, 37, 38]. This approach reduces the number of dilutions of a given test sample that are required for quantitation. However, this advantage is partially offset by the need to add samples to establish a standard curve for each experiment. In our experience, lag phases for a given seed sample can vary between experiments and plate readers, so special care should be taken to run standard curves on each plate using the same sample type. Nonetheless, under properly controlled conditions these kinetics-based approaches to RT-QuIC quantitation should be useful for comparing closely similar sample types.

2.10 Other Practical Considerations

In the development of new or improved applications of RT-QuIC, it is worth considering additional experimental factors that can be influential [39]. N-terminal truncation of $rPrP^{Sen}$ substrate can in some cases (e.g., Syrian hamster PrP^{Sen} to residues 90–231) hasten both prion-seeded and prion-independent reactions while maintaining a clear kinetic distinction between the two. This can result in faster assay times without loss of sensitivity. Raising temperatures or shaking speeds (e.g., to 50–55 °C) can also accelerate RT-QuIC reactions without compromising specificity. For example, when applied to nasal brushings from sCJD patients, higher temperatures accelerated RT-QuIC kinetics, and the use of hamster $rPrP^{Sen}$ (90–231) strengthened RT-QuIC responses. Extension of shaking periods can reduce scrapie-seeded reaction times, but continuous shaking appears to promoted false-positive reactions. Furthermore, under our usual range of conditions, pH 7.4 has promoted more rapid RT-QuIC reactions than more acidic pHs. Additionally, we have found that small variations in SDS concentration can substantially affect assay performance. Finally, when the preferred RT-QuIC assay conditions and appropriate cutoff times for particular sample types have been established, assay throughput and cost-effectiveness can be enhanced by using a multiplate thermoshaker (without fluorescence optics) followed by fluorescence readings at a designated cutoff time in a separate plate reader [39].

Acknowledgments

This work was supported in part by the Intramural Research Program of the NIAID.

References

1. Orru CD, Wilham JM, Vascellari S, Hughson AG, Caughey B (2012) New generation QuIC assays for prion seeding activity. Prion 6(2):147–152
2. Zanusso G, Monaco S, Pocchiari M, Caughey B (2016) Advanced tests for early and accurate diagnosis of Creutzfeldt-Jakob disease. Nat Rev Neurol 12(6):325–333
3. Wilham JM, Orrú CD, Bessen RA, Atarashi R, Sano K, Race B, Meade-White KD, Taubner LM, Timmes A, Caughey B (2010) Rapid end-point quantitation of prion seeding activity with sensitivity comparable to bioassays. PLoS Pathog 6(12):e1001217
4. Atarashi R, Satoh K, Sano K, Fuse T, Yamaguchi N, Ishibashi D, Matsubara T, Nakagaki T, Yamanaka H, Shirabe S, Yamada M, Mizusawa H, Kitamoto T, Klug G, McGlade A, Collins SJ, Nishida N (2011) Ultrasensitive human prion detection in cerebrospinal fluid by real-time quaking-induced conversion. Nat Med 17(2):175–178
5. Vascellari S, Orru CD, Hughson AG, King D, Barron R, Wilham JM, Baron GS, Race B, Pani A, Caughey B (2012) Prion seeding activities of mouse scrapie strains with divergent PrPSc protease sensitivities and amyloid plaque content using RT-QuIC and eQuIC. PLoS One 7(11):e48969
6. Orru CD, Wilham JM, Raymond LD, Kuhn F, Schroeder B, Raeber AJ, Caughey B (2011) Prion disease blood test using immunoprecipitation and improved quaking-induced conversion. MBio 2(3):e00078–e00011
7. Orru CD, Groveman BR, Hughson AG, Zanusso G, Coulthart MB, Caughey B (2015) Rapid and sensitive RT-QuIC detection of human Creutzfeldt-Jakob disease using cerebrospinal fluid. MBio 6(1):e02451–e02414
8. Atarashi R, Wilham JM, Christensen L, Hughson AG, Moore RA, Johnson LM, Onwubiko HA, Priola SA, Caughey B (2008) Simplified ultrasensitive prion detection by recombinant PrP conversion with shaking. Nat Methods 5(3):211–212
9. Atarashi R, Moore RA, Sim VL, Hughson AG, Dorward DW, Onwubiko HA, Priola SA, Caughey B (2007) Ultrasensitive detection of scrapie prion protein using seeded conversion of recombinant prion protein. Nat Methods 4(8):645–650
10. Orrú CD, Wilham JM, Hughson AG, Raymond LD, McNally KL, Bossers A, Ligios C, Caughey B (2009) Human variant Creutzfeldt-Jakob disease and sheep scrapie PrP(res) detection using seeded conversion of recombinant prion protein. Protein Eng Des Sel 22(8):515–521
11. Colby DW, Zhang Q, Wang S, Groth D, Legname G, Riesner D, Prusiner SB (2007) Prion detection by an amyloid seeding assay. Proc Natl Acad Sci U S A 104(52):20914–20919
12. Groveman BR, Dolan MA, Taubner LM, Kraus A, Wickner RB, Caughey B (2014) Parallel in-register intermolecular beta-sheet architectures for prion-seeded prion protein (PrP) amyloids. J Biol Chem 289(35):24129–24142

13. Orru CD, Groveman BR, Raymond LD, Hughson AG, Nonno R, Zou W, Ghetti B, Gambetti P, Caughey B (2015) Bank vole prion protein as an apparently universal substrate for RT-QuIC-based detection and discrimination of prion strains. PLoS Pathog 11(8):e1005117
14. Sano K, Satoh K, Atarashi R, Takashima H, Iwasaki Y, Yoshida M, Sanjo N, Murai H, Mizusawa H, Schmitz M, Zerr I, Kim YS, Nishida N (2013) Early detection of abnormal prion protein in genetic human prion diseases now possible using real-time QUIC assay. PLoS One 8(1):e54915
15. McGuire LI, Peden AH, Orru CD, Wilham JM, Appleford NE, Mallinson G, Andrews M, Head MW, Caughey B, Will RG, Knight RSG, Green AJE (2012) RT-QuIC analysis of cerebrospinal fluid in sporadic Creutzfeldt-Jakob disease. Ann Neurol 72(2):278–285
16. Orru CD, Bongianni M, Tonoli G, Ferrari S, Hughson AG, Groveman BR, Fiorini M, Pocchiari M, Monaco S, Caughey B, Zanusso G (2014) A test for Creutzfeldt-Jakob disease using nasal brushings. New Engl J Med 371(6):519–529
17. Cramm M, Schmitz M, Karch A, Zafar S, Varges D, Mitrova E, Schroeder B, Raeber A, Kuhn F, Zerr I (2015) Characteristic CSF prion seeding efficiency in humans with prion diseases. Mol Neurobiol 51(1):396–405
18. Cramm M, Schmitz M, Karch A, Mitrova E, Kuhn F, Schroeder B, Raeber A, Varges D, Kim YS, Satoh K, Collins S, Zerr I (2016) Stability and reproducibility underscore utility of RT-QuIC for diagnosis of Creutzfeldt-Jakob disease. Mol Neurobiol 53(3):1896–1904. doi:10.1007/s12035-015-9133-2
19. Haley NJ, Van de Motter A, Carver S, Henderson D, Davenport K, Seelig DM, Mathiason C, Hoover E (2013) Prion-seeding activity in cerebrospinal fluid of deer with chronic wasting disease. PLoS One 8(11):e81488
20. Zanusso G, Bongianni M, Caughey B (2014) A test for Creutzfeldt-Jakob disease using nasal brushings. N Engl J Med 371(19):1842–1843
21. Elder AM, Henderson DM, Nalls AV, Wilham JM, Caughey BW, Hoover EA, Kincaid AE, Bartz JC, Mathiason CK (2013) In vitro detection of prionemia in TSE-infected cervids and hamsters. PLoS One 8(11):e80203
22. John TR, Schatzl HM, Gilch S (2013) Early detection of chronic wasting disease prions in urine of pre-symptomatic deer by real-time quaking-induced conversion assay. Prion 7(3):253–258
23. Haley NJ, Siepker C, Hoon-Hanks LL, Mitchell G, Walter WD, Manca M, Monello RJ, Powers JG, Wild MA, Hoover EA, Caughey B, Richt JA (2016) Seeded amplification of chronic wasting disease prions in nasal brushings and recto-anal mucosa associated lymphoid tissues from elk by real time quaking-induced conversion. J Clin Microbiol 54(4):1117–1126. doi:10.1128/JCM.02700-15
24. Haley NJ, Siepker C, Walter WD, Thomsen BV, Greenlee JJ, Lehmkuhl AD, Richt JA (2016) Antemortem detection of chronic wasting disease prions in nasal brush collections and rectal biopsy specimens from white-tailed deer by real-time quaking-induced conversion. J Clin Microbiol 54(4):1108–1116
25. Park JH, Choi YG, Lee YJ, Park SJ, Choi HS, Choi KC, Choi EK, Kim YS (2016) Real-time quaking-induced conversion analysis for the diagnosis of sporadic Creutzfeldt-Jakob disease in Korea. J Clin Neurol 12(1):101–106
26. Zanusso G, Ferrari S, Cardone F, Zampieri P, Gelati M, Fiorini M, Farinazzo A, Gardiman M, Cavallaro T, Bentivoglio M, Righetti PG, Pocchiari M, Rizzuto N, Monaco S (2003) Detection of pathologic prion protein in the olfactory epithelium in sporadic Creutzfeldt-Jakob disease. N Engl J Med 348(8):711–719
27. Bessen RA, Wilham JM, Lowe D, Watschke CP, Shearin H, Martinka S, Caughey B, Wiley JA (2011) Accelerated shedding of prions following damage to the olfactory epithelium. J Virol 86(3):1777–1788
28. Bessen RA, Shearin H, Martinka S, Boharski R, Lowe D, Wilham JM, Caughey B, Wiley JA (2010) Prion shedding from olfactory neurons into nasal secretions. PLoS Pathog 6(4):e1000837
29. Masujin K, Orru CD, Miyazawa K, Groveman BR, Raymond LD, Hughson AG, Caughey B (2016) Detection of atypical H-type bovine spongiform encephalopathy and discrimination of bovine prion strains by real-time quaking-induced conversion. J Clin Microbiol 54(3):676–686
30. Orru CD, Favole A, Corona C, Mazza M, Manca M, Groveman BR, Hughson AG, Acutis PL, Caramelli M, Zanusso G, Casalone C, Caughey B (2015) Detection and discrimination of classical and atypical L-type bovine spongiform encephalopathy by real-time quaking-induced conversion. J Clin Microbiol 53(4):1115–1120
31. Dassanayake RP, Orru CD, Hughson AG, Caughey B, Graca T, Zhuang D, Madsen-Bouterse SA, Knowles DP, Schneider DA (2016) Sensitive and specific detection of classical scrapie prions in the brains of goats by real-time quaking-induced conversion. J Gen Virol 97(3):803–812

32. Henderson DM, Manca M, Haley NJ, Denkers ND, Nalls AV, Mathiason CK, Caughey B, Hoover EA (2013) Rapid antemortem detection of CWD prions in deer saliva. PLoS One 8(9):e74377
33. Haley NJ, Carver S, Hoon-Hanks LL, Henderson DM, Davenport KA, Bunting E, Gray S, Trindle B, Galeota J, LeVan I, Dubovos T, Shelton P, Hoover EA (2014) Detection of chronic wasting disease in the lymph nodes of free-ranging cervids by real-time quaking-induced conversion. J Clin Microbiol 52(9):3237–3243
34. Orru CD, Hughson AG, Race B, Raymond GJ, Caughey B (2012) Time course of prion seeding activity in cerebrospinal fluid of scrapie-infected hamsters after intratongue and intracerebral inoculations. J Clin Microbiol 50(4):1464–1466
35. Gray JG, Graham C, Dudas S, Paxman E, Vuong B, Czub S (2016) Defining and assessing analytical performance criteria for transmissible spongiform encephalopathy-detecting amyloid seeding assays. J Mol Diagn 18(3):454–467
36. Peden AH, McGuire LI, Appleford NE, Mallinson G, Wilham JM, Orru CD, Caughey B, Ironside JW, Knight RS, Will RG, Green AJ, Head MW (2012) Sensitive and specific detection of sporadic Creutzfeldt-Jakob disease brain prion protein using real-time quaking induced conversion. J Gen Virol 93(2):438–449
37. Shi S, Mitteregger-Kretzschmar G, Giese A, Kretzschmar HA (2013) Establishing quantitative real-time quaking-induced conversion (qRT-QuIC) for highly sensitive detection and quantification of PrPSc in prion-infected tissues. Acta Neuropathol Commun 1(1):44–49
38. Henderson DM, Davenport KA, Haley NJ, Denkers ND, Mathiason CK, Hoover EA (2015) Quantitative assessment of prion infectivity in tissues and body fluids by real-time quaking-induced conversion. J Gen Virol 96(Pt_1):210–219
39. Orru CD, Hughson AG, Groveman BR, Campbell KJ, Anson KJ, Manca M, Kraus A, Caughey B (2016) Factors that improve RT-QuIC detection of prion seeding activity. Virus 8(5):140
40. Orru CD, Groveman BR, Hughson AG, Manca M, Raymond GJ, Campbell K, Anson KJ, Kraus A, Caughey B (2017) In: Lawson V (ed) Prion protein protocols, Methods in molecular biology. Springer, in press
41. Schmitz M, Cramm M, Llorens F, Müller-Cramm D, Collins S, Atarashi R, Satoh K, Orrù CD, Groveman BR, Zafar S, Schulz-Schaeffer WJ, Caughey B, Zerr I (2016) The real-time quaking-induced conversion assay for detection of human prion disease and study of other protein misfolding diseases. Nat Protoc 11(11):2233–2242
42. Foutz A, Appleby BS, Hamlin C, Liu X, Yang S, Cohen Y, Chen W, Blevins J, Fausett C, Wang H, Gambetti P, Zhang S, Hughson A, Tatsuoka C, Schonberger LB, Cohen ML, Caughey B, Safar JG (2017) Diagnostic and prognostic value of human prion detection in cerebrospinal fluid. Ann Neurol 81(1):79–92
43. Lattanzio F, Abu-Rumeileh S, Franceschini A, Kai H, Amore G, Poggiolini I, Rossi M, Baiardi S, McGuire L, Ladogana A, Pocchiari M, Green A, Capellari S, Parchi P (2017) Prion-specific and surrogate CSF biomarkers in Creutzfeldt-Jakob disease: diagnostic accuracy in relation to molecular subtypes and analysis of neuropathological correlates of p-tau and Aβ42 levels. Acta Neuropathol 133(4): 559–578
44. Bongianni M, Orrù C, Groveman BR, Sacchetto L, Fiorini M, Tonoli G, Triva G, Capaldi S, Testi S, Ferrari S, Cagnin A, Ladogana A, Poleggi A, Colaizzo E, Tiple D, Vaianella L, Castriciano S, Marchioni D, Hughson AG, Imperiale D, Cattaruzza T, Fabrizi GM, Pocchiari M, Monaco S, Caughey B, Zanusso G (2017) Diagnosis of human prion disease using real-time quaking-induced conversion testing of olfactory mucosa and cerebrospinal fluid samples. JAMA Neurol 74(2):155
45. Groveman BR, Orrú CD, Hughson AG, Bongianni M, Fiorini M, Imperiale D, Ladogana A, Pocchiari M, Zanusso G, Caughey B (2017) Extended and direct evaluation of RT-QuIC assays for Creutzfeldt-Jakob disease diagnosis. Ann Clin Transl Neurol 4(2):139–144
46. Hughson AG, Race B, Kraus A, Sangaré LR, Robins L, Groveman BR, Saijo E, Phillips K, Contreras L, Dhaliwal V, Manca M, Zanusso G, Terry D, Williams JF, Caughey B, Westaway D (2016) Inactivation of prions and amyloid seeds with hypochlorous acid. PLoS Pathog 12(9):e1005914

Chapter 11

Analysis of Charge Isoforms of the Scrapie Prion Protein Using Two-Dimensional Electrophoresis

Elizaveta Katorcha and Ilia V. Baskakov

Abstract

Recent years witnessed extraordinary rise of interest to sialylation and, specifically, to its role in host–pathogen interactions. Prions or PrP^{Sc} are proteinaceous infectious agents that consist of misfolded, aggregated, self-replicating states of a sialoglycoprotein called the prion protein or PrP^{C}. Prions are not conventional pathogens. Nevertheless, due to sialylation of N-linked glycans, prions may use mechanisms similar to those exploited by microbial or viral pathogens in invading a host. Recent studies revealed that sialylation of PrP^{Sc} controls prion infectivity, replication rate, and strain-specific glycoform ratios. As such, sialylation is of paramount importance to prion pathogenesis. For assessing sialylation status of PrP^{Sc}, we developed reliable protocol that involves two-dimensional electrophoresis followed by Western blot (2D). The current chapter describes the procedure for analysis of sialylation status of PrP^{Sc} from various sources including brain, spleen, cultured cells, or Protein Misfolding Cyclic Amplification using 2D.

Key words Prion diseases, Prion proteins, Two-dimensional electrophoresis, Sialylation, Sialic acid, Glycosylation

1 Introduction

Recent years witnessed extraordinary rise of interest to sialylation and, specifically, to its role in host–pathogen interactions and communication between cells of immune system [1, 2]. Prions or PrP^{Sc} are proteinaceous infectious agents that consist of misfolded, aggregated, self-replicating states of a sialoglycoprotein called the prion protein or PrP^{C} [3, 4]. While prions are not conventional pathogens, due to sialylation of N-linked glycans, prions may use mechanisms similar to those exploited by microbial or viral pathogens in invading a host or interacting with cells of immune system. As such, analysis of PrP^{Sc} sialylation is important for the several following reasons. Sialylation of PrP^{Sc} is believed to control the fate of prions in an organism and outcomes of prion infection [5]. Sialylation of prions is enhanced upon their colonization of

Pawel P. Liberski (ed.), *Prion Diseases*, Neuromethods, vol. 129,
DOI 10.1007/978-1-4939-7211-1_11, © Springer Science+Business Media LLC 2017

secondary lymphoid organs; thus, prions may use strategies similar to other pathogens to camouflage themselves from the immune system, facilitating host invasion [6]. In PrP^{Sc}, the glycans are directed outward, with the terminal sialic acid residues creating a negative charge on the surface of prion particles. In fact, electrostatic repulsion between sialic residues creates structural constraints that control prion replication rate and PrP^{Sc} glycoform ratio [7]. Because the structure of prion strains is different, sialylation controls replication rate in a strain-specific fashion [7]. Moreover, due to strain-specific structural constrains, prion strains recruit PrP^{C} sialoglycoforms selectively, i.e., according to their sialylation status rather than their relative expression levels [7]. Analysis of differences in strain-specific sialylation pattern is also important for elucidating prion strain competition for a substrate and strain interference [8, 9].

The fact that the N-linked glycans of PrP^{C} and PrP^{Sc} are sialylated was described more than 30 years ago [10]. Sialic acids are linked to the terminal positions of the two N-linked glycans [11–13] and, upon conversion of PrP^{C} into PrP^{Sc}, are carried over giving rise to sialylated PrP^{Sc} [14, 15]. For assessing sialylation status of PrP^{Sc}, we developed reliable protocol that involves two-dimensional electrophoresis (2D) followed by Western blot [5–7]. In 2D electrophoresis, the first, horizontal dimension (isoelectrofocusing or IEF) separates molecules according to their pI, whereas the second, vertical dimension (SDS-PAGE) separates according to molecular weight. Native PrP^{Sc} particles are multimers of heterogeneous size. For analysis of sialylation status, scrapie material is first treated with proteinase K (PK) to remove PrP^{C} and other proteins. Then PrP^{Sc} multimers are denatured into prion protein (PrP) monomers and separated using 2D according to their charge and molecular weight (Fig. 1). Individual PrP molecules could be un-, mono-, or diglycosylated that are separated in the vertical dimension of 2D according to their glycosylation status (Fig. 1). Each of the two glycans carry up to five terminal sialic acid residues adding negative charges to individual PrP molecules [15]. The distribution of charge isoforms in the horizontal dimension of 2D reflects sialylation status of individual PrP molecules. Heavily sialylated PrPs run toward acidic pH, while weakly sialylated—toward basic pH [5] (Fig. 1). Due to structural heterogeneity of the GPI anchors, unglycosylated PrPs also show several charge isoforms [16] (Fig. 1). Nevertheless, sialylation of glycans shifts the distribution of charge isoform toward acidic pH for monoglycosylated PrPs and even more so for diglycosylated PrPs, when compared to unglycosylated PrPs (Fig. 1).

Separation of PrP charge isoforms by 2D is typically followed by densitometry analysis, which is employed for quantification of isoforms according to their sialylation status. The individual intensity profiles of di- or monoglycosylated isoforms are normal-

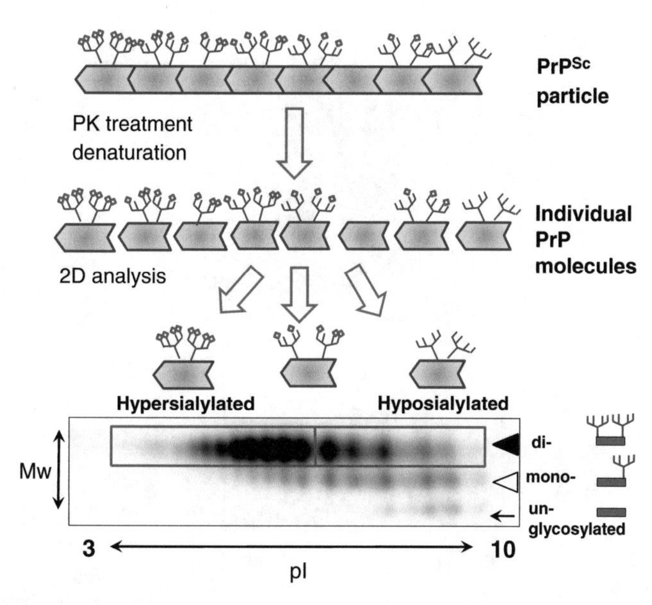

Fig. 1 Schematic diagram illustrating 2D analysis of PrP^{Sc}. Scrapie materials are treated with PK to clear PrP^{C}, denatured into monomers, and then analyzed by 2D. In horizontal dimension of 2D, individual PrP molecules are separated according to their pI. Charge distribution of individual PrP molecules reports on contribution of sialoglycoforms in PrP^{Sc} particles. The charge distribution of monoglycosylated isoforms extends toward acidic pH beyond that of unglycosylated isoforms, and the charge distribution of diglycosylated isoforms extends toward acidic pH beyond that of monoglycosylated isoforms according to the sialylation status of individual PrP molecules. For statistical analysis of sialylation status, diglycosylated charge isoforms are arbitrarily separated into two groups: hypersialylated (on the *left* of pI 7.5) and hyposialylated (on the *right* of pI 7.5). N-linked glycans are shown as *blue lines*, terminal sialic acid residues are shown as *red diamonds*

ized and plotted as a function of pI (Fig. 2a) [17]. This analysis is used for comparing the relative sialylation levels of PrP^{C} or PrP^{Sc} from different sources; however, it does not report on whether the differences between different samples are statistically significant [17]. According to alternative procedure, charged PrP isoforms on individual 2D profiles are separated arbitrarily into two groups. Isoforms located toward acidic pH from pI 7.5 are designated as hypersialylated and those toward basic pH are designated as hyposialylated (Fig. 2b) [7]. The percentage of sum intensities of hypersialylated isoforms relative to the total intensities of all isoforms is used to compare the samples. Using this approach, mean and standard deviations can be calculated when sufficient number of samples within each groups are analyzed (Fig. 2b). In addition, it is possible to plot the percentage of hypersialylation as a function of other parameters, for instance, the percentage of diglycosylated isoforms (Fig. 2b, bottom plot). The current chapter describes the

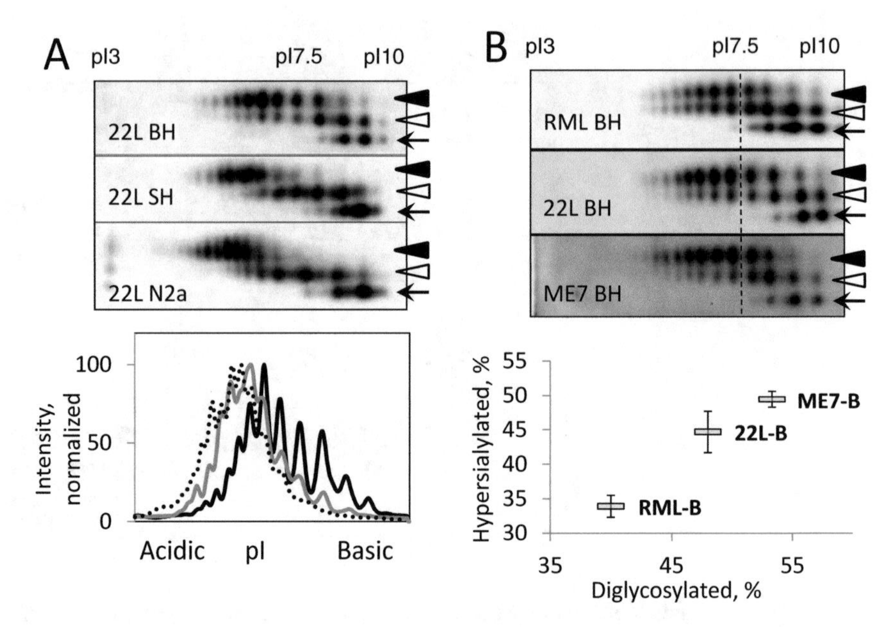

Fig. 2 Analysis of PrP^{Sc} sialylation by 2D analysis. (**a**) Analysis of sialylation status of 22L brain- (BH), spleen- (SH), or N2a-derived material using 2D (*top panels*). Sialylation profiles of diglycosylated isoforms of 22L brain- (*solid line*), spleen- (*gray line*), or N2a- (*dotted line*) derived material (*bottom panel*). Profiles were built using densitometry analysis of 2D Western blots. The highest curve signal value was taken as 100%. (**b**) Analysis of sialylation status of RML, 22L, and ME7 brain-derived material (BH) using 2D (*top panels*). Percentage of hypersialylated isoforms plotted as a function of percentage of diglycosylated glycoforms for brain-derived RML, 22L, and ME7 scrapie material (*bottom panel*). For each strain, at least three values were acquired from independent brain materials; the variations were used to calculate mean and standard deviations. In panels (**a**) and (**b**), *black triangles*, *white triangles*, and *arrows* mark di-, mono-, and non-glycosylated glycoforms, respectively

procedure for analysis of sialylation status of PrP^{Sc} from various sources including brain, spleen, cultured cells, or Protein Misfolding Cyclic Amplification (PMCAb) using 2D electrophoresis followed by Western blot.

2 Materials

Unless otherwise noticed, all reagents were from Sigma-Aldrich (St. Louis, MO). All solutions were prepared with ultrapure water—purified with Barnstead MicroPure Water Purification System (Thermo Scientific, Waltham, MA).

2.1 Scrapie Samples

Scrapie brain and spleen homogenates were prepared from terminally ill hamsters or mice inoculated i.c. or i.p. with brain scrapie material or PMCAb-derived products as previously described [6, 18]. PMCAb samples were produced as described elsewhere [5, 19]. N2a cells infected with 22L strain were collected in PBS.

2.2 Preparation of Scrapie Brain- or Spleen Homogenates or Cell Lysates

1. Ice-cold PBS pH 7.4 (cat # ABA-AB11072-01000, AmericanBio, Natick, MA), supplemented with 5 mM EDTA.
2. Ice-cold conversion buffer (Ca^{2+}-free and Mg^{2+}-free PBS, pH 7.4, supplemented with 0.15 M NaCl, 1.0% Triton, and one tablet of Complete protease inhibitors cocktail (cat# 1836145, Roche, Basel, Switzerland) per 50 mL of conversion buffer).
3. Tissue grinder with pestle, 30 mL (cat # 358049, Wheaton, Milville, NJ).
4. Cordless 12 V compact drill (Ryobi).
5. Tabletop centrifuge with cooling.
6. Thermomixer.
7. Misonix S-4000 microplate horn (Qsonica LLC, Newtown, CT).
8. Proteinase K (PK) (cat # P8107S, New England Biolabs, Ipswich, MA).
9. 1% SDS (cat #351–032-101, Quality Biological, Gaithersburg, MD).
10. 4× LDS Sample Buffer (NP0008, Life Technologies).
11. Water bath.
12. Cell scraper (cat # SAR-83.1832, Sarstedt, Nümbrecht, Germany).
13. Cell media (MEM, cat # 10-010-CV, Corning, Corning, NY).
14. Pre-chilled acetone (Stored at −20 °C).

2.3 Preparation of Samples for 2D

1. Solubilization buffer (8 M Urea, 2% (wt/vol) CHAPS, 5 mM TBP, 20 mM Tris–HCl pH 8.0) (can be prepared and aliquoted in advance, stored at −20 °C).
2. 0.5 M iodoacetamide (freshly prepared).
3. Pre-chilled methanol (stored at −20 °C).
4. Tabletop centrifuge with cooling.
5. Rehydration buffer (7 M urea, 2 M thiourea, 1% (wt/vol) DTT, 1% (wt/vol) CHAPS, 1% (wt/vol) Triton X-100, 1% (vol/vol) ampholyte (cat # ZM0021, Life Technologies, Frederick, MD), trace amount of Bromophenol Blue) (can be prepared and aliquoted in advance, stored at −20 °C).
6. Fixed immobilized pre-cast IPG strips with a linear pH gradient 3–10 (cat. # ZM0018, Life Technologies, Frederick, MD).
7. IPG Runner cassettes (cat. # ZM0003, Life Technologies, Frederick, MD).

2.4 Isoelectrofocusing

1. Power Supply (cat # ZP10001, Life Technologies, Frederick, MD).

2. ZOOM IPG Runner Mini-cell (cat # ZM0001, Life Technologies, Frederick, MD).
3. IPG strips with pI 3–10 linear gradient (cat # ZM0018, Life Technologies, Frederick, MD).

2.5 SDS-Page

1. Provision for equilibration buffers 1 and 2. In a 50 mL tube, combine 10.8 g urea, 7.5 mL Tris pH 8.8 (cat # MBI-21304, Mbiotech), 6 mL glycerol, 6 mL 10% SDS (V6551, Promega), dissolve on rotator. Add 0.3 g amberlite (Mixed Bed Exchanger Amberlite MB-150), incubate on rotator at room temperature for 1 h. Filter to get rid of amberlite particles. Divide into two 15 mL parts. This intermediate preparation can be stored at −20 °C. Defreeze prior to use and add DTT and iodoacetamide as follows.
2. *To make Equilibration buffer 1*: to 15 mL, add 0.3 g DTT. Leave on rotator for 5–10 min (final composition 6 M Urea, 20% (vol/vol) glycerol, 2% SDS, 375 mM Tris–HCl pH 8.8, 130 mM DTT). *To make Equilibration buffer 2*: to 15 mL, add 0.375 g iodoacetamide. Leave on rotator for 5–10 min (final composition 6 M Urea, 20% (vol/vol) glycerol, 2% SDS, 375 mM Tris–HCl pH 8.8, 135 mM iodoacetamide).
3. ZOOM Equilibration Tray (cat. # ZM0007, Life Technologies).
4. Laboratory shaker/rotator.
5. 4–12% Bis-Tris ZOOM SDS-PAGE pre-cast gels (cat. # NP0330BOX, Life Technologies, Frederick, MD).
6. Power Supply 300 Plus 300 V, 4–500 mA (Labnet, Edison, NJ).
7. XCell SureLock Mini-Cell Electrophoresis System (cat. # EI0001, Life Technologies).
8. MES-SDS PAGE buffer, prepared from 20× Bolt MES SDS Running Buffer (cat # B0002-02, Life Technologies, Frederick, MD).
9. 0.5% w/w agarose prepared by melting powdered agarose (cat # 15510, Life Technologies, Frederick, MD, discontinued product) in ultrapure water. Prior to SDS-PAGE, melt 0.5% agarose in a microwave oven and supply with MES-SDS buffer (1 part MES-SDS buffer for 19 parts 0.5% agarose, v/v). Keep at 55–65 °C until used. Each IPG strip requires approximately 400 μL melted agarose.

2.6 Western Blot

1. PVDF membranes (cat # IPVH00010, EMD Millipore, Billerica, MA).
2. Blotting paper.
3. Western Blot Transfer Buffer (100 mL 10× Transfer Buffer (14.5 g Tris; 72.0 g Glycine, dissolve in ultrapure water and

adjust volume to 1 L); 200 mL MeOH; 700 mL ultrapure water).

4. Power Supply 300 Plus 300 V, 4–500 mA (Labnet, Edison, NJ).
5. XCell SureLock Mini-Cell and XCell II Blot Module (cat. # EI0002, Life Technologies).
6. PBST. Supply 1× PBS with 0.1% v/v TWEEN-20.
7. Blocking solution: 5% milk in PBST. Dissolve 2.5 g dry non-fat milk in 50 mL PBST by vortexing, and agitation at room temperature for at least 30 min.
8. Laboratory shaker/rotator.
9. Primary antibodies: 1:10,000 3F4 (BioLegend, San Diego, CA) for hamster PrP^{Sc}, 1:5000 Ab3531 (Abcam, Cambridge, MA) for mouse PrP^{Sc}.
10. Secondary antibodies, HRP-labeled: 1:10,000 Goat-anti-mouse (KPL, Gaithersburg, MD) was used with 3F4; 1:5000 Goat-anti-rabbit (KPL, Gaithersburg, MD) was used with Ab3531.
11. Luminata Forte Western HRP Chemiluminescence Substrate (cat # WBLUF0500, EMD Millipore, Billerica, MA).
12. FluorChem M Western Blot Imaging System and software (ProteinSimple, San Jose, CA).

3 Methods

Unless stated otherwise, all the operations are done at room temperature (20–25 °C). The whole experiment takes at least 2 days, depending on the desired incubation time in primary antibody. Prior to running a 2D, it is recommended to check the quantity of PrP^{Sc} in scrapie material using SDS-PAGE/Western Blot (*See* **Note 1**).

3.1 Sample Preparation

3.1.1 Preparation of Scrapie Brain Samples

1. Prepare 10% brain homogenate (w/v) from scrapie-inoculated animals using ice-cold in PBS pH 7.4, and glass/Teflon tissue grinders cooled on ice and attached to a drill.
2. Dilute 10% brain homogenate tenfold in conversion buffer. Place 100 μL aliquot into thin-wall PCR tubes and sonicate for 30 s at 170 W output in microplate horn.
3. Supplement a 25 μL aliquot from last step with the same amount of PK solution in ultrapure water (final PK concentration of 20 μg/mL) and incubate at 37 °C for 30 min.
4. Take 19 μL from the last step, add 6 μL of 4× LDS, and incubate for 10 min in a boiling water bath.

3.1.2 Preparation of Scrapie Spleen Samples

1. Prepare 10% spleen homogenate (w/v) from scrapie-inoculated animals using ice-cold in PBS pH 7.4, and glass/Teflon tissue grinders cooled on ice and attached to a drill.
2. Dilute 250 μL of 10% (wt/vol) homogenate twofold with PBS. Aliquot into thin-wall PCR tubes and sonicate for 30 s at 170 W output in microplate horn.
3. Combine all aliquots in one centrifuge tube and centrifuge in pre-chilled (4 °C) tabletop centrifuge for 30 min at 16,000 × *g*. Discard the supernatant.
4. Resuspend the pellet in 25 μL of 1% (wt/vol) Triton in PBS and treat with 20 μg/mL PK for 30 min at 37 °C.
5. Take 19 μL from the last step, add 6 μL of 4× LDS, and incubate for 10 min in a boiling water bath.

3.1.3 Preparation of PMCAb Samples

1. Dilute a 10 μL aliquot of PMCAb material tenfold in PBS, place into a thin-wall PCR tube, and sonicate for 30 s at 170 W output in microplate horn.
2. Supplement a 25 μL aliquot from last step with the same amount of PK solution in ultrapure water (final PK concentration of 20 μg/mL) and incubate at 37 °C for 30 min.
3. Take 19 μL from the last step, add 6 μL of 4× LDS, and incubate for 10 min in a boiling water bath.

3.1.4 Preparation of Cultured Cell-Derived Samples

1. Remove media from cultured cells, add 1 mL of fresh media, and scrape cells with a scraper. Collect and spin-down 400 × *g* for 3 min.
2. Resuspend the pellet in 200 μL PBS and supply 1% v/v Triton X-100. Aliquot into thin-wall PCR tubes and sonicate for 30 s at 170 W output in microplate horn.
3. Treat with 10 μg/mL PK for 30 min at 37 °C.
4. Add 1000 μL of pre-chilled acetone. Leave at −20 °C overnight.
5. Discard acetone, let the pellet dry for 15 min. Resuspend the pellet in 25 μL of 1× LDS and incubate for 10 min in a boiling water bath.

3.2 Sample Treatment for 2D

1. Mix 25 μL of LDS-containing sample with 200 μL of solubilization buffer. Incubate for 1 h (*See* **Note 1**).
2. Add 7 μL of freshly prepared 0.5 M iodoacetamide, mix by inverting the tubes a couple of times. Incubate for 1 h in the dark.
3. Add 1160 μL of pre-chilled methanol, mix by inverting the tubes a couple of times. Incubate for at least 2 h at −20 °C.

4. Centrifuge in pre-chilled (4 °C) tabletop centrifuge for 30 min at 16,000 × *g*. Discard the supernatant.
5. Dry the pellet by leaving the tube open for a maximum of 30 min; if needed, dry additionally with an air stream.
6. Resuspend in 160 μL of rehydration buffer by pipetting and, if needed, vortexing.
7. Put a new IPG cassette on a firm leveled surface. Load 155 μL of each sample into a well on from the convex side of the cassette. The samples will immediately start entering into the sample channels. It is not necessary to use all the channels.
8. Peel the IPG strip from the blister with a forceps (*See* **Note 2**). Holding to its (+) marked end, with its printed side facing down, insert a strip into each channel containing a sample. Try not to introduce bubbles (*See* **Note 3**).
9. Seal the sample wells on both ends of the cassette by applying a sticker tape (provided with the cassettes).
10. Leave sample for rehydration overnight, or for at least 16 h.

3.3 Isoelectrofocusing

1. Remove sealing tape with plastic sample loading wells from both sides of cassette thus exposing adhesive surface.
2. Make sure that portions of the blue-colored gel on IPG strips are exposed on both ends of the cassette; adjust with forceps if needed (*See* **Note 4**).
3. Place electrode wick (provided with cassettes) over the adhesive. Use the black alignment marks on the cassette to properly place the wicks.
4. Evenly apply 600 μL ultrapure water to each wick.
5. Assemble the IPGRunner sandwich with the help of the gel dummy (if using two cassettes at a time, the dummy is not needed). The electrode wicks must come in contact with the electrodes of the ZOOM IPG RunnerTM Core.
6. Place the assembled module into the Mini-Cell Chamber and secure with a gel wedge.
7. Fill outer chamber of the Mini-Cell with deionized or ultrapure water (*See* **Note 5**). Be careful not to spill water into the inner chamber.
8. Place the ZOOM IPGRunner Cell Lid on the ZOOM IPGRunner Core. The lid can only be positioned in one position, with the (−) electrode on the right.
9. Connect the electrode cords to the power supply.
10. Turn on the power supply and choose the IEF current conditions. We find that the standard scheme suggested by the manufacturer for broad range IPG strips works for our purposes. Specifically, 175 V for 15 min, then 175–2000 V linear gradient for 45 min, then 2000 V for 30 min (*See* **Note 6**).

11. After the IEF program has finished, turn off the power supply, disconnect the cables, and remove the lid.
12. Carefully discard the water from the outer chamber.
13. Disassemble the module by first taking out the gel wedge and then the cassette with the Runner Core.
14. Put the cassette on a paper towel with IPG wells facing up. Blot any extra liquid. Trying not to disturb the IPG strips, remove the film cover from the cassette (*See* **Note 7**).

3.4 SDS-Page

1. Peel off adhesive liner from ZOOM Equilibration Tray. Carefully, trying not to touch the IPG strips, place the Equilibration Tray on the cassette: their outline shapes are similar and the cassette has protruding ribs which must come into the indentations in the edges of the tray. After placing the tray, secure it by applying pressure to the adhesive contacts of the tray.
2. Add 15 mL of Equilibration Buffer 1 into the orifice of the tray. Incubate the strips with gentle agitation for 15 min. Discard Equilibration Buffer 1 by gently turning the cassette/tray assembly upside down (*See* **Note 8**).
3. Add 15 mL of Equilibration Buffer 2 into the orifice of the tray. Incubate the strips with gentle agitation for 15 min. Discard Equilibration Buffer 2 by gently turning the cassette/tray assembly upside down.
4. Meanwhile, prepare the SDS-PAGE cells: unpack ZOOM gels, peel off the tapes, take out the combs. Rinse the wells with ultrapure water to remove any excess of polyacrylamide and blot excess water with blotting paper. If the marker well is bent it can be straightened with a help of a fine plastic tip or a forceps. Insert the gels into Mini Cell(s) with their Buffer Core and dummies, if needed. Secure with gel wedges.
5. Peel off the tray from the cassette. Blot any excess liquid.
6. Take each IPG strip and cut plastic ends. Do not cut off gel portions (*See* **Note 9**).
7. Place the IPG into the gel with the gel part of the strip facing toward the outer chamber of the Mini-Cell. We use the position of the (+) part of the strip toward the marker lane; however, the reverse position is also possible. Align the strip horizontally, trying to avoid bubbles.
8. Seal the strip in the well by adding approximately 400 μL of agarose (heated to 55–65 °C) supplied with MES SDS-PAGE running buffer. Let the agarose set for a couple of minutes.

9. Add SDS-PAGE running buffer into the inner and outer chambers of the cell. If needed, add the marker into the marker lane (*See* **Note 10**).
10. Close the lid of the Mini-Cell and connect the electrode cords to the power supply.
11. Turn on the power supply and choose the SDS-PAGE current conditions. We use constant voltage of 170 V for 60 min.
12. After the SDS-PAGE program has finished, turn off the power supply, disconnect the cables, remove the lid, and take out the gels.

3.5 Western Blot

1. Pre-soak four pads in Transfer Buffer.
2. Prepare PVDF membrane: rinse marked PVDF membrane, in methanol for 20 s. Discard methanol, incubate in ultrapure water for 2 min, then incubate in Transfer Buffer.
3. Extract ZOOM gel, briefly rinse with ultrapure water.
4. Assemble blot cell. Conduct transfer at 33 V for 1 h on ice. After the program is finished, disassemble the cell and incubate membrane in 20 mL blocking solution for 30–60 min with gentle agitation.
5. Discard the blocking solution, incubate membrane in the primary antibody for either 1 h at room temperature or overnight at 4 °C.
6. Decant primary antibody (it can be reused during 5–7 days if kept at 4 °C).
7. Wash membrane in PBST twice for 15 min with gentle agitation.
8. Discard the PBST, incubate the membrane in the secondary antibody for 45–60 min with agitation.
9. Wash membrane in PBST three times for 10–15 min with gentle agitation.
10. Discard PBST, incubate the membrane briefly (10–15 s) in the developing solution, and immediately proceed to developing. The developing solution can be used for several membranes during 1 h.

3.6 Analysis of 2D Images

1. Open the acquired images in the AlphaView software window.
2. To generate individual sialylation profiles for graphical representation, use the "Lane Profile" function. Select the lane of interest and follow the instructions in the dialog box. The resulting curve can be transferred to Excel or other graph building software for further analysis (see Fig. 2a, bottom plot).
3. For quantification of intensities of spot(s) of interest, use "Multiplex band analysis" option. First, draw a vertical line

through carefully horizontally aligned 2D images at pI of ~7.5 (dash line Fig. 2b); this line is used to arbitrarily separate charge isoforms into hypersialylated (to the left from the line) and hyposialylated (to the right). In the "Multiplex band analysis" dialog box, choose a rectangle, circle, or other shape and place it on the digitized blot to confine the spots of interest; subtract intensity of an equal background area from the same blot. The intensities of hyper- and hyposialylated isoforms combined are counted as 100% for each sample, and the percentage of hypersialylated can be easily derived from each 2D blot. After calculating mean and standard deviation from multiple repeats, a plot can be generated in Excel software. Figure 2b shows the percentage of hypersialylated molecules versus the percentage of diglycosylated, as previously reported for three mouse strains [7].

4 Notes

1. It is important that an appropriate amount of sample is taken for 2D. Both overloading and underloading should be avoided. In Fig. 3, examples of inadequate sample loading are presented: the results of overloading are shown in the higher and middle rows, whereas an example of underloading is shown on the lower left panel. The following artifacts appear due to overloading: (1) smearing or "tailing", (2) poor resolution of spots, or (3) both effects. Underloading results in a high background and loss of a signal. All mentioned types of artifacts make the images unsuitable for profile building or calculations.
2. It may be advisable to write down the numbers of the IPG strips and the samples they carry for convenience. Each IPG strip has its own unique number printed on its back. The strips can be thus easily identified afterwards.
3. If a bubble is formed, try retrieving the strip from the channel and inserting it again, this time trying to push the bubble from the other end of the channel. If the bubbles are forming due to low sample volume, try adding some more rehydration buffer. It will not significantly dilute the sample.
4. The water is only providing a stable temperature, so its purity is of minor importance
5. It is important that each strip has contact with electrode wick for current to pass efficiently. Moreover, it is advisable that all strips are exposed on both ends and aligned similarly to simplify later analysis and comparison.
6. Note that the Bromophenol Blue contained in the rehydration buffer will start moving slowly toward the (+) electrode. If that is not observed in 10–15 min, check the current on the screen

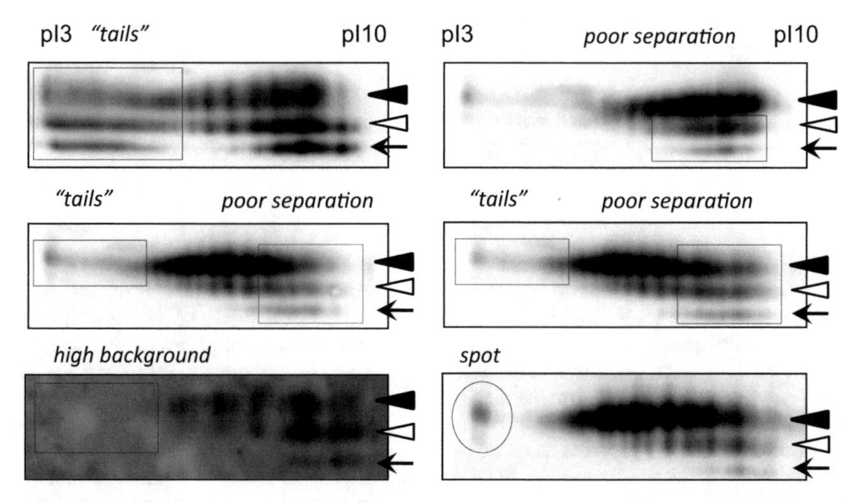

Fig. 3 Artifacts related to inadequate sample loading on 2D. Inadequate sample loading on 2D results in "tailing" effect, poor spot separation, high background, or spots at acidic pI. *Black triangles*, *white triangles*, and *arrows* mark di-, mono-, and non-glycosylated glycoforms, respectively

of the power supply. Check if there is water in the inner chamber. If so, stop the current, take off the lid and carefully discard the water from both the outer and inner chambers. Disassemble the module, blot the inner chamber surfaces with kimwipes, replace the soaked electrode wicks with the new ones, then start again from step 5.

7. According to the manufacturer, at this stage IPG strips can be frozen and kept at −80 °C in a sealed container. However, we have never done this and do not know whether this may interfere with PrP sialoform analysis.
8. If the tray/cassette contact is leaking, it is possible to incubate strips in any clean container of an appropriate volume given that the strip numbers were taken to prevent subsequent confusion.
9. We have noticed that sometimes, due to high amount of material in the strip and/or the position of the strip relative to the electrodes, a spot may appear on either ((−) or (+)) side of the pI range on the Western Blot (Fig. 2, lower right panel). We advise, on subsequent runs, to decrease protein load on the strip and/or to verify its position in the cassette so that neither of its ends is past the respective electrode. In the latter case the (+) end of the strip will be colored with Bromophenol Blue after IEF.
10. In the case of using the same sample in the marker lane, we advise to dilute initial 2D sample (step 1 from Sect. 3.2) 10–25 times in the SDS sample loading buffer prior to loading. It will thus not overpower the weaker signal from the IEF.

Acknowledgments

This work was supported by the National Institute of Health grant R01 NS045585.

References

1. Varki A (2008) Sialic acids in human health and disease. Trends Mol Med 14(8):351–360
2. Varki A (2010) Uniquely human evolution of sialic acid genetics and biology. Proc Natl Acad Sci U S A 107 (Supplement_2): 8939–8946
3. Prusiner SB (1982) Novel proteinaceous infectious particles cause scrapie. Science 216(4542):136–144
4. Legname G, Baskakov IV, Nguyen HOB, Riesner D, Cohen FE, DeArmond SJ, Prusiner SB (2004) Synthetic mammalian prions. Science 305(5684):673–676
5. Katorcha E, Makarava N, Savtchenko R, D'Azzo A, Baskakov IV (2014) Sialylation of prion protein controls the rate of prion amplification, the cross-species barrier, the ratio of PrPSc glycoform and prion infectivity. PLoS Pathog 10(9):e1004366
6. Srivastava S, Makarava N, Katorcha E, Savtchenko R, Brossmer R, Baskakov IV (2015) Post-conversion sialylation of prions in lymphoid tissues. Proc Natl Acad Sci U S A 112(48):E6654–E6662
7. Katorcha E, Makarava N, Savtchenko R, Baskakov IV (2015) Sialylation of the prion protein glycans controls prion replication rate and glycoform ratio. Sci Rep 5(1):16912
8. Makarava N, Savtchenko R, Baskakov IV (2015) Two alternative pathways for generating transmissible prion disease de novo. Acta Neuropathol Commun 3(1):69–13
9. Makarava N, Savtchenko R, Baskakov IV (2013) Selective amplification of classical and atypical prions using modified protein misfolding cyclic amplification. J Biol Chem 288(1):33–41
10. Bolton DC, Meyer RK, Prusiner SB (1985) Scrapie PrP 27-30 is a sialoglycoprotein. J Virol 53(2):596–606
11. Turk E, Teplow DB, Hood LE, Prusiner SB (1988) Purification and properties of the cellular and scrapie hamster prion proteins. Eur J Biochem 176(1):21–30
12. Endo T, Groth D, Prusiner SB, Kobata A (1989) Diversity of oligosaccharide structures linked to asparagines of the scrapie prion protein. Biochemistry 28(21):8380–8388
13. Stimson E, Hope J, Chong A, Burlingame AL (1999) Site-specific characterization of the N-linked glycans of murine prion protein by high-performance liquid chromatography/electrospray mass spectrometry and exoglycosidase digestions. Biochemistry 38(15):4885–4895
14. Stahl N, Baldwin MA, Teplow DB, Hood L, Gibson BW, Burlingame AL, Prusiner SB (1993) Structural studies of the scrapie prion protein using mass spectrometry and amino acid sequencing. Biochemistry 32(8):1991–2002
15. Rudd PM, Endo T, Colominas C, Groth D, Wheeler SF, Harvey DJ, Wormald MR, Serban H, Prusiner SB, Kobata A et al (1999) Glycosylation differences between the normal and pathogenic prion protein isoforms. Proc Natl Acad Sci U S A 96(23):13044–13049
16. Stahl N, Baldwin MA, Hecker R, Pan KM, Burlingame AL, Prusiner SB (1992) Glycosylinositol phospholipid anchors of the scrapie and cellular prion proteins contain sialic acid. Biochemistry 31(21):5043–5053
17. Katorcha E, Klimova N, Makarava N, Savtchenko R, Pan X, Annunziata I, Takahashi K, Miyagi T, Pshezhetsky AV, D'Azzo A et al (2015) Knocking out of cellular neuraminidases Neu1, Neu3 or Neu4 does not affect sialylation status of the prion protein. PLoS One 10(11):e0143218
18. Makarava N, Kovacs GG, Savtchenko R, Alexeeva I, Budka H, Rohwer RG, Baskakov IV (2012) Stabilization of a prion strain of synthetic origin requires multiple serial passages. J Biol Chem 287(36):30205–30214
19. Gonzalez-Montalban N, Makarava N, Ostapchenko VG, Savtchenko R, Alexeeva I, Rohwer RG, Baskakov IV (2011) Highly efficient protein misfolding cyclic amplification. PLoS Pathog 7(2):e1001277

Chapter 12

Exosomes in Prion Diseases

Alexander Hartmann, Hermann Altmeppen, Susanne Krasemann, and Markus Glatzel

Abstract

Dementias are characterized by generation and tissue deposition of proteins altered in their secondary or tertiary structure. Prion diseases are prominent and well-studied examples of these diseases. Initiation of prion disease is associated to the conversion of the cellular prion protein (PrP^C) to its pathogenic isoform (PrP^{Sc}). Spread of PrP^{Sc} throughout the central nervous system leads to disease progression and is achieved by cell-to-cell transfer, axonal or nanotube-mediated transport or exosomes.

In this chapter we describe how to isolate, purify, and quality control exosomes, and provide helpful notes for practical guidance and troubleshooting in these techniques.

Key words Exosomes, Prion protein, PrP, Neurotoxicity, Neurodegeneration, Protein aggregation, Protocol

1 Introduction

The vast majority of dementias are characterized by generation and tissue deposition of proteins altered in their secondary or tertiary structure [1]. Prominent examples of these diseases include Alzheimer's disease, frontotemporal dementia, and prion diseases. Although altered proteins and the pattern of tissue deposition differ and the clinical presentations of these diseases is distinct, shared neuropathological features at end stage of the disease include disturbed synaptic function, activation of glial cells, and loss of neurons [1, 2]. Initiation of disease is closely linked to a disturbed equilibrium where generation of abnormally folded proteins exceeds its degradation [3]. Progression of disease requires spread of abnormally folded proteins throughout the central nervous system [4, 5]. Prion diseases are model diseases for abovementioned entities and owing to the excellent quality of murine and cell culture-based prion disease models, our knowledge on the initiation and progression of prion disease is vastly superior to other dementias, thus we will focus on these diseases for the rest of the

Pawel P. Liberski (ed.), *Prion Diseases*, Neuromethods, vol. 129,
DOI 10.1007/978-1-4939-7211-1_12, © Springer Science+Business Media LLC 2017

chapter [6]. Prion diseases occur when the cellular prion protein (PrP^C), a glycosylphosphatidylinositol (GPI)-anchored membrane protein converts to its pathogenic isoform (PrP^{Sc}) [7]. Although PrP^{Sc} is only one component of the infectious agent (the "prion"), generation and tissue deposition of PrP^{Sc} correlates with neurodegeneration [8]. How the spread of abnormally folded proteins, in our example, PrP^{Sc}, from neuron to neuron is established has only been partially resolved. To date, at least four main routes are experimentally corroborated. These are: (a) direct neuron-to-neuron transfer, (b) transport along axons, (c) transfer via membrane nanotubes, and (d) vesicle (exosome) bound transport. It is likely that more modes exist and that the four established modes are not mutually exclusive [9–16]. Direct neuron-to-neuron transfer may occur when neurons have physical contact but also when abnormally folded proteins are released from affected neurons, i.e., by ectodomain shedding of PrP^{Sc} [17]. Transport of PrP^{Sc} along axons exploits axonal transport mechanisms such as antero/retrograde transport and may bridge the synaptic cleft enabling spread along neuronal tracts [9, 18]. Nanotubes consisting of membrane protrusions (tunnelling nanotubes) directly connect cells with each other and enable rapid signal transduction or exchange of organelles. PrP^{Sc} but also prion infectivity exploits this transport pathway enabling efficient cell-to-cell spread [19].

1.1 Exosomes and Prions

There is an ongoing debate on the definition of exosomes and some authors prefer to avoid the term exosomes in favor of a broader terminology such as microvesicles or extracellular vesicles. We are aware of the limitations in the definition of exosomes by size, running behavior in sucrose density gradients, protein composition, and shape [20]. Nevertheless, we will use the term exosome to make this chapter more reader friendly and to facilitate integration of recent literature in the field of exosomes and dementia research.

Exosomes are small (~120 nm) membrane vesicles that are released into the extracellular space from multivesicular bodies, which fuse with the plasma membrane [21]. They have diverse functions including removal of proteins, intercellular communication by transfer of microRNAs, transfer of pathogens, and antigen presentation [22, 23]. Exosomes are released by a variety of cell types including reticulocytes, antigen presenting cells, neoplastic cells, and neurons [24–26]. Exosomes exert their function by release of protein or RNA (i.e., shedding of superfluous proteins in reticulocytes or cell-to-cell transfer of RNA), or interaction with defined receptors (i.e., antigen-presentation in immune cells) [27, 28].

Exosomes derived from numerous donor cells contain abundant PrP^C, possibly in an N-terminally modified form [29, 30]. Furthermore, PrP^{Sc} and prion infectivity is associated with exosomes and neurons can transmit prion disease via exosomes [14, 31]. Although it is not established if PrP^{Sc} and/or prion infectivity are contained within the lumen of exosomes or rather associate with membranes on the outside of exosomes, the way by which PrP^{Sc} and/or prion infectivity is transported from cell to cell by exosomes is sometimes compared to a Trojan horse mechanism. Exosomes have also been implied in the transport of beta-amyloid in AD [32, 33]. In this instance, beta-amyloid is mainly associated to the outer leaflet of the exosomal membrane, the majority being bound PrP^{C} [30].

2 Methods

Exosomes may be used for experimental purposes but also for diagnostic purposes. For this, exosomes have to be isolated and purified from cell culture supernatant or from bodily fluids such as cerebrospinal fluid (Fig. 1). Once isolation and purification is achieved, a number of quality control measures have to be performed to establish purity of exosomes and to quantify amounts of exosomes (Fig. 2).

2.1 Isolation and Purification of Exosomes from Cell Culture Supernatants

2.1.1 Protocol 1: Serial Differential Centrifugation

There are numerous protocols and commercially available kits to isolate and purify exosomes. These comprise differential ultracentrifugation protocols, designed to exploit the physical properties of exosomes [34]. Commercially available kits (Total Exosome Isolation Kit, Life Technologies, USA; ExoSpin Exosome Purification Kit, Cell Guidance Systems, USA) precipitate vesicles, such as exosomes, with polyether compounds, for example, polyethylene glycol, to allow for exosome-isolation using low-speed centrifugation. Recently methods using immunoaffinity capture with antibody-coated magnetic beads have been established [35]. Yet others and we consider the differential ultracentrifugation protocol followed by extensive quality control measures as a good and reliable protocol to harvest sufficient amounts of exosomes with a reasonable purity from cell culture supernatants [35].

1. Like all cell culture-based techniques, growing and maintain high-quality cells is essential for obtaining highest quality exosomes.

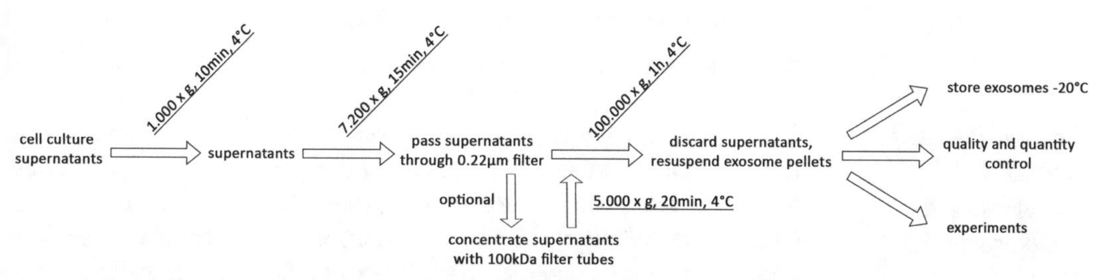

Fig. 1 Flow chart for the isolation of exosomes by serial differential centrifugation. This flow chart illustrates the workflow of exosome isolation. The optional concentration step before ultracentrifugation is only necessary if the amount of supernatant has to be scaled down

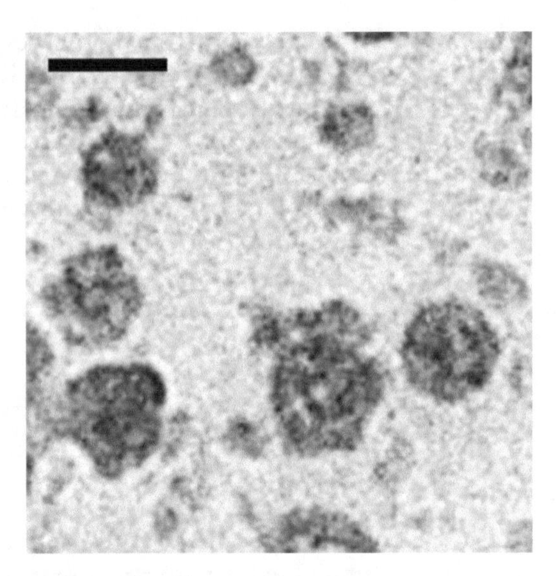

Fig. 2 Ultrastructural image of an exosome. Shown are exosomes isolated from cell culture supernatant using above-described protocols. Note the typical cup-shaped appearance of exosomes in EM analysis. Scale bar is 100 nm

2. Cells have to be incubated for at least 48 h in media using exosome-depleted FBS (Atlas Biologicals).
3. For all following steps supernatants have to be stored on ice.
4. To get rid of death cells and cell debris supernatants have to be centrifuged at 1000 × *g* for 10 min at 4 °C. Supernatants are transferred in a new tube followed by another centrifugation at 7500 × *g* for 15 min at 4 °C.
5. Supernatants have to be passed through a 0.22 μm filter into a new tube to remove cell debris, microvesicles, and apoptotic bodies.
6. Ultracentrifugation of supernatants at 100,000 × *g* for 1 h at 4 °C will pellet exosomes. The supernatant can be discarded (**Note 1**).
7. Resuspend exosome pellets in PBS containing EDTA-free protease inhibitor cocktail (**Notes 2** and **3**).
8. Exosomes can be stored at −20 °C.

2.2 Isolation and Purification of Exosomes from Cerebrospinal Fluid (CSF)

2.2.1 Protocol 1: Serial Differential Centrifugation

There principles of exosome isolation are the same as the protocol used to isolate exosomes from cell culture supernatants. The limitations of the protocol are also comparable. Since we are using a naturally occurring fluid, the composition of extracellular vesicles is even more heterogeneous than the population of extracellular vesicles obtained from cell culture supernatants; thus, special attention has to be given to quality control measures.

CSF samples should be processed as fast as possible to avoid changes in quality. Since the viscosity of the CSF samples is higher than that of cell culture supernatant and sample volume is generally limited, CSF is diluted with PBS to increase volume and avoid sample loss.

1. Centrifuge sample at 2000 × *g* for 30 min at 4 °C to get rid of death cells and cell debris. Sample supernatants are transferred in a new tube and may be stored at −80 °C if samples cannot be processed for direct exosome isolation at this time point. The cell pellet may be kept for further analysis, e.g., cytospins, Western blotting.
2. Dilute precleared CSF (1–2 mL) with an equal volume of cold, filtered PBS (**Note 3**).
3. For all following steps, supernatants have to be stored on ice.
4. CSF samples have to be centrifuged at 2000 × *g* for 30 min at 4 °C. Cell-free CSF sample is cleared by centrifugation at 4500 × *g* for 10 min. Supernatants are further purified at 10,000 × *g* for 30 min at 4 °C.
5. Dilute sample supernatants again with an equal volume of PBS and pass through a 0.22 μm filter into a new tube to reduce contamination by cell debris, microvesicles, and apoptotic bodies.
6. Rinse filter by passing another 2 mL of cold PBS and collect together with the sample to avoid loss of material.
7. Harvest exosomes by ultracentrifugation of supernatants at 110,000 × *g* for 70 min at 4 °C. Discard the supernatant.
8. Resuspend CSF exosome pellet in PBS (**Note 3**) and collect exosomes by another ultracentrifugation at 110,000 × *g* for 70 min at 4 °C to get rid of residual protein contamination.
9. Resuspend CSF exosome pellets in PBS containing EDTA-free protease inhibitor cocktail for further analysis (**Note 3**).
10. Exosomes can be stored at −20 °C.

2.3 Quality Control Measures to Ensure Enrichment of Exosomes

For further investigations, it is necessary to ensure selective enrichment of exosomes in the obtained exosome suspension after exosome isolation. Apoptotic vesicles or contaminating proteins can be part of the suspension after exosome isolation.

2.3.1 Protocol 1: Western Blotting for Marker Proteins

1. Mix the desired amount of exosomes (exosome concentrations can be measured by (Nanoparticle tracking analysis, see below) 5× sample buffer (0.5 M Tris–HCl, pH 6.8, 15% SDS, 50% glycerol, 25% ß-mercaptoethanol, 0.01% bromophenol blue).
2. Incubate the samples at 95 °C for 5 min.
3. Load the samples on 10% Bis/Tris SDS gels. For protein separation the run is performed in 1% SDS page running buffer at 100–120 V for 90 min.

4. Proteins are transferred to nitrocellulose membranes using 1% blotting buffer with an electric current of 400 mA for 70 min.
5. Exosomal marker proteins (e.g., CD63, Flottilin-1, Tsg101, and Alix) are detected using monoclonal antibodies (there are various companies selling high-quality antibodies to these proteins, this can be assessed at www.antibodypedia.com) and horseradish peroxidase-labeled anti-mouse secondary antibody (1:2500) (**Note 4**).

2.3.2 Protocol 2: Electron Microscopy (EM) to Assess Morphology of Exosomes

Exosome pellets are usually very small, if not invisible after purification. Therefore, standard electron microscopy protocols are difficult to apply to analyze exosomes by EM. Here, we describe a method to directly mount purified exosomes on precoated EM grids.

1. Resuspend the purified 100,000 × *g* exosome pellet in freshly prepared 2% paraformaldehyde (PFA) (**Note 5**). Depending on the source of exosomes (cell culture supernatant or CSF), use about 20–80 μL for resuspension. Exosomes may be stored for several days at 4 °C at this point.
2. Cut a piece of Parafilm and place it with the clean side upwards into a chamber. Carefully deposit 2–3 Formvar-carbon coated EM grids (Formvar-carbon coated 200 mesh copper grids; #PYFC200-CU; Science Services) per sample with the bluish side up onto the Parafilm (the coppery, glossy side should be upside down). For this, use a pointy forceps, but take care not to puncture the Formvar coat.
3. Deposit a 5 μL drop of resuspended exosomes on the grid surface and cover the chamber to let the membrane adsorb the exosomes for 20 min.
4. Cut new Parafilm and again place it with the clean side upwards. Add a 100 μL drop of PBS per sample on Parafilm and carefully transfer the grid with the sample side down onto the PBS drop using a clean forceps.
5. Add 100 μL drops (one per grid) of 0.3% glutaraldehyde (dilute EM grade glutaraldehyde 25% aqueous solution in 0.1 M sodium phosphate buffer, pH 7.4) (**Note 6**) on the Parafilm and subsequently transfer the grids upside down and incubate for 10 min.
6. Wash the grids eight times in distilled water. Therefore, place eight 100 μL drops of water per sample on the Parafilm and transfer the grids from drop to drop incubating for 2 min on every drop (take care not to wet the glossy side of the grid).
7. To increase contrast of exosomes, the grids are treated with uranyl acetate (add 100 mg uranyl acetate to 2.5 mL 70% ethanol; thoroughly mix and spin down; transfer supernatant to

fresh tube, store at 4 °C; this solution is stable for 4–5 days only). For this, a 50 μL drop of uranyl acetate is place on a small fresh piece of Parafilm and the grid is placed sample side down onto the drop. Incubate for 5 min at RT. Wash grid as described above with two consecutive drops of water. Uranyl acetate is both radioactive and toxic and requires special handling. See your institutional Radiation Safety Office/EM facility for guidelines concerning proper handling and disposal.

8. Blot excess water on a small piece of Whatman filter paper and let the grid air dry for 10 min. Store in appropriated grid storage box.
9. Samples can be observed under a transmission electron microscope operating at 80 kV and images can be recorded using specialized digital cameras and appropriate software.

2.3.3 Protocol 3: Sucrose Gradient to Establish Buoyancy of Exosomes

This protocol is designed for exosome gradient isolation using OptiPrep™ solution (Axis-Shield PoC, Norway) [34].

1. Prepare a discontinuous iodixanol gradient of 40% (w/v), 20% (w/v), 10% (w/v), and 5% (w/v) using the OptiPrep™ stock solution with 0.25 M sucrose, 10 mM Tris, and a pH of 7.5.
2. The gradient is built in a 14 × 89 polyallomer tube (Beckman Coulter).
3. Add 3 mL of 40% iodixanol solution and carefully layer 3 mL each of 20% and 10% solutions and 2.5 mL of 5% the solution.
4. On top of the gradient the exosome-containing solution (500 μL, 1.5 mg protein) is added.
5. Centrifuge at 100,000 × *g* for 18 h at 4 °C.
6. Collect individual 1 mL samples of the gradient fractions and dilute them with 2 mL PBS.
7. Centrifuge at 100,000 × *g* for 3 h at 4 °C.
8. Wash the pellet carefully with 1 mL PBS and resuspend pellets in 50 μL PBS.
9. The fractions can be used for characterization via western blots or other methods for quantification.

2.4 Measuring Sizes and Amounts of Exosomes

For further experiments, it is important to characterize and quantify the isolated exosomes. Exosomes are vesicles originating from cells. The amounts and sizes of released exosomes in cell culture can differ due to a variety of factors (temperature, cell culture media, density of cells, etc.). To ensure that the isolated vesicles are in the range of exosomes and to determine the concentration of isolated exosomes further characterizations and quantifications are necessary.

2.4.1 Protocol 1: DSL

The following protocol is designed for the use of a SpectroLight 600 in-situ dynamic light scattering instrument (DLS 600) (XtalConcepts, Hamburg, Germany) with a laser wavelength of 660 nm and an output of 100 mW. The detection of the scattered light is obtained at an angle of 142°.

1. Prepare a sample solution containing 0.1875×10^{10} particles/mL in a volume of 70 μL, 5 μL of this solution were used for the DLS measurements.
2. Overlay a Nunc plate (Nunc™ MicroWell™ Minitrays) with paraffin oil. Sample volume is pipetted in the wells under the paraffin oil.
3. Adjust the laser beam to your sample solution to obtain an optimal scattering profile.
4. Samples are measured at 293 K and each measurement is performed for 30 s.
5. Size ranges of exosomes are visualized by SpectroLight 600 software.

2.4.2 Protocol 2: Nanoparticle Tracking

The following protocol is designed for the NanoSight LM14 (NanoSight NTA 2.3 Build 0033, Malvern Instruments) equipped with a 638 nm laser and a Marlin F-033B IRF camera.

1. Exosome solution has to be diluted to a concentration of 8×10^8–2×10^9 particles/mL.
2. Clean the sample chamber (LM10 unit) with 70% EtOH. Make sure that the Viton Gasket (O-Ring) on the underside of the top plate is adjusted correctly in its channel, tighten the screws only with finger-tight pressure.
3. Inject at least 0.3 mL exosome solution using a syringe (1 mL) into the sample chamber and switch on the laser. An automated syringe pump facilitates injection into the chamber.
4. Adjust the microscope objective (position and height) to obtain a clear image of particles within the beam.
5. Record five (or more) videos of 10 s–1 min length (dependent on the particle concentration of the exosome solution (i.e., for <10 particles per visual field, 1 min is optimal)) at a camera intensity of 16. For every video inject a little bit more of your sample solution.
6. Via the batch processing function of the NanoSight NTA software the videos are used to calculate exosome size and particle concentration.
7. Clean the sample chamber (see point 2 above) before measuring the next sample.

3 Conclusions

The field of exosome research is a relatively new and dynamic area. Methods are constantly being refined and it is impossible to include and discuss all methods in a singular book chapter. We have tried to focus on methods which are relevant for dementia, especially prion research, and that have proven reliable and realistic in an academic research laboratory setting. This hopefully facilities and stimulates further research in this field and will help to deepen our understanding on the role that these vesicles play for the spread of neurotoxicity, for the diagnostics for these diseases, and for potential future therapeutics.

4 Notes

1. After ultracentrifugation exosome pellets mostly are not visible. Mark the tube before ultracentrifugation at the side where the pellet is expected to be sure you can resuspend it in even a small volume of PBS containing EDTA-free protease inhibitor cocktail.
2. For ultracentrifugation it is essential to use polyallomer tubes (Beckmann Coulter) to ensure the exosome pellet tightly sticks to the bottom of the tube. In contrast ultra-clear tubes (Beckmann Coulter) are not suited since the pellet might easily detach and get lost with the supernatant. Ultra-clear tubes are more suitable for gradient centrifugation.
3. PBS may form aggregates/complexes and these could influence further analysis. Thus 0.9% NaCl or HEPES have been suggested as alternative buffer for resuspension of exosomes.
4. For western blotting for exosomal marker proteins it is recommendable to check for additional marker proteins which could label possible contaminations of the obtained exosome solution after exosome isolation. To exclude the abundance of apoptotic vesicles in the preparation, antibodies against Histone-H3 can be used, since exosomes do not contain this protein. This marker could also be used for internal quality control, since it's presents is indicating that the exosome producing cell culture might not be in optimal density or condition. Furthermore, antibodies against Calnexin and GM130, which are marker proteins for the endoplasmic reticulum and the Golgi complex, respectively, can be used to exclude the abundance of intracellular vesicles originating from broken cells. Additionally not all exosomal marker proteins are working with every exosome due to their originating cell type (e.g., Alix is not working perfectly for exosomes derived from neurons).

5. Dissolve 2 g of PFA powder in 100 mL PBS while heating up to 65 °C and stirring; PFA solution can be aliquoted and stored at −20/−80 °C; quickly thaw in 37 °C water bath before using; discard after using.
6. To prepare a 10× phosphate buffer add 13.11 g of $NaH_2PO_4 \cdot H_2O$ and 72.09 g of $Na_2HPO_4 \cdot 2H_2O$ to 900 mL Aqua bidest; dissolve; adjust pH to 7.4 and fill up to 1000 mL. Before usage, prepare a 1× solution with Aqua bidest (containing 0.3% sucrose).

References

1. Aguzzi A, Lakkaraju AK (2016) Cell biology of prions and prionoids: a status report. Trends Cell Biol 26(1):40–51
2. Glatzel M, Stoeck K, Seeger H, Luhrs T, Aguzzi A (2005) Human prion diseases: molecular and clinical aspects. Arch Neurol 62(4):545–552
3. Hachinski V, Sposato LA (2013) Dementia: from muddled diagnoses to treatable mechanisms. Brain 136(9):2652–2654
4. Schipanski A, Lange S, Segref A, Gutschmidt A, Lomas DA, Miranda E et al (2013) A novel interaction between ageing and ER overload in a protein conformational dementia. Genetics 193(3):865–876
5. Glatzel M, Linsenmeier L, Dohler F, Krasemann S, Puig B, Altmeppen HC (2015) Shedding light on prion disease. Prion 9(4): 244–256
6. Moreno-Gonzalez I, Soto C (2012) Natural animal models of neurodegenerative protein misfolding diseases. Curr Pharm Des 18(8): 1148–1158
7. Geissen M, Krasemann S, Matschke J, Glatzel M (2007) Understanding the natural variability of prion diseases. Vaccine 25(30):5631–5636
8. Colby DW, Prusiner SB (2011) Prions. Cold Spring Harb Perspect Biol 3(1):a006833
9. Glatzel M, Aguzzi A (2000) PrP(C) expression in the peripheral nervous system is a determinant of prion neuroinvasion. J Gen Virol 81(11):2813–2821
10. Magalhaes AC, Baron GS, Lee KS, Steele-Mortimer O, Dorward D, Prado MA et al (2005) Uptake and neuritic transport of scrapie prion protein coincident with infection of neuronal cells. J Neurosci 25(21):5207–5216
11. Porto-Carreiro I, Fevrier B, Paquet S, Vilette D, Raposo G (2005) Prions and exosomes: from PrPc trafficking to PrPsc propagation. Blood Cells Mol Dis 35(2):143–148
12. Alais S, Simoes S, Baas D, Lehmann S, Raposo G, Darlix JL et al (2008) Mouse neuroblastoma cells release prion infectivity associated with exosomal vesicles. Biol Cell 100(10): 603–618
13. Wadia JS, Schaller M, Williamson RA, Dowdy SF (2008) Pathologic prion protein infects cells by lipid-raft dependent macropinocytosis. PLoS One 3(10):e3314
14. Vella LJ, Sharples RA, Nisbet RM, Cappai R, Hill AF (2008) The role of exosomes in the processing of proteins associated with neurodegenerative diseases. Eur Biophys J 37(3):323–332
15. Fevrier B, Vilette D, Archer F, Loew D, Faigle W, Vidal M et al (2004) Cells release prions in association with exosomes. Proc Natl Acad Sci U S A 101(26):9683–9688
16. Leblanc P, Alais S, Porto-Carreiro I, Lehmann S, Grassi J, Raposo G et al (2006) Retrovirus infection strongly enhances scrapie infectivity release in cell culture. EMBO J 25(12): 2674–2685
17. Altmeppen HC, Prox J, Puig B, Dohler F, Falker C, Krasemann S et al (2013) Roles of endoproteolytic alpha-cleavage and shedding of the prion protein in neurodegeneration. FEBS J 280(18):4338–4347
18. Kunzi V, Glatzel M, Nakano MY, Greber UF, Van Leuven F, Aguzzi A (2002) Unhampered prion neuroinvasion despite impaired fast axonal transport in transgenic mice overexpressing four-repeat tau. J Neurosci 22(17): 7471–7477
19. Gousset K, Schiff E, Langevin C, Marijanovic Z, Caputo A, Browman DT et al (2009) Prions hijack tunnelling nanotubes for intercellular spread. Nat Cell Biol 11(3):328–336
20. Varga Z, Yuana Y, Grootemaat AE, van der Pol E, Gollwitzer C, Krumrey M et al (2014) Towards traceable size determination of

extracellular vesicles. J Extracell Vesicles 3(1):23298. doi:10.3402/jev.v3.23298

21. Lakkaraju A, Rodriguez-Boulan E (2008) Itinerant exosomes: emerging roles in cell and tissue polarity. Trends Cell Biol 18(5):199–209
22. Silverman JM, Clos J, de'Oliveira CC, Shirvani O, Fang Y, Wang C et al (2010) An exosome-based secretion pathway is responsible for protein export from Leishmania and communication with macrophages. J Cell Sci 123(6):842–852
23. Testa JS, Apcher GS, Comber JD, Eisenlohr LC (2010) Exosome-driven antigen transfer for MHC class II presentation facilitated by the receptor binding activity of influenza hemagglutinin. J Immunol 185(11):6608–6616
24. Faure J, Lachenal G, Court M, Hirrlinger J, Chatellard-Causse C, Blot B et al (2006) Exosomes are released by cultured cortical neurones. Mol Cell Neurosci 31(4):642–648
25. Skog J, Wurdinger T, van Rijn S, Meijer DH, Gainche L, Sena-Esteves M et al (2008) Glioblastoma microvesicles transport RNA and proteins that promote tumour growth and provide diagnostic biomarkers. Nat Cell Biol 10(12):1470–1476
26. Kramer-Albers EM, Bretz N, Tenzer S, Winterstein C, Mobius W, Berger H et al (2007) Oligodendrocytes secrete exosomes containing major myelin and stress-protective proteins: trophic support for axons? Proteomics Clin Appl 1(11):1446–1461
27. Kogure T, Lin WL, Yan IK, Braconi C, Patel T (2011) Intercellular nanovesicle-mediated microRNA transfer: a mechanism of environmental modulation of hepatocellular cancer cell growth. Hepatology 54(4):1237–1248
28. Simons M, Raposo G (2009) Exosomes—vesicular carriers for intercellular communication. Curr Opin Cell Biol 21(4):575–581
29. Vella LJ, Sharples RA, Lawson VA, Masters CL, Cappai R, Hill AF (2007) Packaging of prions into exosomes is associated with a novel pathway of PrP processing. J Pathol 211(5):582–590
30. Falker C, Hartmann A, Guett I, Dohler F, Altmeppen H, Betzel C et al (2016) Exosomal PrP drives fibrillization of amyloid beta and counteracts amyloid beta-mediated neurotoxicity. J Neurochem 137(1):88–100. doi:10.1111/jnc.13514
31. Raposo G, Fevrier B, Vilette D, Archer F, Loew D, Faigle W et al (2004) Cells release prions in association with exosomes. Proc Natl Acad Sci U S A 101:9683–9688
32. Rajendran L, Honsho M, Zahn TR, Keller P, Geiger KD, Verkade P et al (2006) Alzheimer's disease beta-amyloid peptides are released in association with exosomes. Proc Natl Acad Sci U S A 103(30):11172–11177
33. Yuyama K, Sun H, Sakai S, Mitsutake S, Okada M, Tahara H et al (2014) Decreased amyloid-beta pathologies by intracerebral loading of glycosphingolipid-enriched exosomes in Alzheimer model mice. J Biol Chem 289(35):24488–24498
34. Tauro BJ, Greening DW, Mathias RA, Ji H, Mathivanan S, Scott AM et al (2012) Comparison of ultracentrifugation, density gradient separation, and immunoaffinity capture methods for isolating human colon cancer cell line LIM1863-derived exosomes. Methods 56(2):293–304
35. Greening DW, Xu R, Ji H, Tauro BJ, Simpson RJ (2015) A protocol for exosome isolation and characterization: evaluation of ultracentrifugation, density-gradient separation, and immunoaffinity capture methods. Methods Mol Biol 1295:179–209

Chapter 13

Synthetic Mammalian Prions

Fabio Moda, Edoardo Bistaffa, Joanna Narkiewicz, Giulia Salzano, and Giuseppe Legname

Abstract

We describe a detailed method to generate different synthetic prions characterized by defined abnormal structures, which confer to each isolate specific infectious properties. When challenged in vitro or in vivo some of these isolates were able to cause illness and produced pathological features similar to that observed in animals or human with naturally occurring prion diseases, including the sporadic form of the Creutzfeldt–Jakob disease. Thus, synthetic prions are of fundamental importance to shed light on the intricate molecular events leading to the misfolding of the normal prion protein. Understanding prion conversion mechanism could allow designing therapeutic strategies aimed at blocking this process.

Key words Synthetic prions, Prion protein, Fibrillization, Aggregation, Amyloid

1 Introduction

1.1 Historical Background About Prions and the "Protein Only" Hypothesis

Prion diseases, or Transmissible Spongiform Encephalopathies (TSEs), are a large group of neurodegenerative diseases that afflict mammalian species. Animal TSEs include scrapie in sheep, goats, and mufflons [1], transmissible mink encephalopathy (TME) in ranch-reared mink [2], chronic wasting disease (CWD) of mule deer and elk [3], bovine spongiform encephalopathy (BSE) in cow [4], and feline spongiform encephalopathy (FSE) in domestic cats [5]. Human prion disorders comprise kuru [6], Creutzfeldt–Jakob disease (CJD) [7], Gerstmann–Sträussler–Scheinker (GSS) syndrome [8, 9], and Fatal Familial Insomnia (FFI) [9]. The first report of prion disease dates back to 1732 in Lincolnshire after an outbreak of scrapie in British farms [10]. However, the real scrapie origin is unknown and some writers traced its origin somewhat earlier. For approximately one century the causes of this disease lacked any explanation. In the late 1800s, the hypothetical viral nature of the infectious agent was postulated by Besnoit. The way to look at this disease changed in 1936 when two French

Pawel P. Liberski (ed.), *Prion Diseases*, Neuromethods, vol. 129,
DOI 10.1007/978-1-4939-7211-1_13, © Springer Science+Business Media LLC 2017

veterinary, Jean Cullie and Paul-Louis Chelle, while working on a vaccine against a common sheep virus (made of formalin-fixed brain tissue from scrapie-infected animals), accidentally transmitted the scrapie agent to treated sheep. After this episode, William Gordon (Moredun Institute in Edinburgh) tried to repeat the experiment of scrapie transmission (accidentally discovered by Cullie and Chelle) by challenging 697 animals of which over 200 developed scrapie pathology after 2 years [11]. Scrapie was then transmitted from sheep to mice by Roger J. Morris and Daniel Carleton Gajdusek [12] and from goats to mice by Richard Chandler [13], thus creating a prion strain adapted to common laboratory animals (mice) showing that scrapie can be transmissible. This accomplishment opened the way to numerous studies that would have been almost impossible to test in sheep. After these observations, many hypothesis on the nature of the infectious agent were put forward: (1) a self-replicating membrane [14, 15], (2) a sub-virus [15], (3) a viroid and a spiroplasma or a retrovirus-like element [16] but none of these were exhaustively shown. In 1957, a young virologist and pediatrician, Daniel Carleton Gajdusek, moved to Papua New Guinea to investigate the cause of an atypical encephalitis-like pathology that afflicted the local population [17]. Today we know that such disease (kuru) is a prion disorder horizontally transmitted by cannibalistic rituals. The key event that provided the first link between scrapie and kuru came from an observation of William Hadlow (a veterinarian and neuropathologist). He was invited to a conference where he saw photographs of kuru pathology and recognized strong similarities with the scrapie pathology in goat. Hadlow sent a letter to Gajdusek proposing that the same infectious agent may be responsible for both kuru and scrapie. Shortly after that episode, Gajdusek started experiments to test whether kuru could be transmitted to primates. During this time, under the supervision of Dr. Clarence Joseph Gibbs, Gajdusek performed the first transmission of kuru to chimpanzee [6]. These studies provided an important step forward in understanding the infectious properties of prions, but because of the long incubation period of the disease, the infectious agent was initially thought to be a slow virus [18]. In 1967, Tikvah Alper and colleagues observed that the infectious agent associated to scrapie had a very strong resistance to UV treatment and ionizing radiation [19]. Moreover, the authors reported the low minimum molecular weight necessary for the infectivity (being at around 2×10^5 Da), thus ruling out the presence of complex infectious agents such as viruses. These studies supported the theory of Alan G. Dikinson and George Outram [20] suggesting that the agent of infection could be a virino, a small nucleic acid tightly complexed and protected by a protein coat that preserves the integrity of the genetic material from conventional denaturing agents. After the 30 years that followed these initial observations no one was able to show the association of TSE infectivity with the presence of nucleic acid. In 1967 John S. Griffith, a mathematician, proposed an innovative theory based on experimental observations that the pathological agent of prion disorder could be a protein able to self-replicate in the body of the host [21]. This is the so-called "protein only hypothesis" that was largely supported by Stanley B. Prusiner that provided a substantial amount of experimental evidence showing that the infectious agent was indeed a protein [22]. Subsequently, Prusiner coined the term "prions" to indicate this novel proteinaceous infectious agent and went to win the Nobel Prize in 1997 for his discovery of prions, a new biological principle of infection. In 1982 Prusiner's group purified the protease-resistant prion protein (PrP) from infectious material [23] and described the direct correlation between the amount of purified material and the infectivity titer of this agent. Particularly, they discovered that the level of infectivity could be reduced using (1) agent that destroy all the protein structure or (2) antibodies against PrP [24]. This data further supported the protein only hypothesis. This theory is now well accepted in the prion field and many authors have provided important contribution in support of this theory. In 1985, Bruce Chesebro and colleagues identified the specific gene that encodes for prion protein (*PRNP*) [25]. The work led to the conclusion that the physiological cellular prion protein (PrP^{C}) and the pathological related scrapie prion protein (PrP^{Sc}) have the same properties [26] and the conversion mechanism seems to be dictate by a conformational rearrangement of PrP^{C} into PrP^{Sc}. Genetic studies on familial prion disorders showed a direct association between mutation on the *PRNP* gene and prion disease. The first genetic form of prion disease discovered in human was the Gerstmann–Sträussler–Scheinker (GSS) syndrome characterized by a point mutation (P102L) in the *PRNP* gene [27]. Additional studies revealed that there are several additional mutations of the *PRNP* gene that are associated with different genetic form of prion disorders [28]. In 1993, Charles Weissmann developed a transgenic mouse model devoid of PrP. These mice were not susceptible to prion disease [29]. This observation provided an important step forward in showing that PrP^{C} is necessary and sufficient to sustain prion pathology.

1.2 First Attempts to Generate Infectivity In Vitro

The first effort towards understanding the phenomenon of prion conversion have been described by Byron Caughey and coworkers in a seminal experiment showing that infectious PrP (PrP^{Sc}) could be generated in vitro by incubating normal PrP^{C} with the brain-derived PrP^{Sc} [30]. Following these original observations, similar studies showed that the incubation of PrP^{C} with two different strains of hamster prions (Hyper and Drowsy) produced two distinct types of Proteinase K (PK)-resistant PrP [31]. Likewise, a chimeric hamster/mouse PrP^{C} (named MH2M) incubated with a

PrP^{Sc} collected from the brain of Sc237-infected hamsters resulted in PK-resistant misfolded PrP (and named PrP^{res}) that after inoculation in wild-type mice (Swiss CD-1) and hamsters failed to induce prion disease in both animal species. Thus, the acquisition of proteinase resistance by PrP^{C} was not sufficient for the propagation of infectivity [32]. Subsequently, many other attempts to generate prion infectivity from misfolded recombinant PrP have been unsuccessful [33–35]. The first evidence that synthetic prion could acquire infectious properties came from an important study by Kiotoshi Kaneko and collaborators [36]. This experiment was based on the use of transgenic mice expressing low levels of the mutant P101L PrP (Tg196), which spontaneously develop mild central nervous system (CNS) alterations after more than 600 days [37]. These animals were intracerebrally inoculated with a synthetic peptide (55 residues) of recombinant mouse PrP carrying the P101L mutation refolded into a β-sheet structure that was named MoPrP(89–143, P101L). Two hundred days after the infection, mice began to show severe neuropathological changes typical of prion diseases, while the injection of different peptides with the same sequence but different conformations (not folded into β-rich structures) failed to induce any pathological alteration. The de novo generated prion strain revealed different properties and neurological signs from the typical RML mouse prion strain. Interestingly, PrP^{Sc} deposits found in the brain were not PK resistant [38]. This peptide-induced disease retained its infectious properties when passaged to Tg196 mice while the infectivity was lost when passaged into wild-type animals [39].

Starting from these findings, several physical-chemical approaches to induce misfolding of recombinant PrP (recPrP) into β-sheet rich conformation were explored. These studies were aimed to address the question whether PrP^{C} alone is sufficient for the spontaneous formation of prions without the presence of any exogenous agent. Unfortunately, most the experiments failed to produce infectivity in vivo or their potential infectivity still needs to be assessed in animal models [40].

1.3 Generation of the First Mammalian Synthetic Prion (MoSP1)

Under pathological conditions, the normal form of the prion protein (PrP^{C}) misfolds to the pathological isoform, PrP^{Sc}, which is rich in β-sheet structures. Proteolytic digestion of PrP^{Sc} leads to the formation of a PK-resistant core, PrP27–30, which can form amyloid fibrils. In 2002, Ilia Baskakov and collaborators studied the kinetic pathways of amyloid formation in vivo using the unglycosylated recombinant PrP corresponding to the PrP27–30 (residues 89–230). In details, under specific denaturing conditions two misfolded forms were adopted: β-oligomers (in the presence of acidic pH and urea) or fibrillar structures (neutral pH and low concentration of urea), which developed to amyloids as measured by Thioflavin-T (ThT) binding [33]. An important discovery achieved in this experiment was the fact that the addition of preformed amyloid assemblies (seed) to the reaction considerably reduced the time of the fibrillization process, thus showing that the formation of recPrP fibrils can be promoted by seeding.

Starting from these findings, one of us verified whether these fibrils were infectious when challenged in mice [41]. In particular, transgenic mice (Tg9949) which overexpress N-terminally truncated PrP (residues 89–230) at 16–32 times normal levels were intracerebrally inoculated with (1) pre-folded amyloid fibrils (unseeded) and (2) seeded amyloid fibrils. In both cases, animals succumbed to prion disease. Particularly, the first group of mice exhibited longer incubation time than the second one, 474 vs. 382 days post infection, respectively. Moreover, both groups were characterized by unique neuropathological changes (vacuolation profile and pattern of PrP^{Sc} deposition). Biochemically, the animals inoculated with seeded amyloid fibrils were characterized by the presence of more protease-resistant PrP^{Sc} than the brain of the unseeded amyloid-inoculated mice. Prions in the brains of Tg9949 mice inoculated with seeded amyloid fibrils were designated as MoSP1 and, conversely to MoPrP(89–143, P101L), the new strain could be passaged to wild-type mice and was protease-resistant.

Accordingly [42, 43], a direct correlation was found between the stability of a specific prion isolate and the length of incubation time after inoculation in these animals. Basically, the most stable amyloids produces the most stable prion strain that exhibits the longest incubation times. These findings suggest that labile prion isolates obtained in these mice accumulate more rapidly and kill the host faster.

1.4 Generation of New Synthetic Mammalian Prions

As follow-up of the first synthetic prion experiment [41] a series of recPrP amyloid fibers were produced by the same laboratory and challenged in vivo. In particular, David Colby and coworkers were able to produce 11 PrP amyloids with different conformations by varying the biochemical conditions used for their formation, including pH, urea concentration, and temperature [43]. These amyloid inocula were then injected in the CNS of transgenic mice overexpressing (4–8 times) the wild-type mouse prion protein (Tg4053). Of the 11 amyloids, 10 caused disease with unique pattern of neuropathological alterations and PrP deposition in the brain, suggesting that each inoculum was a unique strain. Moreover, the conformational stability of each prion strain was closely correlated to the conformational stability of the amyloid preparation used to generate it. The amyloid that was not able to produce any prion-like pathology was characterized by the lowest conformational stability and, likely, it was cleared readily by interstitial proteases. Furthermore, this work established that the infectivity resides mostly within the polypeptide chain, and conversely to what is reported in other studies [44–47] the N-linked glycosylation could have a marginal role. These observations raised a crucial question in the prion field: which factors influence the conformation diversity of prion strains and which is the specific role of the primary sequence of PrP in determining this phenomenon? These issues are of fundamental importance for the development of efficient strategies to prevent intra- and interspecies prion spread. Notably, the effectiveness of anti-prion compounds seems to be linked to particular strains of PrP^{Sc}, as no universal inhibitor exists for all strains [48, 49]. Beside generation of new strains, there is another event that should be taken into account, the so-called evolution of prions. A wealth of evidence has shown that prions have a tendency for adaptation through a Darwinian process: the most aggressive strains generally dominate [50–53]. Thus, prions are capable of altering their molecular properties in order to adapt to new hosts and environment. This phenomenon is often observed when prions are transmitted to different host species. Particularly, two mechanisms might take part in this process: (1) prions are a mixture of strains with a distribution of incubation periods and *adaptation* appears when the strain with the shortest incubation time amplifies to sufficient titers to appear within the population or (2) the adaptation may results from slight conformational changes during (or after) prion replication.

Because the cloning of prion strains requires repeated passaging in inbred mice, the first synthetic prion generated by Legname et al. [41], named MoSP1, was repeatedly passaged in an inbred host. In such a scenario, MoSP1 was found to be composed at least of two different prion isolates: MoSP1(1) and MoSP1(2) [54, 55]. These results argued that a cellular machinery take part in the propagation of different PrP^{Sc} conformers, each of which enciphers a distinct biological phenotype. Additional studies on MoSP1(1) and MoSP1(2) revealed that these two strains were characterized by two different molecular mass of the protease-resistant core, 21 kDa and 19 kDa, respectively [56]. When cloned in cells MoSP1(1) and MoSP1(2) could breed with fidelity; however when present as a mixture, MoSP1(1) preferentially proliferated leading to the disappearance of MoSP1(2). Additionally, the inoculation of mice with MoSP1(2) resulted in the formation and accumulation of small amount of MoSP1(1). Thus, MoSP1(1) replicates faster than MoSP1(2) and this isolate might be more effective at interacting with PrP^{C} (or other factors) denying MoSP1(2), the resources required for its propagation. MoSP1(1) prion strain has a selective advantage over MoSP1(2). Moreover, recent experiments have shown that mammalian [51, 57] and yeast prions [58] can evolve conformationally in response to environmental pressure.

1.5 Other Attempts to Generate Infectious Synthetic Prion

After Legname et al. work, several groups started to focus on the generation of synthetic prions using different approaches. In 2010, Natallia Makarava and coworkers [59] were able to convert full-length hamster PrP into amyloid fibrils. These fibrils were incubated with hamster normal brain homogenate (NBH) and subjected to five cycles of 1 min incubation at 80 °C followed by 1 min incubation at 37 °C (annealing). The resulting homogenate was analyzed by Western blotting and a PK-resistant band migrating at around 16 kDa was detected. The same material was intracerebrally inoculated in hamsters and after serial transmissions produced a new type of prion disease characterized by a very long incubation time. The newly generated synthetic prion was named SSLOW (Synthetic Strain Leading to OverWeight) and its conformational stability was similar to that of 263 K, even though these two strains were characterized by different incubation times. Another strategy to generate synthetic prion was based on protein misfolding cyclic amplification (PMCA), an approach developed by Claudio Soto's group [60]. In this case PrP^{Sc} is mixed with an excess of PrP^{C} (obtained from healthy brain homogenate). After an incubation period (generally 30 min) the material is sonicated in order to disaggregate PrP^{Sc} fibrils, thus generating different PrP^{Sc} seeds that can recruit and convert normal PrP^{C}. The entire process in repeated in a cyclic manner. After 48 h, the amplified product is further diluted in healthy brain homogenates and additional cycles of amplification (rounds) are performed. In this way, PMCA generates millions of infectious units (PrP^{Sc}), starting with the equivalent to one PrP^{Sc} oligomer. Thus, the sensitivity of PMCA is similar to that of PCR technique, which is used to amplify DNA. By taking advantage of this technique, Nathan Deleault et al. [61] showed that using a purified PrP^{C} preparation, efficient PMCA propagation of infectious prions required the presence of polyanions, such as nucleic acids, glycosaminoglycans, phospholipid-rich membranes, and chaperone proteins. Although polyanions may not be absolutely required to form an infectious prion, they may increase the efficiency of prion conversion or modify some biochemical properties of PrP^{Sc} [62]. Different polyanions, in particular RNA [63], synthetic lipid in combination with purified RNA [64] or only synthetic lipids [65, 66] were found to facilitate PrP^{C} conversion in vitro and promote the generation of prion de novo. In 2013, Zhihong Zhang et al. [67] were able to perform serial PMCA using recombinant PrP in a laboratory that has never been exposed to any prion. After several rounds of amplification, two types of PK-resistant and self-perpetuating recombinant PrP conformers ($rPrP^{res}$) were generated de novo. Both PrP^{res} were characterized by two different PK-resistant cores of 17 kDa and 14 kDa, respectively. Only $rPrP^{res}$ (17 kDa) was highly infectious when injected in mice, causing prion disease with an average survival time of about 172 days. In contrast, $rPrP^{res}$ (14 kDa) completely failed to induce any disease. In 2009, Marcelo A. Barria et al. [68] were able to modify PMCA and produce PrP^{Sc} in the absence of preexisting PrP^{Sc}. The de novo generated PrP^{Sc} was infectious when inoculated into wild-type hamsters, producing a new disease phenotype with unique neuropathological features. Considering all these findings,

Fei Wang et al. [69] were able to use PMCA to produce a new prion strain starting from full-length mouse recombinant PrP (recMoPrP) using different combination of synthetic anionic phospholipid POPG (1-palmitoyl-2-oleoylphosphatidylglycerol) and RNA isolated from mouse liver. The generated PrP^{Sc} was infectious and the infectivity was maintained after serial passages in wild-type mice. The same group in 2012 [70] showed that infectious prions can be generated by replacing the total RNA with a synthetic polyriboadenylic acid [poly(rA)], thus excluding the possibility that a conjectured informational RNA present in the tissue of the host could have facilitated the conversion of PrP^C in its infectious form. These data show that synthetic prion produced in vitro by different methods is able to cause disease in both transgenic and wild-type mice and provide important evidence in support of the prion-only hypothesis.

1.6 Generation of Synthetic Prions Able to Infect Cells and Wild-Type Animals

We have recently showed that, under controlled and well-defined biophysical and biochemical conditions, full-length recombinant mouse PrP (recMoPrP(23–231)) converted to amyloid fibrils (without any seeding factor) and acquired different conformational structures [71]. When injected in cells, most of these fibrils were able to induce endogenous PrP^C to misfold and aggregate at membrane and cytosolic compartments. Converted PrP acquired partially resistance to PK treatment, which was maintained during cellular passages. Some of these amyloids were also able to induce PrP^C conversion in PMCA. After inoculation in animals, PMCA amplified fibrils were able to induce sever prion pathology. Fibrils that were not amplified by PMCA were not able to induce prion pathology when challenged animals (first passage). Serial transmission passages in animals might be required to generate stable and infectious material.

2 Materials

2.1 Expression and Purification of Recombinant MousePrP (recMoPrP(23–231))

2.1.1 Summary

Mouse recombinant PrP protein (from residue 23 to 231) without any "tag" was expressed in *Escherichia coli* Rosetta2(DE3) cells using the pET expression system (Novagen). PrP protein was extracted from inclusion bodies and purified in denaturing conditions by Nickel-affinity chromatography using its octarepeat sequence as a natural affinity tag, followed by Reverse Phase chromatography. The purified protein was freeze-dried and stored at −80 °C.

Materials

- Nickel-affinity chromatography column HisTrap FF crude (GE Healthcare)

- Reverse Phase Chromatography column Jupiter C4, 300A (Phenomenex)
- Vacuum filter units 0.45 μm (Millipore)
- Millex-GS Syringe Filter Unit, 0.22 μm, mixed cellulose esters, 33 mm (Merck Millipore cod. SLGS033SS)

Reagents

- Competent BL21Rosetta2(DE3) cells *Escherichia coli* (Stratagene)
- Bactoagar (Sigma cod. A5306)
- $CaCl_2$ (Sigma cod. C3881)
- Glycerol (Sigma cod. G6279)
- Bactotryptone (Sigma cod. T9419)
- Yeast extract (Sigma cod. Y1625)
- Ampicillin (AppliChem cod. A0839)
- Chloramphenicol (Sigma cod. 23275)
- Isopropyl β-D-1-thiogalactopyranoside (IPTG) (Sigma cod. I6758)
- **Phenylmethylsulfonyl fluoride** (PMSF) (Sigma cod. P7626)
- Trizma base (Sigma cod. T1503)
- **Ethylenediaminetetraacetic acid** (EDTA) (Sigma cod. E9884)
- Triton (Sigma cod. T9284)
- Imidazole (Merck millipore cod. UN3263)
- Guanidine Hydrochloride (Sigma cod. 50950)
- Sodium Chloride (NaCl) (Sigma cod. S7653)
- Trifluoroacetic acid (TFA) cod. 91700)

Equipment

- Autoclave (Fedegari Group mod. FE_FVGBK)
- Benchtop centrifuge (Eppendorf)
- Heated Incubator Shaker (Innova, New Brunswick)
- Homogenizer Panda Plus 2000 (GeaNiroSoavi) cooled by MultiTemp III unit (Amersham Biosciences)
- Centrifuge Avanti J-26 XP (Beckman Coulter)
- Rotor JLA-8100 with 1 L polypropylene bottles (Beckman Coulter)
- Rotor JA-18 (Beckman Coulter)
- AKTA Purifier protein purification system (GE Healthcare)
- Freeze Dry System FreeZone 2.5 (Labconco)

- Ultrasonicbath (Branson)
- SDS-PAGE equipment (Mini-Protean Tetra Cell System, Biorad)

Buffers Preparation

- $CaCl_2$: Dissolve $CaCl_2$ at 0.1 M final concentration into H_2O. Sterilize by autoclaving.
- Glycerol: Prepare glycerol at 20% final concentration. Sterilize by autoclaving.
- Luria Bertani (LB) medium: Dissolve 10 g bactotryptone, 5 g yeast extract and 10 g NaCl into 1 L H_2O. Adjust pH to 7.8. Sterilize by autoclaving.
- LB agar plates: Add 15 g agar to 1 L LB medium before autoclaving. After autoclaving, prepare plates allowing medium to cool until bottle can be held in hands without burning. Add antibiotics (the final concentration of ampicillin should be 100 μg/mL and the final concentration of chloramphenicol should be 30 μg/mL) and pour 30 mL into each sterile Petri dish (100 mm diameter).
- Ampicillin: Prepare a stock solution of 100 mg/mL in H_2O and filter-sterilize. Store at −20 °C.
- Chloramphenicol: Prepare a stock solution of 30 mg/mL in 100% ethanol and filter-sterilize. Store at 4 °C.
- IPTG: Prepare a 1 M stock solution in H_2O and filter-sterilize. Store at 4 °C.
- PMSF: Prepare a 0.2 M stock solution in isopropanol. Store at 4 °C.
- Buffer A (*Protein expression and purification*): 25 mM Tris-base, 5 mM EDTA, pH 8.0.
- Buffer B (*Protein expression and purification*): 25 mM Tris-base, 5 mM EDTA, 0.8% Triton X 100, pH 8.0.
- Buffer C (*Protein expression and purification*): 25 mM Tris-base, 8 M Gdn-HCl, pH 8.0.
- Binding/Washing buffer (*Metal ion affinity chromatography):* 20 mM Tris–HCl pH 8.0, 0.5 M NaCl, 10 mM imidazole, 2 M guanidine hydrochloride (Gdn-HCl).
- Elution buffer (*Metal ion affinity chromatography)*: 20 mM Tris–HCl pH 8.0, 0.5 M NaCl, 500 mM imidazole, 2 M Gdn-HCl.
- Buffer A (*Reverse Phase Chromatography*): 0.1% trifluoroacetic acid (TFA) in H_2O.
- Buffer B (*Reverse Phase Chromatography*): 0.1% TFA in 100% acetonitrile.

2.2 Fibrillization of Recombinant MousePrP (recMoPrP(23–231))

2.2.1 Summary

Different amyloid fibrils of recMoPrP (23–231) were prepared following controlled and well-defined biochemical conditions using only recPrP and common chemicals. All the conditions described in this protocol are characterized by denaturant condition (Gdn-HCl and Urea) that might be able to induce the generation of misfolded PrP amyloids. However, the infectivity of these preparations required to be checked using several in vitro or in vivo conditions.

Materials

- Millex-GS Syringe Filter Unit, 0.22 μm, mixed cellulose esters, 33 mm (Merck Millipore cod. SLGS033SS)
- Syringe 10 mL (InJ/Light Rays cod. 1021CM38)
- Nunc™ MicroWell™ 96-Well Optical-Bottom Plates (Thermo Fisher cod. 265301)
- Nunc™ Sealing Tapes (Thermo Fisher cod. 232702)
- Glass beads (Sigma cod. Z143928)

Reagents

- Guanidine hydrochloride (≥98% Sigma cod. 50950)
- Sodium hydroxide (Sigma cod. S8045)
- Thioflavin T (Sigma cod. T3516)
- Phosphate buffer 10× (Gibco cod. 14200-067)
- Dithiothreitol (Sigma cod. D0632)
- Sodium acetate trihydrate (Sigma cod. 71188)
- Urea (Sigma cod. U5378)
- NaCl (Sigma cod. 71376)
- Acetic acid (Sigma cod. 537020)
- MilliQ water

Equipment

- Eppendorf Thermomixer R (mod. 5355)
- BMG FLUOstar OPTIMA
- Ultracentrifuge Beckman Coulter Optima MAX

Buffers Preparation

- Acetate buffer pH 5.0: Prepare acetate buffer 0.2 M as following: Sodium acetate trihydrate: 54.43 g, Glacial acetic acid: 6 mL and MilliQ water to 1 L. Dilute the solution at required concentration and adjust the pH to 5.0 with NaOH 10 N.
- Guanidine Hydrochloride: Dissolve Guanidine Hydrochloride (Gdn-HCl) in PBS or acetate buffer at different concentrations (see Table 1 for details) and filter the final solution.

Table 1
Condition for recMoPrP (23–231) fibrillization

Buffer	Denaturant	Concentration	pH	[MoPrP]	DTT	NaCl
50 mM Acetate	Gdn-HCl	0.25	5	100	−	−
		1			+	+
					+	−
				200	−	+
		2	5	100	+	+
					+	−
					−	−
		3	5	100	+	+
					+	−
					−	−
		4	5	100	+	+
					+	−
					−	−
	Urea	4	3.5	100	−	−
			5	100	−	−
			6	100	−	−
			5	100	−	+
				200	−	+
PBS	Gdn-HCl	0.25	7.5	100	−	−
		0.5	7.5	100	−	−
		1	7.5	100	−	−
					+	+
					+	−
		2	7.5	100	+	−
					+	+
				200	−	−
		3	7.5	100	−	−
					+	−
					+	+
		4	7.5	100	−	−
					+	−
					+	+

- *Note 1*: To avoid any protein contamination, we recommend only glassware for the preparation of buffers and solutions.
- *Note 2*: Stock buffer solutions can be stored at +4° for several months but before the reuse the correct pH should be carefully checked.
- Thioflavin T preparation: Perform all the passages of Thioflavin T (ThT) preparation avoiding any exposure to ambient light. Dissolve ThT powder in MilliQ water, incubate at 37 °C for 20 min, and then filter the solution using 0.22 μm filter. Store aliquots of stock solution at −20 °C.
- *Note*: ThT aliquots should be used only one time when thaw out for guaranteeing the efficiency of the reagent.
- Dithiothreitol preparation: Dissolve dithiothreitol (DTT) powder in MilliQ water, filter, and store at −20 °C. As indicated for ThT reagent the aliquots of DTT can be used only once when thaw.
- Sodium chloride preparation: Dissolve sodium chloride (NaCl) powder in MilliQ water and filter the solution with 0.22 μm syringe filter unit. NaCl solution can be conserved at room temperature.

Note: Prepare all buffers in sterile conditions with MilliQ water (13 MΩ/cm), adjust the pH when necessary, and filter all the solutions using 0.22 μm syringe filter unit. Perform all the procedures of preparation of buffers and proteins in a prion-free laboratory using only decontaminated equipment and consumables. However, after the fibrillization process the products should be handled for biochemical investigations in a biosafety class 2 laboratory.

3 Methods

3.1 Expression and Purification of Recombinant MousePrP (recMoPrP(23–231))

Preparation of Competent Cells

1. Spread BL21 Rosetta2 (DE3) *E. coli* cells stock on an LB agar plate with chloramphenicol.
2. Inoculate single colony of BL21 Rosetta2 (DE3) *E. coli* cells in 5 mL LB medium, incubate at 37 °C overnight (O/N).
3. Inoculate 1 mL of the O/N culture into 60 mL of LB medium in a 250 mL flask and incubate at 37 °C to OD_{600} = 0.5.
4. Stop the growth at 4 °C for 30 min and centrifuge the culture at 2700 × *g*, 4 °C, 5 min.
5. Resuspend the pellet in 30 mL of cold $CaCl_2$ 0.1 M and incubate in ice for 30 min.
6. Centrifuge the suspension at 2700 × *g*, 4 °C, 5 min.

7. Resuspend pellet with 2 mL of cold $CaCl_2$, 0.1 M, glycerol 20%.
8. Incubate at 4 °C O/N, and then aliquot cells in sterile tubes (50 μL each) and store at −80 °C.

Protein Expression

1. Transform 50 μL competent BL21 Rosetta2 (DE3) cells with pET11a(MoPrP23–231) and spread 300 μL on an agar plate containing 100 μg/mL ampicillin and 30 μg/mL chloramphenicol. Incubate the plate O/N at 37 °C.
2. Inoculate 100 mL LB medium in a 500 mL flask with a single colony. Shake O/N at 200 rpm at 37 °C.
3. Add 25 mL of the O/N culture to 2 L LB medium containing 100 μg/mL ampicillin and 30 μg/mL chloramphenicol in a fermenter 2 L vessel. Shake the culture at 250 rpm at 37 °C until the cells reach mid-log phase (OD_{600} = 0.5).
4. Shift the temperature to 30 °C and then add IPTG to a final concentration of 1 mM. Continue shaking for 16 h.
5. Collect the cells and centrifuge at 12,000 × *g* for 30 min at 4 °C. Discard the supernatant and collect the bacterial paste.

Cell Disruption and Isolation of Inclusion Bodies

1. Add 15 mL of Buffer A for each gram of bacterial paste, resuspend the pellet, and centrifuge at 12,000 × *g* for 30 min, 4 °C.
2. Add 20 mL of Buffer B for each bacterial paste gram and resuspend the pellet.
3. Lyse the cell suspension using the 4 °C-cooled down homogenizer. Load in the funnel the resuspended sample (max volume 500 mL, min volume 100 mL). Perform homogenization four times at around 1500 bar (22,000 psi). After this process the sample appears clear. Add PMSF to a final concentration of 1 mM. Centrifuge at 12,000 × *g* for 30 min, 4 °C. After centrifugation, a solid amber-colored pellet corresponding to the inclusion bodies (IBs) is visible. Discard the supernatant.
4. Add 15 mL of Buffer B for each gram of IBs and resuspend the pellet. Centrifuge at 12,000 × *g* for 30 min, 4 °C. Discard the supernatant.
5. Add 15 mL of MilliQ water for each gram of IBs and resuspend the pellet. Centrifuge at 12,000 × *g* for 30 min, 4 °C. Discard the supernatant.

   Note: Repeat passage 5 if bacterial debris are still present (Optional).
6. Add 5 mL of Buffer C for each gram of IBs and resuspend the IBs. Keep the IBs at 37 °C at 200 rpm O/N to dissolve completely the IBs.

7. Centrifuge at 12,000 × *g* for 30 min, 4 °C, to remove any debris. Collect the supernatant.

3.1.1 Metal Affinity Chromatography (Protein Purification)

Inclusion bodies containing recPrP were solubilized in 8 M Gdn-HCl with shaking overnight at 37 °C. The solution was then centrifuged (20,000 × rpm, 30 min), filtered through 0.22 μm syringe filter, and diluted with buffer: 80 mM Tris-base, 2 M NaCl, pH 8.0 (to obtain final concentration: 6 M Gdn-HCl, 20 mM Tris–HCl pH 8.0, 500 mMNaCl). The HisTrap column was equilibrated with 10 column volumes (CV) of binding/washing buffer before protein loading. The protein solution was loaded onto the column at a flow rate of 1–2 mL/min. The unbound proteins were washed (4–5 mL/min) with buffer A until the baseline absorbance at A280 was reached. RecPrP was eluted with linear imidazole gradient (buffer B) for 30 min at flow rate 5 mL/min and 5 mL fractions were collected. Small aliquots of elution fractions were resolved by SDS-PAGE and fractions containing PrP were combined in a 50 mL tube.

Note 1: all solutions should be filtered (0.45 μm vacuum filter units) and degassed (ultrasonic bath) before chromatography.

Note 2: Gdn-HCl present in samples is not compatible with SDS-PAGE, so the buffer should be exchanged before the run. The samples are precipitated by methanol/chloroform precipitation procedure and resuspended in Laemmli sample buffer.

3.1.2 Protein Precipitation with Methanol Chloroform

Four hundred microliters of of methanol was added to 100 μL of protein sample and vortexed. Hundred microliters of chloroform and 300 μL of H_2O are added into the mixture following vortexing after every step. The solution was centrifuged for 1 min at 14,000 × *g*. Top aqueous layer was removed without disturbing the interface (contains proteins). Four hundred microliters of methanol was then added into bottom layer, mixed, and centrifuged as before to pellet the protein. Supernatant was removed, the pellet was left to dry, and sample buffer was added.

3.1.3 Reverse Phase Chromatography (Protein Purification)

Before every run the blank gradient (cleaning program) was performed to ensure that no protein elute from the reverse phase (RP) column. The column was washed with several CV of A buffer, followed by a gradient to 95% of buffer B for several CV, again gradient to A buffer, and finally equilibrated with several CV with A buffer. If some peaks elute from the column during cleaning program, the procedure should be repeated.

The solution containing recPrP protein after HisTrap purification was filtered (0.22 μm syringe filter) and loaded into RP column at a flow rate 1 mL/min. The column was washed with A buffer until the baseline absorbance at A280 was reached and the protein was eluted with linear gradient (till 95% of buffer B for 10 CV, flow rate 5 mL/min). Fractions of 5 mL were collected during elution step. Twenty microliters of samples from each fraction were lyophilized and resolved by SDS-PAGE. Fraction containing PrP were pooled, lyophilized, and stored at −80 °C until future use.

3.2 Fibrillization of Recombinant Mouse PrP (recMoPrP(23–231))

Fibrillization Protocol

1. Dissolve lyophilized recMoPrP(23–231) in the appropriate denaturing buffer at 10 mg/mL, aliquot, and store at −80 °C.

   Note: The protein can be kept for several months at −80 °C but for longer storage we recommend to conserve the protein as lyophilized powder.
2. Thaw recMoPrP(23–231) immediately before the preparation of the reaction buffer.

   Note: When necessary, recombinant protein can be stored at +4 °C for few hours.
3. To form amyloids with dithiothreitol incubate the recombinant proteins with 100 mM DTT at 37 °C for 1 h before the preparation of the reaction buffers.
4. Prepare the reaction buffer under sterile hood using dedicated pipette set and sterile tips. The final concentrations of variable reagents are reported in Table 1. All the fibrillization mix contains Thioflavin T 10 μM as fluorescent dye. When indicated, the final concentration of NaCl is 0.4 M.
5. Split the reaction buffer after gently mixing the 96-wells plates at final volume of 200 μL/well.

   Note: Fibrillization reactions are highly variable among replicates. To reduce the variability, at least 10 replicates for each condition of fibrillization are prepared.
6. Add 3 mm glass bead to each well to increase the kinetic of fibrillization reaction.
7. Seal the plate with sealing tape and place it on Eppendorf thermomixer.

   Note: To avoid the contact with light, wrap the plate with aluminum foil.
8. Place the thermomixer and the plate into chemical oven at 37 °C. Set up the shacking at 900 rpm, continuously.
9. Measure the ThT fluorescence signal every hour in the first phase of the fibrillization (lag phase) and every 15 min during the log phase with BMG FLUOstar Optima (444 nm excitation and 485 nm emission).

   Note: The duration of the lag phase is highly variable; we recommend following the reaction, at least, for 10–20 h. By contrast, the log phase is very fast and if you need a precise fibrillization curve the fluorescence should be check continuously until reach the *plateau*.

10. After fibrillization reaction collect the prion amyloids by ultracentrifugation at 100,000 × *g*, 1 h at 4 °C. Resuspend the amyloids in the appropriate buffer for further analysis.

4 Conclusion

We have recently performed a study for elucidating the conformational diversity of pathological recPrP amyloids and their biological activities implicated in prion conversion and propagation. In order to explore this aspect we have generated infectious materials that possess different conformational structures by using a precise mixture of common chemicals. Our methodology for prion conversion of recPrP required only purified rec full-length mouse (Mo) PrP. Neither infected brain extracts nor amplified PrP^{Sc} were used. Using atomic force microscopy (AFM) we observed that some of our preparations showed the typical amyloid shape with well-characterized fibrils while other structures appeared to be more similar to β-oligomers with spherical conformation. Mouse hypothalamic GT1 and neuroblastoma N2a cell lines were infected with these amyloid preparations to characterize their infectious properties [72]. Remarkably, a large number of amyloid preparations were able to induce the conformational conversion of endogenous PrP^{C} to a proteinase-resistant PrP conformer. In order to assess the ability of our preparation to induce prion pathology in vivo, we have inoculated these amyloids in the brain of wild-type mice. None of the animals showed any prion-like pathology (either biochemically or histologically) and were culled at the end of their life span. After PMCA, we were able to amplify and detect a PK-resistant PrP from the CNS of a group mice injected with one of our amyloid preparation. Compared to well-known prion strains, this PK-resistant PrP showed a unique biochemical feature (e.g., glycoform ratio and electrophoretical mobility). The amplified product was subsequently injected to wild-type animals which developed an aggressive prion pathology characterized by very short incubation and survival time. Interestingly, by means of PMCA we were able to detect the prion isolate in the blood (plasma fraction) of symptomatic animals at 140 days post inoculation. Immunohistochemical analyses of the brains revealed typical lesion associated with prion pathology such as spongiform changes, PrP^{Res} deposition, and astrogliosis. The most important information that we have gathered from this experiment is the possibility to generate putative infectious materials in vitro using solely recPrP, common chemicals, and controlled and well-defined biophysical and biochemical conditions. Remarkably, we were able to generate infectious prion without employing any external source of PrP^{Sc}. Surprisingly, the amyloid preparation produced in vitro maintained the same biochemical features when amplified directly from (1) the synthetic preparation (2), the infected cell lysates, and (3) from the brain of infected mouse. This work supports the protein only hypothesis previously discussed and, in this case, provides the first evidence that following a specific and defined set of chemicals is possible to achieve condition to synthetize and produce specific prion isolates characterized by unique conformation and infectious properties.

Working with natural prions is not always so simple since they are complex and heterogeneous. The synthetic ones are instead easier to control, homogeneous, and structurally defined. And yet they still show the same consequences as biological ones. These features are of fundamental importance to identify the mechanisms, which can block the pathogenic effect, in order to develop personalized treatments for each type of prion disease.

References

1. Wood JL, Lund LJ, Done SH (1992) The natural occurrence of scrapie in moufflon. Vet Rec 130(2):25–27
2. Barlow RM (1972) Transmissible mink encephalopathy: pathogenesis and nature of the aetiological agent. J Clin Pathol Suppl 6:102–109
3. Williams ES, Young S (1982) Spongiform encephalopathy of Rocky Mountain elk. J Wildl Dis 18(4):465–471
4. Wells GA, Scott AC, Johnson CT, Gunning RF, Hancock RD, Jeffrey M et al (1987) A novel progressive spongiform encephalopathy in cattle. Vet Rec 121(18):419–420
5. Wyatt JM, Pearson GR, Smerdon TN, Gruffydd-Jones TJ, Wells GA, Wilesmith JW (1991) Naturally occurring scrapie-like spongiform encephalopathy in five domestic cats. Vet Rec 129(11):233–236
6. Gajdusek DC, Gibbs CJ, Alpers M (1966) Experimental transmission of a Kuru-like syndrome to chimpanzees. Nature 209(5025):794–796
7. Gibbs CJ Jr, Gajdusek DC, Asher DM, Alpers MP, Beck E, Daniel PM et al (1968) Creutzfeldt-Jakob disease (spongiform encephalopathy): transmission to the chimpanzee. Science 161(3839):388–389
8. Masters CL, Gajdusek DC, Gibbs CJ Jr (1981) Creutzfeldt-Jakob disease virus isolations from the Gerstmann-Straussler syndrome with an analysis of the various forms of amyloid plaque deposition in the virus-induced spongiform encephalopathies. Brain 104(3):559–588
9. Medori R, Tritschler HJ, LeBlanc A, Villare F, Manetto V, Chen HY et al (1992) Fatal familial insomnia, a prion disease with a mutation at codon 178 of the prion protein gene. N Engl J Med 326(7):444–449
10. Liberski PP (2012) Historical overview of prion diseases: a view from afar. Folia Neuropathol 50(1):1–12
11. Gordon WS (1946) Louping ill, tickbome fever and scrapie. Vet Rec 47:516–520
12. Millson GC et al (1976) The physico-chemical nature of the scrapie agent. In: Kimberlin RH (ed) Slow virus diseases of animals and man. North-Holland Publishing Company, Amsterdam, pp 243–266
13. Chandler RL (1961) Encephalopathy in mice produced by inoculation with scrapie brain material. Lancet 1(7191):1378–1379
14. Gibbons RA, Hunter GD (1967) Nature of the scrapie agent. Nature 215(5105):1041–1043
15. Adams DH (1970) The nature of the scrapie agent. A review of recent progress. Pathol Biol 18(9):559–577
16. Bastian FO (2005) Spiroplasma as a candidate agent for the transmissible spongiform encephalopathies. J Neuropathol Exp Neurol 64(10):833–838
17. Liberski PP (2009) Kuru and D. Carleton Gajdusek: a close encounter. Folia Neuropathol 47(2):114–137
18. Cho HJ (1976) Is the scrapie agent a virus? Nature 262(5567):411–412
19. Alper T, Cramp WA, Haig DA, Clarke MC (1967) Does the agent of scrapie replicate without nucleic acid? Nature 214(5090): 764–766
20. Kimberlin RH (1982) Scrapie agent: prions or virinos? Nature 297(5862):107–108
21. Griffith JS (1967) Self-replication and scrapie. Nature 215(5105):1043–1044
22. Prusiner SB (1982) Novel proteinaceous infectious particles cause scrapie. Science 216(4542):136–144

23. Bolton DC, McKinley MP, Prusiner SB (1982) Identification of a protein that purifies with the scrapie prion. Science 218(4579):1309–1311
24. Gabizon R, McKinley MP, Groth D, Prusiner SB (1988) Immunoaffinity purification and neutralization of scrapie prion infectivity. Proc Natl Acad Sci U S A 85(18):6617–6621
25. Chesebro B, Race R, Wehrly K, Nishio J, Bloom M, Lechner D et al (1985) Identification of scrapie prion protein-specific mRNA in scrapie-infected and uninfected brain. Nature 315(6017):331–333
26. Stahl N, Baldwin MA, Teplow DB, Hood L, Gibson BW, Burlingame AL et al (1993) Structural studies of the scrapie prion protein using mass spectrometry and amino acid sequencing. Biochemistry 32(8):1991–2002
27. Hsiao K, Baker HF, Crow TJ, Poulter M, Owen F, Terwilliger JD et al (1989) Linkage of a prion protein missense variant to Gerstmann-Straussler syndrome. Nature 338(6213):342–345
28. Collinge J (2001) Prion diseases of humans and animals: their causes and molecular basis. Annu Rev Neurosci 24:519–550
29. Bueler H, Aguzzi A, Sailer A, Greiner RA, Autenried P, Aguet M et al (1993) Mice devoid of PrP are resistant to scrapie. Cell 73(7):1339–1347
30. Kocisko DA, Come JH, Priola SA, Chesebro B, Raymond GJ, Lansbury PT et al (1994) Cell-free formation of protease-resistant prion protein. Nature 370(6489):471–474
31. Bessen RA, Kocisko DA, Raymond GJ, Nandan S, Lansbury PT, Caughey B (1995) Non-genetic propagation of strain-specific properties of scrapie prion protein. Nature 375(6533):698–700
32. Hill AF, Antoniou M, Collinge J (1999) Protease-resistant prion protein produced in vitro lacks detectable infectivity. J Gen Virol 80(Pt 1):11–14
33. Baskakov IV, Legname G, Baldwin MA, Prusiner SB, Cohen FE (2002) Pathway complexity of prion protein assembly into amyloid. J Biol Chem 277(24):21140–21148
34. Baskakov IV, Legname G, Prusiner SB, Cohen FE (2001) Folding of prion protein to its native alpha-helical conformation is under kinetic control. J Biol Chem 276(23): 19687–19690
35. Baskakov IV, Aagaard C, Mehlhorn I, Wille H, Groth D, Baldwin MA et al (2000) Self-assembly of recombinant prion protein of 106 residues. Biochemistry 39(10):2792–2804
36. Kaneko K, Ball HL, Wille H, Zhang H, Groth D, Torchia M et al (2000) A synthetic peptide initiates Gerstmann-Straussler-Scheinker (GSS) disease in transgenic mice. J Mol Biol 295(4):997–1007
37. Hsiao KK, Groth D, Scott M, Yang SL, Serban H, Rapp D et al (1994) Serial transmission in rodents of neurodegeneration from transgenic mice expressing mutant prion protein. Proc Natl Acad Sci U S A 91(19):9126–9130
38. Nazor KE, Kuhn F, Seward T, Green M, Zwald D, Purro M et al (2005) Immunodetection of disease-associated mutant PrP, which accelerates disease in GSS transgenic mice. EMBO J 24(13):2472–2480
39. Tremblay P, Ball HL, Kaneko K, Groth D, Hegde RS, Cohen FE et al (2004) Mutant PrPSc conformers induced by a synthetic peptide and several prion strains. J Virol 78(4):2088–2099
40. Benetti F, Legname G (2009) De novo mammalian prion synthesis. Prion 3(4):213–219
41. Legname G, Baskakov IV, Nguyen HO, Riesner D, Cohen FE, DeArmond SJ et al (2004) Synthetic mammalian prions. Science 305(5684):673–676
42. Peretz D, Scott MR, Groth D, Williamson RA, Burton DR, Cohen FE et al (2001) Strain-specified relative conformational stability of the scrapie prion protein. Protein Sci 10(4):854–863
43. Colby DW, Giles K, Legname G, Wille H, Baskakov IV, DeArmond SJ et al (2009) Design and construction of diverse mammalian prion strains. Proc Natl Acad Sci USA 106(48):20417–20422
44. Collinge J, Clarke AR (2007) A general model of prion strains and their pathogenicity. Science 318(5852):930–936
45. Collinge J, Sidle KC, Meads J, Ironside J, Hill AF (1996) Molecular analysis of prion strain variation and the aetiology of 'new variant' CJD. Nature 383(6602):685–690
46. DeArmond SJ, Sanchez H, Yehiely F, Qiu Y, Ninchak-Casey A, Daggett V et al (1997) Selective neuronal targeting in prion disease. Neuron 19(6):1337–1348
47. Tuzi NL, Cancellotti E, Baybutt H, Blackford L, Bradford B, Plinston C et al (2008) Host PrP glycosylation: a major factor determining the outcome of prion infection. PLoS Biol 6(4):e100
48. McKenzie D, Kaczkowski J, Marsh R, Aiken J (1994) Amphotericin B delays both scrapie agent replication and PrP-res accumulation early in infection. J Virol 68(11):7534–7536
49. Supattapone S, Wille H, Uyechi L, Safar J, Tremblay P, Szoka FC et al (2001) Branched polyamines cure prion-infected neuroblastoma cells. J Virol 75(7):3453–3461

50. Bartz JC, Bessen RA, McKenzie D, Marsh RF, Aiken JM (2000) Adaptation and selection of prion protein strain conformations following interspecies transmission of transmissible mink encephalopathy. J Virol 74(12):5542–5547
51. Li J, Browning S, Mahal SP, Oelschlegel AM, Weissmann C (2010) Darwinian evolution of prions in cell culture. Science 327(5967): 869–872
52. Makarava N, Baskakov IV (2013) The evolution of transmissible prions: the role of deformed templating. PLoS Pathog 9(12):e1003759
53. Baskakov IV (2014) The many shades of prion strain adaptation. Prion 8(2)
54. Legname G, Nguyen HO, Baskakov IV, Cohen FE, Dearmond SJ, Prusiner SB (2005) Strain-specified characteristics of mouse synthetic prions. Proc Natl Acad Sci U S A 102(6): 2168–2173
55. Legname G, Nguyen HO, Peretz D, Cohen FE, DeArmond SJ, Prusiner SB (2006) Continuum of prion protein structures enciphers a multitude of prion isolate-specified phenotypes. Proc Natl Acad Sci U S A 103(50):19105–19110
56. Ghaemmaghami S, Watts JC, Nguyen HO, Hayashi S, DeArmond SJ, Prusiner SB (2011) Conformational transformation and selection of synthetic prion strains. J Mol Biol 413(3):527–542
57. Ghaemmaghami S, Ahn M, Lessard P, Giles K, Legname G, DeArmond SJ et al (2009) Continuous quinacrine treatment results in the formation of drug-resistant prions. PLoS Pathog 5(11):e1000673
58. Chen B, Bruce KL, Newnam GP, Gyoneva S, Romanyuk AV, Chernoff YO (2010) Genetic and epigenetic control of the efficiency and fidelity of cross-species prion transmission. Mol Microbiol 76(6):1483–1499
59. Makarava N, Kovacs GG, Bocharova O, Savtchenko R, Alexeeva I, Budka H et al (2010) Recombinant prion protein induces a new transmissible prion disease in wild-type animals. Acta Neuropathol 119(2):177–187
60. Saborio GP, Permanne B, Soto C (2001) Sensitive detection of pathological prion protein by cyclic amplification of protein misfolding. Nature 411(6839):810–813
61. Deleault NR, Harris BT, Rees JR, Supattapone S (2007) Formation of native prions from minimal components in vitro. Proc Natl Acad Sci U S A 104(23):9741–9746
62. Gonzalez-Montalban N, Lee YJ, Makarava N, Savtchenko R, Baskakov IV (2013) Changes in prion replication environment cause prion strain mutation. FASEB J 27(9):3702–3710
63. Geoghegan JC, Valdes PA, Orem NR, Deleault NR, Williamson RA, Harris BT et al (2007) Selective incorporation of polyanionic molecules into hamster prions. J Biol Chem 282(50):36341–36353
64. Wang F, Yang F, Hu Y, Wang X, Wang X, Jin C et al (2007) Lipid interaction converts prion protein to a PrPSc-like proteinase K-resistant conformation under physiological conditions. Biochemistry 46(23):7045–7053
65. Deleault NR, Walsh DJ, Piro JR, Wang F, Wang X, Ma J et al (2012) Cofactor molecules maintain infectious conformation and restrict strain properties in purified prions. Proc Natl Acad Sci U S A 109(28):E1938–E1946
66. Deleault NR, Piro JR, Walsh DJ, Wang F, Ma J, Geoghegan JC et al (2012) Isolation of phosphatidylethanolamine as a solitary cofactor for prion formation in the absence of nucleic acids. Proc Natl Acad Sci U S A 109(22): 8546–8551
67. Zhang Z, Zhang Y, Wang F, Wang X, Xu Y, Yang H et al (2013) De novo generation of infectious prions with bacterially expressed recombinant prion protein. FASEB J 27(12):4768–4775
68. Barria MA, Mukherjee A, Gonzalez-Romero D, Morales R, Soto C (2009) De novo generation of infectious prions in vitro produces a new disease phenotype. PLoS Pathog 5(5):e1000421
69. Wang F, Wang X, Yuan CG, Ma J (2010) Generating a prion with bacterially expressed recombinant prion protein. Science 327(5969):1132–1135
70. Wang F, Zhang Z, Wang X, Li J, Zha L, Yuan CG et al (2012) Genetic informational RNA is not required for recombinant prion infectivity. J Virol 86(3):1874–1876
71. Moda F, Le TN, Aulic S, Bistaffa E, Campagnani I, Virgilio T et al (2015) Synthetic prions with novel strain-specified properties. PLoS Pathog 11(12):e1005354
72. Bosque PJ, Prusiner SB (2000) Cultured cell sublines highly susceptible to prion infection. J Virol 74(9):4377–4386

Chapter 14

Biomarkers in Cerebrospinal Fluid

Joanna Gawinecka, Matthias Schmitz, and Inga Zerr

Abstract

Cerebrospinal fluid (CSF) is the main component of the brain extracellular space and participates in the exchange of many biochemical products in the central nervous system (CNS). Consequently, CSF contains a dynamic and complex mixture of proteins, which reflects physiological or pathological state of CNS. Changes in CSF proteome have been described in various neurodegenerative disorders. These alterations are also discussed to reflect pathological changes in the brain and thus contribute to a better understanding of the pathophysiology of the underlying disorder.

Proteomics offer a new methodology for the analysis of pathological changes and mechanisms occurring in neurodegenerative processes and provide a possibility of novel biomarker discovery to supplement faster, earlier, and precise diagnosis. In general, following criteria have to be applied in order to qualify a protein or a gene as a potential biomarker: the selected parameters have to be sensitive (able to detect the abnormalities in early stage of disease), specific (allow the differential diagnosis), reproducible across different laboratories, with high positive predictive value (indicate the disease when test is positive) and allow the disease monitoring as well as a potential therapy response. In Creutzfeldt–Jakob disease, two major lines of approaches have been followed aiming to detect the pathological form of prion protein (PrP^{Sc}) in various peripheral tissues on one hand, but also looking for surrogate parameters as a consequence of the underlying neurodegenerative process. While the amounts of the abnormal disease-related PrP^{Sc} in CSF and blood in human TSEs seem to be extremely low, the development of PrP^{Sc}-based biomarker was hampered by technical problems and the detection limits. On the other hand, a variety of other proteins have been investigated in the CSF and recently a variety of potential biomarkers have been reported, which contribute to the clinical diagnosis. Already established biomarkers are 14-3-3, β-amyloid, tau-protein and phosphorylated tau isoforms, S100b, as well as neuron-specific enolase (NSE). Since some of these markers display certain limitations, the search continues. The review summarizes the current knowledge of the biomarker development in prion diseases and discusses perspectives for new approaches.

Key words TSE, CJD, 14-3-3, Tau, CSF, Biomarker, Proteomics, RT QuIC

The book chapter is based on the article published in Future Neurology by Joanna Gawinecka and Inga Zerr in January 2006, Title Cerebrospinal fluid biomarkers in human prion diseases.

Pawel P. Liberski (ed.), *Prion Diseases*, Neuromethods, vol. 129,
DOI 10.1007/978-1-4939-7211-1_14,

1 Introduction

Creutzfeldt–Jakob disease (CJD) is a rare, fatal, rapidly progressive neurodegenerative disorder that belongs to the family of transmissible spongiform encephalopathies (TSE). CJD can be categorized in three groups: sporadic, genetic, and iatrogenic. The most common form is sporadic CJD (sCJD) and consists of about 85% of all cases. Approximately 10–15% of all cases account for genetic TSE (gTSE), which is associated with number of mutations in the gene encoding PrP (PRNP). The most frequent forms of gTSE are Gerstmann–Sträussler–Scheinker syndrome (GSS), fatal familial insomnia (FFI), and gCJD caused by E200K mutation ($CJD^{E200K\text{-}129M}$). Acquired forms such as Kuru, iatrogenic CJD (iCJD), and new variant CJD (vCJD) are distinguished after discovery that CJD could be accidentally transmitted [1]. In the case of iCJD, transmission mostly occurred after central nervous tissue transplants of CJD-contaminated source or cadaver-extracted hormones [2]. The variant CJD is linked with bovine spongiform encephalopathy (BSE) epidemic in Europe [3].

World Health Organization diagnostic criteria for these diseases allow the classification of patients into probable CJD or possible CJD during life. The criteria include the presence of typical clinical features such as dementia, pyramidal signs, myoclonus, cerebellar signs, a characteristic EEG pattern with periodic sharp and slow wave complexes, and, finally, the presence of 14-3-3 protein in the cerebrospinal fluid (CSF) [4]. Recently, it has been shown that MRI can strongly support the diagnosis of the disease and diagnostic criteria were amended [5]. The diagnosis of definite CJD still depends on the neuropathological examination of the brain tissue [6].

The biochemical analysis of CSF proteins provides an easily accessible and robust platform to test various markers in TSE patients. Some of them, such as 14-3-3, β-amyloid ($A\beta_{1\text{-}42}$), tau protein and its phosphorylated isoforms, S100B, or neuron-specific enolase (NSE), are already used to support the clinical diagnosis and are crucial to different TSE from other forms of rapid progressive dementia. However, recent advances in proteomic methodologies offer a huge potential for discovery of new biomarkers and protein analysis in CSF of CJD and other dementia.

In this review, our main aims are first to give a detailed overview of currently available markers, which can be used to support the diagnosis of CJD; and second to describe current proteomic methodologies that can be applied for the discovery of novel CSF targets.

1.1 Routine CSF Tests

The CSF analysis plays an important role in the clinical diagnosis of CJD to exclude an inflammatory disease on one hand, but also to confirm the clinical suspicion of CJD on the other.

The routine examination of CSF from patients with CJD or GSS usually reveals normal results. An unspecific increase in total protein, the presence of oligoclonal IgG bands, or raised cell count was observed in less than 10% of the cases [7, 8]. In general, inflammatory CSF findings in rapid progressive dementia exclude the diagnosis of a prion disease.

2 Biomarker in CSF in CJD

2.1 14-3-3 Protein

In 1986 Harrington et al. reported for the first time the presence of two 30 kDa proteins designated as 130 and 131 proteins in CSF of CJD patients [9]. Ten years later these proteins were identified as proteins belonging to 14-3-3 family [10].

14-3-3 proteins were initially described as an abundant, acidic brain proteins and their name is derived from the combination of its fraction number on DEAE-cellulose chromatography and migration position in the subsequent starch-gel electrophoresis [11]. The 14-3-3 family belongs to highly conserved, abundant neuronal proteins and consists of seven very similar isoforms (β, ε, γ, η, τ, ζ, and σ) [12, 13]. Among them, four isoforms (β, ε, γ and η) could be detected in CSF of sCJD patients [14–16]. The 14-3-3 proteins are multifunctional and through phosphorylated serine motif (RSxpSxP) bind a large number of partners including phosphatases, kinases, and transmembrane receptors [17]. The interaction with their effector proteins modulate diverse biological processes including neuronal development, cell cycle and death, cell growth control, as well as signal transduction. The binding of 14-3-3 to Raf-1 promotes the stabilization of its active conformation and therefore regulates Ras-Raf-dependent mitogenic signaling pathway [18, 19]. 14-3-3 can also interact with Cdc25 phosphatases, which play a pivotal role in cell cycle progression by activating cyclin-dependent kinases [20]. Moreover, 14-3-3 forms a complex with Bad preventing the interaction of Bad with Bcl-X_L and Bcl-2 and, thus, Bad-induced cell death [21]. 14-3-3 is also rate-limiting enzyme in the biosynthesis of neurotransmitters, such as dopamine and serotonin [22]. The involvement of 14–3-3 in CJD pathogenesis still has to be elucidated. However, 14-3-3 ζ isoform was detected in plaque-like PrP^{Sc} deposits in some forms of sCJD and 14-3-3 immunoreactivity was found in glial cells of sCJD-affected brain [23, 24].

Despite only limited knowledge about the pathology behind the elevated level of 14-3-3 in CJD, the detection of 14-3-3 protein in CSF is part of clinical diagnostic criteria for probable sCJD (Fig. 1) [25]. The large number of studies proved that in the appropriate clinical circumstances a positive 14-3-3 is highly sensitive and specific for sCJD diagnosis. The 14-3-3 detection correlated with clinical diagnosis in 85–94% (Table 1) [26–37]. However, the occurrence of 14-3-3 is influenced by both the form and the subtype of CJD, but also by the stage of disease [15, 27, 38, 39]. The 14-3-3 protein is elevated in more than 90% of probable or definite sCJD cases or gCJD (E200K and V210I mutations). Only around 50% of GSS, vCJD, and iCJD

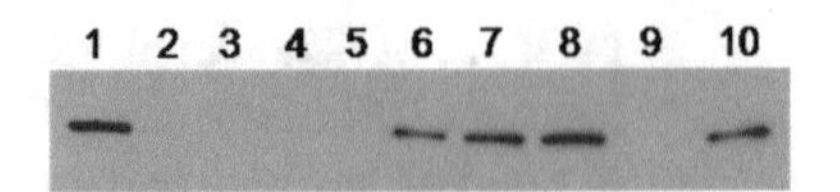

Fig. 1 Detection of 14-3-3 in CSF from patients suspected for TSE. *Lane 1*, CSF from patient with confirmed sCJD (positive control); *Lane 2*, CSF from patient suffered from other confirmed neurological disorder (negative control); *Lanes 3–10*, patients undergoing TSE diagnosis

Table 1
Reported sensitivities and specificities of 14-3-3 in the diagnosis of sCJD

Study	Method	Sensitivity (%)	Specificity in sCJD subtypes (%)						Specificity (%)
			MM1	MM2	MV1	MV2	VV1	VV2	
Castellani et al. [27]	WB	87	94	70	100	57	100	84	
Sanchez-Juan et al. [49]	WB	85	92	78	91	65	100	90	85
Collins et al. [26]	WB	88	91	61	86	71	90	95	
Gmitterova et al. [38]	ELISA (cutoff 290 pg/mL)	94	100	75	89	89	100	100	
Hsich et al. [10]	WB	96							96
Zerr et al. [32]	WB	94							93
Zerr et al. [31]	WB	94							84
Beaudry et al. [90]	WB	60							100
Lemstra et al. [30]	WB	97							97
Van Everbroeck et al. [41]	WB	100							92
		94							96
Green et al. [33]	WB	89							97
Green et al. [147]	Capture assay	82							94
Kenney et al. [37]	ELISA (cutoff 8,3 ng/mL)	93							98
Aksamit et al. [32]	sICMA (cutoff 8 ng/ mL)	61							100
	(cutoff 4 ng/ mL)	94							49
Sanchez-Juan et al. [39][a]	WB	88–91							
Pennington et al. [36][b]	WB	96							67

[a]Sensitivity measured in different disease stages
[b]Early stage of sCJD, parameters measured within 6 weeks from disease onset

Table 2
Sensitivity of 14-3-3 in different form of human TSEs (Adapted [31])

Disease	Total number of cases	14-3-3 positive	Sensitivity (%)
Definite or probable sCJD	413	376	91
Possible sCJD	127	79	62
vnCJD	11	5	45
iCJD	10	6	60
gCJD			
E200K	13	13	100
V210I	15	15	100
GSS	5	2	40
FFI	15	0	0
Controls (other neurological diseases)	392	34	–

display detectable 14-3-3 level in CSF and almost all FFI cases are negative (Table 2) [31, 40]. As far as 14-3-3 protein is not 100% specific for CJD and is a surrogate marker of the disease, it is not suited for the general screening of patients with dementia. The increased 14-3-3 concentration is found in cerebral hypoxia, inflammatory CNS disorders, intracerebral hemorrhage, and blood-contaminated CSF samples [30, 31]. However, these diseases are clinically easy to distinguish from CJD. The relevant differential diagnosis of CJD, such as Alzheimer's disease, Parkinson's disease, frontotemporal dementia, or Hashimoto encephalitis, only rarely display elevated 14-3-3 in CSF.

The 14-3-3 level in CSF is influenced by various biological parameters. The 14-3-3 test shows a lower sensitivity in CJD cases with longer disease duration, with the Met/Val heterozygosity at codon 129 of prion protein gene or type 2 of protease-resistant core of scrapie prion protein (PrP^{Sc} type 2) [27, 38–41]. The most appropriate time point for the highest test sensitivity still remains questionable. Although the test was often found positive at the onset of the first neurological symptoms, the higher sensitivity was reported in the middle stage or the late stage of the disease [29, 41–43]. Moreover, in the terminal stage of disease 14-3-3 level might be decreased in CSF, but this observation is based on case reports and might reflect extremely long disease duration [42, 44]. In sequential studies of iatrogenic CJD, 14-3-3 was rarely positive in early stage, but was always detectable after 7 months when the

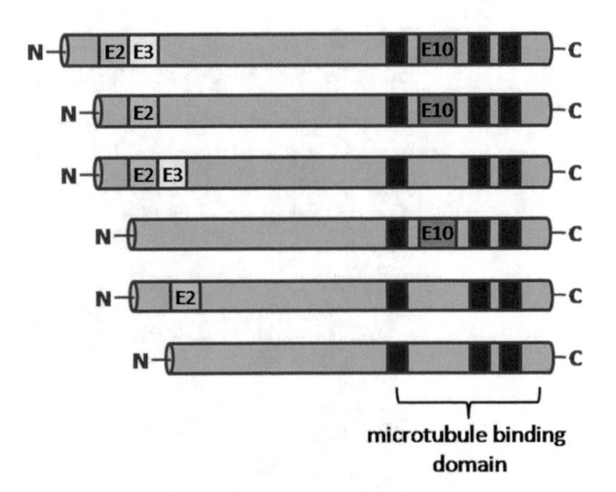

Fig. 2 Schematic representation of six tau isoforms formed by alternative mRNA splicing of a single gene. The isoforms differ in the presence or absence of inserts coded by exon 2 (E2), exon (E3), and exon 10 (E10). Microtubule binding domains within tau molecule are indicated in *bark blue*

disease aggravated and dementia appeared [28]. The substantial increase of the sensitivity and specificity of 14-3-3 test was observed when the second lumbar puncture was performed in the later stage of disease. It can strongly support diagnosis of difficult, unclear cases [39, 40]. Usually, in CJD, the increase of 14-3-3 level is observed, whereas the level returns to undetectable in CSF in the other acute neuronal disorders.

2.2 Tau Protein

Tau protein is a microtubule-associated protein, which interacts with tubulin and promotes microtubule assembly and stability; it is also involved in neurogenesis, axonal maintenance, and axonal transport. There are six different tau isoforms present in the human adult brain, which are generated by an alternative mRNA splicing from a single gene (Fig. 2) [45]. Tau is a phosphoprotein, with 79 putative serine or threonine phosphorylation sites on the longest tau isoform [46] (Fig. 2). The hyperphosphorylated tau has a reduced affinity for microtubules and reduced ability to promote their assembly [44]. In the tauopathies like Alzheimer's disease, tau detaches from microtubules and aggregates in paired helical filaments (PHFs). Tau isolated from these aggregates is found to be about four times more phosphorylated than tau isolated from nondemented individuals [47, 48]. The elevated CSF level of nonphosphorylated and phosphorylated tau is one of the AD hallmarks [49–54].

Tau concentration in CSF of CJD cases is highly increased and its quantitative analysis with generally accepted 1300 pg/mL cutoff is a good diagnostic tool for CJD. The specificity and sensitivity of tau determination was found to be very similar as for 14-3-3 test (Table 3) [33, 36, 40, 41, 55–60]. It was further observed that the highest elevation of tau concentration occurred in the middle stage

Table 3
Reported sensitivities and specificities of tau, NSE, and S100B in the diagnosis of sCJD

Study	Tau		NSE		S100B	
	Sensitivity (%)	Specificity (%)	Sensitivity (%)	Specificity (%)	Sensitivity (%)	Specificity (%)
Sanchez-Juan et al. [40]	86		73	95	82	76
Sanchez-Juan et al. [39][a]	81–91		73–94		83	
Gloeckner et al. [60]	88	89				
Goodall et al. [76]	84	91				
Pennington et al. [36][b]	98	82				
Van Everbroeck et al. [41]	87	97				
Zerr et al. [89]			80	92		
Beaudry et al. [100]			80	92	94	85
Otto et al. [75, 92]	100	95			84	91

[a]Sensitivity measured in different disease stages
[b]Early stage of sCJD, parameters measured within 6 weeks from disease onset

of the disease. The disease duration longer than 1 year as well as the Met/Val or Val/Val genotype was associated with relatively lower increase of tau level in CSF [41]. Moreover, increased tau was shown to be more sensitive for vCJD than 14-3-3, but using 500 pg/ml cutoff for tau testing (80% sensitivity) [33]. The high tau level was detected in gCJD with different type of mutations or inserts in the PRNP gene. However, similar to 14-3-3, it was not the case for GSS and FFI [61]. Concerning the phosphorylated tau isoforms in CSF of CJD, tau phosphorylated at threonine 181 (p-tau) was significantly raised in sCJD as well as in vCJD. Interestingly, tau concentration was lower in vCJD when compared to sCJD, whereas p-tau concentration was much higher in vCJD than in sCJD. It is believed that tau is released into CSF as a result of neuronal damage and rapid progress of sCJD, which could explain the higher mean concentration of tau in sCJD. From

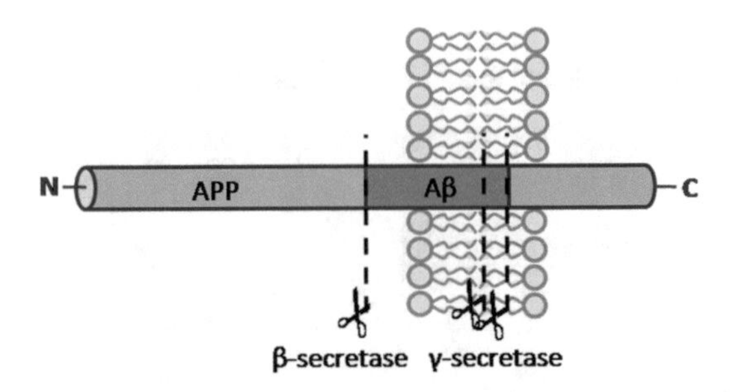

Fig. 3 Schematic representation of β-amyloids formation from amyloid precursor protein (APP). The cleavage of APP by β- and γ-secretase leads to generation of two major Aβ peptides: $A\beta_{1-40}$ and slightly larger $A\beta_{1-42}$

the other hand side, the raised level of p-tau could suggest yet unexplained factors which influence tau phosphorylation in vCJD [62]. Moreover, the ratio between p-tau and total tau was found to be very helpful for the discrimination of CJD patients from the other kind of dementia [63].

2.3 β-Amyloid $A\beta_{1-42}$

β-Amyloids (Aβs) are generated by a sequential cleavage of amyloid precursor protein (APP) by β- and γ-secretase (Fig. 3). Aβs are small hydrophobic peptides existing mainly in two lengths: $A\beta_{1-40}$ and $A\beta_{1-42}$. It was initially assumed that the production of Aβs occurs only under pathological conditions. This concept was confuted when Aβ was shown to be constitutively released from APP and secreted to blood and CSF [64–66]. Aβ peptides contribute in the oxidative modification of proteins and lipids and are also involved in the inhibition of several glial and neuronal transport systems including glutamate and glucose transporters, GTP-coupled transmembrane signaling proteins, as well as ion motive ATPases [67–69]. In AD, Aβ peptides are involved in pathological processes and accumulate in the brain as amyloid or senile plaques [70, 71]. There is clear correlation between Aβ levels in CSF and plaque depositions in the brain coupled with the concept of casual involvement of APP and Aβs in the pathogenesis of AD. The concentration of Aβs is thought to reflect disease-associated changes and is widely applied as a diagnostic biomarker of AD [72–75].

The reduced level of $A\beta_{1-42}$ was also reported in other dementia including CJD, and increased $A\beta_{1-42}$ concentration was found in control patients with false-positive 14-3-3 test [41, 60, 76–78].

2.4 S-100B and Neuron-Specific Enolase (NSE)

S100B is a calcium-binding low molecular weight protein predominantly produced in astrocytes and exists as a disulfide-linked homodimer [79, 80]. S100B modulates the integrity of the cytoskeleton via the inhibition of the microtubule and type III intermediate filament assembly [81, 82]. Nanomolar levels of S100B protect against the cell death and enhance neuronal survival [83, 84]. From the other hand side, at micromolar concentration, S100B may have a neurotoxic effect caused by the interaction with RAGE and the stimulation of increased production of reactive oxygen species (ROS) [85]. S100B can also induce the upregulation of expression of several genes implicated in apoptosis such as c-fos, c-jun, bax, and bcl-x [86].

Neuron-specific enolase (NSE) is a glycolytic enzyme which is mainly found in the cytoplasm of neurons and cells of neuroendocrine origin [87, 88]. NSE and S100B are known markers of neuronal and astroglial cell death, respectively [89].

Due to their comparable low sensitivity and specificity, both NSE and S100B displayed limited usefulness in supporting the diagnosis of human TSE (Table 3) [40, 42, 90–95]. The S100B protein was elevated only in around 80% cases of CJD including sporadic, genetic, iatrogenic, and variant form. The increase in concentration of NSE proteins was observed in approximately 70%, 60%, and 50% cases of sCJD, gCJD, and vCJD, respectively [33, 40, 76].

2.5 Other Proposed Diagnostic Marker

Besides common TSE markers, several other proteins have been proposed as possibly useful in the diagnosis of the human TSE. However, they were tested only in small subsets of patients and so far remain irrelevant for the standard diagnostic procedure. For instance, the cystatin C, LDH-1, GFAP, H-FABP, and prostaglandin E(2) were reported to be elevated in CSF of CJD-affected patients [96–100]. Moreover, the pro-inflammatory factor TGF-β was found to be significantly reduced, whereas the anti-inflammatory IL-4, IL-8, and IL-10 were found to be elevated in CJD [101]. Furthermore, the CJD-specific alterations in the glycosylation pattern of acetylcholinesterase were reported [102]. All the suggested candidates for CJD-markers are presented in Table 4.

2.6 Prion Protein

The prion protein (PrP) is an evolutionary highly conserved glycoprotein, which is predominantly located in lipid rafts on the outer surface of neuronal and glial cells, lymphocytes B, as well as platelets [103–108]. The cellular PrP (PrP^C) undergoes endocytosis and recycling stimulated by copper ions [109–111]. The physiological role of PrP^C is not fully understood. However, many evidences suggest that PrP is a multifunctional protein involved in several cellular processes, such as the regulation of cell death, copper metabolism, protection against oxidative stress, and the modulation of signal transduction [109, 112–119]. In prion disease, PrP^C is posttranslationally misfolded into a protease-resistant and aggregating scrapie isoform (PrP^{Sc}) [120]. According to the “protein only” hypothesis, an interaction between the endogenous PrP and its scrapie form is sufficient to cause the template-driven conversion of PrP^C to PrP^{Sc} [121].

Table 4
Other reported CSF candidates for marker of human prion diseases

Study	Proposed CJD-marker	Level in CSF
Sanchez et al. [98] Piubelli et al. [140]	Cystatin C	Elevated
Choe et al. [143][a]	ApoE	Elevated
Guillaume et al. [100]	H-FABP	Elevated
Schmidt [99]	LDH-1	Elevated
Jesse et al. [96]	GFAP	Elevated
Stoeck et al. [101]	TGF-β	Reduced
Stoeck et al. [101]	IL-4, IL-8, and IL-10	Elevated
Silveyra et al. [102]	Acetylcholinesterase	Altered glycosylation pattern
Gloeckner et al. [78]	Transthyretin	Normal
Bleich et al. [148]	Homocysteine	Normal
Mollenhauer et al. [149]	Alpha-synuclein	Elevated
Kettlun et al. [150]	Matrix metalloproteinase	Elevated
Holsinger et al. [151]	BACE1	Increased activity
Cartier et al. [152]	Fibronectin, Thrombospondin, Heparan sulfate proteoglycan	Elevated
Manaka et al. [153]	Ubiquitin	Elevated
Alberti et al. [154]	Neurofilament heavy subunit	Elevated
Albrecht et al. [155]	Beta-nerve growth factor	Elevated
Minghetti et al. [97]	Prostaglandin E(2)	Elevated

[a]More specific for vCJD

The ultimate diagnostic tool for human TSEs would be to identify the scrapie agent rather than a surrogate marker in CSF or plasma. However, so far it has been impossible to detect differences either in the protease-resistance or the conformation of PrP in CJD- and non-CJD CSF by both immunoblotting and conformation-dependent assay [122]. Moreover, even using ultrasensitive method called Scanning for Intense Fluorescent Targets (SIFT) with the femtomolar detection range, PrP aggregates in CSF could be detected only in 20% sCJD cases [123]. The detection of the pathological PrP is hampered by potentially its very low amounts or it aggregates in CSF and blood. Thus, the development of a diagnostic test requires either a novel and extremely sensitive detection techniques, additional PrP^{Sc} isolation steps, filtration assays, or amplification methods such as protein misfolding cycling amplification (PMCA) [124, 125].

Nevertheless, it should be kept in mind that there are cases of disease transmission via blood in animals and humans (such as very well documented vCJD transmission via blood transfusion), which strongly suggests that blood may contain infectious agent and thus be a platform for a test development in CJD [126–129].

Whereas most research on biomarker discovery in CJD was directed towards pathological PrP, only limited data are available on PrP^C in CJD. For the first time, Tagliavini et al. showed the presence of soluble form of PrP in human CSF [130]. A lower PrP^C concentration was found in CJD, but also in patients suffering from other kind of neurodegenerative disorders, when compared to age-related, healthy controls. However, due to poor discrimination between CJD and other neurodegenerative conditions the analysis of the physiological PrP in CSF cannot be used as a diagnostic test [131–133].

3 Aggregation Assays (RT QuIC)

The adaptation of in vitro amplification systems for the detection of aggregated PrPSc in human CSF was an innovation for pre-mortem diagnostic because it enables for the first time to study the conversion processes of PrPSc in vitro. Real-time quaking-induced conversion (RT-QuIC) analysis uses recombinant prion protein (recPrP) as a substrate to amplify small amounts of a PrPSc seed in human CSF to a detectable limit (in principle comparable to a PCR for proteins). Converted and aggregated PrP can be monitored in real time by fluorescence dye analysis using a fluorescent plate reader. This method has already been applied to brain tissue or cerebrospinal fluid (CSF) and has already been established in various laboratories. It exhibited a specificity of 99% and a sensitivity of 85% [156–160]. Furthermore, a high reproducibility and stability of RT-QuIC across various CSF storage conditions, such as short-term CSF storage at different temperatures, long-term storage, repeated freezing, and thawing cycles, had already been demonstrated [159, 160].

The technique bears a huge diagnostic and analytical potential. The seeding activity in CSF depends, among other factors, on the type of the disease (genetic versus sporadic). Further studies are urgently needed to determine the potential of the technique for strain-specific disease diagnosis (Fig. 4).

4 Proteomics and Biomarker Discovery

Emerging technologies in the proteomic field offer a great promise for the discovery of new biomarkers with improved sensitivity and

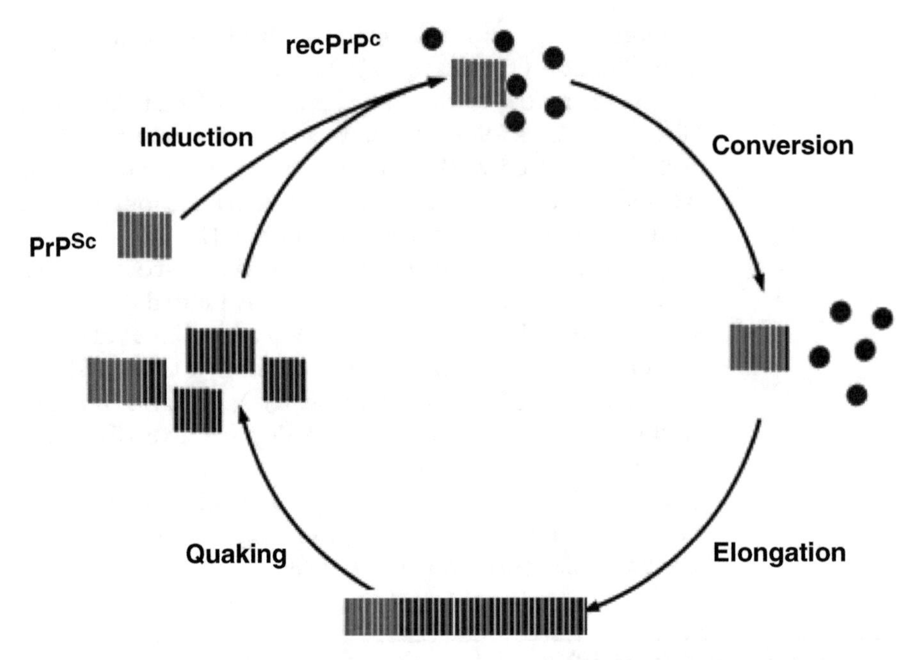

Fig. 4 Schematic diagram of the RT-QuIC assay. A PrP^{Sc} seed and recPrPC-substrate are mixed and incubated, inducing the conversion and incorporation of the substrate molecules producing misfolded β-strand-enriched PrP polymers. Cycles of shaking and incubation of β-strand-enriched PrP polymers breaks and generates more seeds inducing an exponential amplification reaction

specificity. The fundamental goal of biomarker discovery is to find a distinctive molecular signal with clear-cut clinical value for fast and an efficient diagnostics, an improved patient stratification, or a better monitoring of therapeutic intervention. The key role played by proteins in both disease etiology and treatment make them prime biomarker candidates.

Proteomic technology combine high-resolution separation techniques applied to complex mixture of proteins with identification methods such as mass spectrometry (MS). It is generally accepted that none of the existing separation-identification methodologies can give a full account of the protein composition in a complex mixture such as biological body fluids or tissue extracts. However, this limitation has not prevented the application of current proteomic technology to provide valuable information on protein state which can be correlated with disease or health state. At present, two different approaches are used in the search for a biomarker. The first one is the direct search in the body fluids where the concentration of marker is expected to be relatively low and nowadays detection methods do not allow identifying it in most of the cases. The second one is the search for a biomarker in affected tissue where marker is presumably present at much higher concentration. However, a risk emerges that a biomarker candidate cannot be later detected in peripheral fluids such as blood or urine [134].

Since CSF is in close proximity to the brain tissue, it reflects physiological and pathological processes occurring in the central nervous system (CNS). It can serve as rich source of candidates for biomarker of neurodegenerative disorders. However, the first problem to overcome when choosing CSF as analyzed protein mixture is the high dynamic range of protein abundances. The estimated range of protein concentrations in CSF spans 12 orders of magnitude [135]. Moreover, the high abundant proteins such as albumin, immunoglobulins, transferrin, haptoglobin, and several others consist up to 90–95% of total protein concentration. Therefore, the depletion of the well-known and characterized high abundant proteins from CSF is an indispensible step in the direct approach for the biomarker discovery.

In tissue-based approach, the enrichment of target protein pool can be achieved by either laser capture microdissection (LCM) or subcellular fractionation. LCM is the most specific enrichment of cells or tissue section involved in disease process. The subcellular fractionation by differential centrifugation steps allows analyzing proteome of particular organelles such as mitochondria or plasma membrane.

The standard workflow in biomarker discovery can be broken down into following phases: (1) the isolation and separation of the protein mixture; (2) the identification of biomarker candidates; (3) and the validation of the biomarker specificity and sensitivity (Fig. 5).

In many proteomic studies, 2D gel electrophoresis (2-DE) is still a method of choice for high-resolution protein separation. 2-DE applied two distinct properties to resolve proteins. In the first dimension proteins are separated according to their net charge and according to their molecular weight in the second dimension. This technique is laborious, not-automate and suffered from some limitations. For instance, several classes of proteins including very basic or acidic, small and hydrophobic cannot be efficiently separated by 2-DE. An exciting advance in 2-DE technology is the fluorescence difference gel electrophoresis (DIGE) [136], which utilizes fluorescent tagging of two protein samples and an internal standard with three different dyes. The tagged proteins are resolved on the same 2-D gel and postrun imaging is used to create three images. The introducing of DIGE efficiently decreased gel-to-gel variations and significantly increased accuracy and reproducibility of studies based on 2-DE. The limitations of 2-DE inspired a number of approaches to bypass gel electrophoresis. The multidimensional liquid chromatography coupled with tandem MS (MudPIT) is powerful alternative for protein separation [137]. In this approach, not intact proteins, but peptides generated by specific enzymes such as trypsin are analyzed. Therefore, even very large, hydrophobic, very basic or acidic proteins can be sufficiently investigated.

MS provides protein structural information like amino acid sequence and peptide mass for the protein identification by

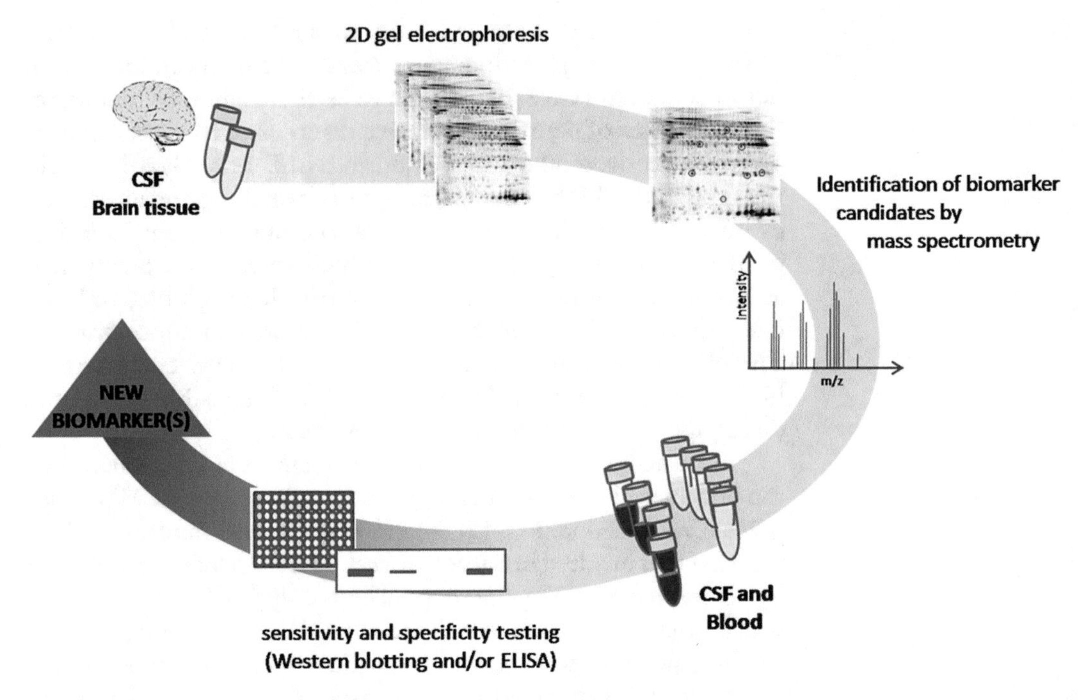

Fig. 5 Schematic workflow of biomarker discovery using gel-based proteomic approach

searching protein databases such as ProFound, MASCOT, or Findmod. It can also enable information about posttranslational modifications of proteins.

It has to be taken under consideration that the assumption that a single protein can specify a disease and serve as a biomarker may be an oversimplification and an unrealistic goal. For this reason, the application of a multi-parameter biomarker would be more effective and specific [138].

The gel-based proteomic approach has been already applied for the investigation of CSF proteome changes in CJD and the searching for novel biomarkers [9, 74, 139–143]. Actually, the first biomarker reported in CJD were proteins 130 and 131, later known as 14-3-3, and they were detected by 2D gel electrophoresis [9]. Piubelli et al. [140] analyzed native CSF from MM1 subtype of sCJD and found seven upregulated and six downregulated CJD-associated proteins, among them ubiquitin, gelsolin, apolipoproteins, and transferrin. However, they are also altered in AD and thus seem to be rather involved in general neurodegenerative process [144–146]. Only Cystatin C appears to be more specific for CJD and was already previously proposed as potential CJD biomarker [97]. Then Cepek et al. [141] compared CSF patterns between CJD, AD, and controls and found 5 protein spots only present in CJD, but without performing further characterization of these proteins. Finally, Brechlin et al. [142] applied DIGE technology to investigate CSF in CJD. Using fluorescent labeling, depleting albumin and IgG and including other dementia as controls significantly increased specificity of obtained results. Unfortunately, neither specific nor promising candidate for biomarker was detected in these studies. Choe et al. [143] compared native CSF from sCJD and vCJD to CSF from other dementia and found seven protein spots with different abundance in two CJD forms. Among them, ApoE showed significantly higher level in vCJD comparing to sCJD. In conclusion, the gel-based proteomic approach has been proven to be capable to identify biomarkers and further studies involving appropriate controls are needed to fully explore this option in CJD and other dementia.

Appendix: Protocol for 14-3-3 Detection by ELISA

The measurement of 14-3-3 protein gamma in CSF was performed by using the CircuLex 14-3-3 Gamma ELISA Kit (BIOZOL Diagnostica Vertrieb GmbH, Echin, Germany).

Materials

1. CircuLex 14-3-3 Gamma ELISA Kit (BIOZOL Diagnostica Vertrieb GmbH, Echin, Germany).
2. 450 nm filter and spectrophotometric microplate reader for example 1420 Multilabel Counter Victor 2 (PerkinElmer, Massachusetts, USA)
3. Microplate washer or multi-channel pipette

Reagent Setup

CSF sample preparation. Prepare CSF sample at RT. Centrifuge native CSF at 720 × *g* for 15 min to remove insoluble impurities, using a centrifuge 5415C (or an equivalent). Remove supernatant and discard pellet. Transfer supernatant within 1.5 mL tubes and keep it on ice until usage. Before use dilute CSF samples 1:5 using the dilution buffer from the kit.

Solutions

Wash Buffer: Prepare 1× Wash Buffer by adding 100 mL of 10× Buffer concentrate to 900 mL of deionized (distilled) water.

Standard preparation: Reconstitute 14-3-3 standard with 1 mL of sample/standard diluent buffer to 64,000 AU/mL (master standard). Use the master mix to produce as dilution series. Mix each tube before next transfer. Prepare standard 1 by diluting 150 μL master mix in 450 mL dilution buffer. Prepare standard 2 by diluting 300 μL of standard 1 in 300 μL dilution buffer. Standard 3 by diluting 300 μL of standard 2 in 300 μL dilution buffer and so on until standard 7. Standard 8 is a blank control consisting of dilution buffer.

Antibody dilutions: Prepare a working solution for 1× 14-3-3 gamma detection antibody by 100 fold diluting the 100× antibody with detection antibody dilution buffer, e.g. 10 μL antibody in 990 μL diluent.

Prepare 1× HRP conjugated Anti-IgG by 100-fold diluting the 100× HRP conjugated Anti-IgG antibody with conjugate buffer in the same manner.

Assay Procedure

1. Pipette 100 μL of standard 1–7 and blank control into the appropriated wells.
2. Pipette 100 μL of samples in duplicates into the appropriated wells.
3. Incubate the wells for 1 h at 25 °C.
4. Wash four times in 1× Wash buffer
5. Add 100 μL of 1× primary antibody (14-3-3 gamma) into each well and incubate for 1 h at 25 °C.
6. Wash four times in 1× Wash buffer
7. Add 100 μL of HRP conjugated anti-IgG into each well and incubate for 1 h at 25 °C.
8. Wash four times in 1× Wash buffer
9. Add 100 μL substrate reagent into each and incubate the wells for 10–30 min at 25 °C.
10. Add 100 μL Stop solution to each well.
11. Measure the absorbance by using a spectrophotometric microplate reader at wavelength 450 nm. Wells must be read within 30 min of adding the Stop solution.

References

1. Will RG, Matthews WB (1982) Evidence for case-to-case transmission of Creutzfeldt-Jakob disease. J Neurol Neurosurg Psychiatry 45(3):235–238
2. Lang CJ, Heckmann JG, Neundorfer B (1998) Creutzfeldt-Jakob disease via dural and corneal transplants. J Neurol Sci 160(2):128–139
3. Will RG, Ironside JW, Zeidler M, Cousens SN, Estibeiro K, Alperovitch A, Poser S, Pocchiari M, Hofman A, Smith PG (1996) A new variant of Creutzfeldt-Jakob disease in the UK. Lancet 347(9006):921–925
4. WHO (1999) WHO manual for strengthening diagnosis and surveillance of Creutzfeldt-Jakob disease. World Health Organisation Communicable Disease Surveillance and Response
5. Meissner B, Kallenberg K, Sanchez-Juan P, Collie D, Summers DM, Almonti S, Collins SJ, Smith P, Cras P, Jansen GH, Brandel JP, Coulthart MB, Roberts H, Van Everbroeck B, Galanaud D, Mellina V, Will RG, Zerr I (2009) MRI lesion profiles in sporadic Creutzfeldt-Jakob disease. Neurology 72(23):1994–2001
6. Kretzschmar HA, Ironside JW, DeArmond SJ, Tateishi J (1996) Diagnostic criteria for sporadic Creutzfeldt-Jakob disease. Arch Neurol 53(9):913–920
7. Jacobi C, Arlt S, Reiber H, Westner I, Kretzschmar HA, Poser S, Zerr I (2005) Immunoglobulins and virus-specific antibodies in patients with Creutzfeldt-Jakob disease. Acta Neurol Scand 111(3):185–190
8. Green A, Sanchez-Juan P, Ladogana A, Cuadrado-Corrales N, Sánchez-Valle R, Mitrová E, Stoeck K, Sklaviadis T, Kulczycki J, Heinemann U, Hess K, Slivarichová D, Saiz A, Calero M, Mellina V, Knight R, van Duijn CM, Zerr I (2007) CSF analysis in patients with sporadic CJD and other transmissible spongiform encephalopathies. Eur J Neurol 14(2):121–124
9. Harrington MG, Merril CR, Asher DM, Gajdusek DC (1986) Abnormal proteins in the cerebrospinal fluid of patients with Creutzfeldt-Jakob disease. N Engl J Med 315(5):279–283
10. Hsich G, Kenney K, Gibbs CJ Jr, Lee KH, Harrington MG (1996) The 14-3-3 brain protein in cerebrospinal fluid as a marker for

transmissible spongifrom encephalopathies. N Engl J Med 335(13):924–930
11. Moore BWPV (1967) Specific acidic proteins of the nervous system. In: Carslon FD (ed) Physiological and biochemical aspects of nervous system integration. Prentice-Hall, Englewood Cliffs, NJ, pp 343–359
12. Boston PF, Jackson P, Thompson RJ (1982) Human 14-3-3 protein: radioimmunoassay, tissue distribution, and cerebrospinal fluid levels in patients with neurological disorders. J Neurochem 38:1475–1482
13. Ichimura T, Isobe T, Okuyama T, Takahashi N, Araki K, Kuwano R, Takahashi Y (1988) Molecular cloning of cDNA coding for brain-specific 14-3-3 protein, a protein kinase-dependent activator of tyrosine and tryptophan hydroxylases. PNAS 85:7084–7088
14. Wiltfang J, Otto M, Baxter HC, Bodemer M, Steinacker P, Bahn E, Zerr I, Kornhuber J, Kretzschmar HA, Poser S et al (1999) Isoform pattern of 14-3-3 proteins in the cerebrospinal fluid of patients with Creutzfeldt-Jakob disease. J Neurochem 73(6):2485–2490
15. Shiga Y, Wakabayashi H, Miyazawa K, Kido H, Itoyama Y (2006) 14-3-3 protein levels and isoform patterns in the cerebrospinal fluid of Creutzfeldt-Jakob disease patients in the progressive and terminal stages. J Clin Neurosci 13(6):661–665
16. Ries L, Melbert D, Krapcho M, Mariotto A, Miller BA, Feuer EJ, Clegg L, Horner MJ, Howlader N, Eisner MP, Reichman M, Edwards BK (eds) (2007) SEER Cancer Statistics Review, 1975–2004. National Cancer Institute, http://seer.cancer.gov/csr/1975_2004/, based on November 2006 SEER data submission, posted to the SEER web site: SEER Cancer Statistics Review, 1975–2004
17. Aitken A, Collinge DB, Van Heusden BP, Isobe T, Roseboom PH, Rosenfeld G, Soll J (1992) 14-3-3 proteins: a highly conserved, widespread family of eukaryotic proteins. Trends Biochem Sci 17:498–501
18. Beimling P, Niehof M, Radziwill G, Moelling K (1994) The aminoterminus of c-Raf-1 binds a protein kinase phosphorylating Ser259. Biochem Biophys Res Commun 204(2):841–848
19. Muslin AJ, Tanner J, Allen PM, Shaw AS (1996) Interaction of 14-3-3 with signaling proteins is mediated by the recognition of phosphoserine. Cell 84(6):889–897
20. Conklin DS, Galaktionov K, Beach D (1995) 14-3-3 proteins associate with cdc25 phosphatases. PNAS 92(17):7892–7896
21. Zha J, Harada H, Yang E, Jockel J, Korsmeyer SJ (1996) Serine phosphorylation of death agonist BAD in response to survival factor results in binding to 14-3-3 not BCL-X(L). Cell 87:619–628
22. Ichimura T, Uchiyama J, Kunihiro O, Ito M, Horigome T, Omata S, Shinkai F, Kaji H, Isobe T (1995) Identification of the site of interaction of the 14-3-3 protein with phosphorylated tryptophan hydroxylase. J Biol Chem 270(48):28515–28518
23. Richard M, Biacabe AG, Streichenberger N, Ironside JW, Mohr M, Kopp N, Perret-Liaudet A (2003) Immunohistochemical localization of 14.3.3 zeta protein in amyloid plaques in human spongiform encephalopathies. Acta Neuropathol (Berl) 105(3):296–302
24. Kawamoto Y, Akiguchi I, Jarius C, Budka H (2004) Enhanced expression of 14-3-3 proteins in reactive astrocytes in Creutzfeldt-Jakob disease brains. Acta Neuropathol (Berl) 108:302–308
25. WHO (1998) Human transmissible spongiform encephalopathies. Wkly Epidemiol Rec 47:361–365
26. Collins SJ, Sanchez-Juan P, Masters CL, Klug GM, van Duijn C, Poleggi A, Pocchiari M, Almonti S, Cuadrado-Corrales N, de Pedro-Cuesta J et al (2006) Determinants of diagnostic investigation results across the clinical spectrum of sporadic CJD. Brain 129:2278–2287
27. Castellani RJ, Colucci M, Xie Z, Zou W, Li C, Parchi P, Capellari S, Pastore M, Rahbar MH, Chen SG et al (2004) Sensitivity of 14-3-3 protein test varies in subtypes of sporadic Creutzfeldt-Jakob disease. Neurology 63:436–442
28. Brandel JP, Peoc'h K, Beaudry P, Welaratne A, Bottos C, Agid Y, Laplanche JL (2001) 14-3-3 protein cerebrospinal fluid detection in human growth hormone-treated Creutzfeldt-Jakob disease patients. Ann Neurol 49(2):257–260
29. Zerr I, Bodemer M, Gefeller O, Otto M, Poser S, Wiltfang J, Windl O, Kretzschmar HA, Weber T (1998) Detection of 14-3-3 protein in the cerebrospinal fluid supports the diagnosis of Creutzfeldt-Jakob disease. Ann Neurol 43(1):32–40
30. Lemstra AW, van Meegen MT, Vreyling JP, Meijerink PH, Jansen GH, Bulk S, Baas F, van Gool WA (2000) 14-3-3 testing in diagnosing Creutzfeldt-Jakob disease: a prospective study in 112 patients. Neurology 55(4):514–516
31. Zerr I, Pocchiari M, Collins S, Brandel JP, de Pedro CJ, Knight RSG, Bernheimer H, Cardone F, Delasnerie-Lauprêtre N, Cuadrado Corrales N et al (2000) Analysis of EEG and CSF 14-3-3 proteins as aids to the

diagnosis of Creutzfeldt-Jakob disease. Neurology 55:811–815
32. Zerr I, Bodemer M, Westermann R, Schröter A, Jacobi C, Arlt S, Otto M, Poser S (2000) 14-3-3 proteins in neurological disorders. J Neurol 247(Suppl 3):III/14
33. Green AJ, Thompson EJ, Stewart GE, Zeidler M, McKenzie JM, MacLeod M-A, Ironside JW, Will RG, Knight RS (2001) Use of 14-3-3 and other brain-specific proteins in CSF in the diagnosis of variant Creutzfeldt-Jakob disease. J Neurol Neurosurg Psychiatry 70:744–748
34. Aksamit AJ Jr, Preissner CM, Homburger HA (2001) Quantitation of 14-3-3 and neuron-specific enolase proteins in CSF in Creutzfeldt-Jakob disease. Neurology 57(4):728–730
35. Sanchez-Valle R, Saiz A, Graus F (2002) 14-3-3 protein isoforms and atypical patterns of the 14-3-3 assay in the diagnosis of Creutzfeldt-Jakob disease. Neurosci Lett 320:69–72
36. Pennington C, Chohan G, Mackenzie J, Andrews M, Will R, Knight R, Green A (2009) The role of cerebrospinal fluid proteins as early diagnostic markers for sporadic Creutzfeldt-Jakob disease. Neurosci Lett 455(1):56–59
37. Kenney K, Brechtel C, Takahashi H, Kurohara K, Anderson P, Gibbs CJ Jr (2000) An enzyme-linked immunosorbent assay to quantify 14-3-3 proteins in the cerebrospinal fluid of suspected Creutzfeldt-Jakob disease patients. Ann Neurol 48(3):395–398
38. Gmitterová K, Heinemann U, Bodemer M, Krasnianski A, Meissner B, Kretzschmar HA, Zerr I (2009) 14-3-3 CSF levels in sporadic Creutzfeldt-Jakob disease differ across molecular subtypes. Neurobiol Aging 30(11):1842–1850
39. Sanchez-Juan P, Sanchez-Valle R, Green A, Ladogana A, Cuadrado-Corrales N, Mitrová E, Stoeck K, Sklaviadis T, Kulczycki J, Hess K, Krasnianski A, Equestre M, Slivarichová D, Saiz A, Calero M, Pocchiari M, Knight R, van Duijn CM, Zerr I (2007) Influence of timing on CSF tests value for Creutzfeldt-Jakob disease diagnosis. J Neurol 254(7):901–906
40. Sanchez-Juan P, Green A, Ladogana A, Cuadrado-Corrales N, Sanchez-Valle R, Mitrova E, Stoeck K, Sklaviadis T, Kulczycki J, Hess K et al (2006) Cerebrospinal fluid tests in the differential diagnosis of CJD. Neurology 67(4):637–643
41. Van Everbroeck B, Quoilin S, Boons J, Martin JJ, Cras P (2003) A prospective study of CSF markers in 250 patients with possible Creutzfeldt-Jakob disease. J Neurol Neurosurg Psychiatry 74:1210–1214
42. Jimi T, Wakayama Y, Shibuya S, Nakata H, Tomaru T, Takahashi Y, Kosaka K, Asano T, Kato K (1992) High levels of nervous system-specific proteins in cerebrospinal fluid in patients with early stage Creutzfeldt-Jakob disease. Clin Chim Acta 211(1–2):37–46
43. Giraud P, Biacabe AG, Chazot G, Later R, Joyeuy O, Moene Y, Perret-Liaudet A (2002) Increased detection of 14-3-3 protein in cerebrospinal fluid in sporadic Creutzfeldt-Jakob disease during the disease course. Eur Neurol 48:218–221
44. Mollenhauer B, Serafin S, Zerr I, Steinhoff BJ, Otto M, Scherer M, Schul-Schaeffer W, Poser S (2003) Diagnostic problems during late course in Creutzfeldt-Jakob disease. J Neurol 250:629–630
45. Goedert M, Spillantin MG, Potier MC, Ulrich J, Crowther RA (1989) Cloning and sequencing of the cDNA encoding an isoform of microtubule-associated protein tau containing four tandem repeats: differential expression of tau protein mRNAs in human brain. EMBO J 8(2):393–399
46. Lindwall G, Colle JG (1984) Phosphorylation affects the ability of tau protein to promote microtubule assembly. J Biol Chem 259(8):5301–5305
47. Köpke E, Tung Y, Shaikh S, Alonso A, Iqbal K, Grundke-Iqbal I (1993) Microtubule-associated protein tau: abnormal phosphorylation of a non-paired helical filament pool in Alzheimer's disease. J Biol Chem 268:24374–24384
48. Alonso A, Zaidi T, Novak M, Grundke-Iqbal I, Iqbal K (2001) Hyperphosphorylation induces self-assembly of tau into tangles of paired helical filaments/straight filaments. PNAS 98(12):6923–6928
49. Andreasen N, Minthon L, Clarberg A, Davidson P, Gottfries J, Vanmechelen E, Vanderstichele H, Winblad B, Blennow K (1999) Sensitivity, specificity and stability of CSF-tau in AD in a community-based patient sample. Neurology 7:1488–1494
50. Galasko D, Clark C, Chang L, Miller B, Green RC, Motter R, Seubert P (1997) Assessment of CSF levels of tau protein in mildly demented patients with Alzheimer's disease. Neurology 48(3):632–635
51. Arai H, Morikawa Y, Higuchi M, Matsui T, Clark CM, Miura M, Machida N, Lee VM, Trojanowski JQ, Sasaki H (1997) Cerebrospinal fluid tau levels in neurodegenerative diseases with distinct tau-related pathology. Biochem Biophys Res Commun 236:262–264
52. Mecocci P, Cherubini A, Bregnocchi M, Chionne F, Cecchetti R, Lowenthal DT,

Senin U (1998) Tau protein in cerebrospinal fluid: a new diagnostic and prognostic marker in Alzheimer disease? Alzheimer Dis Assoc Disord 12:211–214
53. Ishiguro KOH, Arai H, Yamaguchi H, Urakami K, Park JM, Sato K, Kohno H, Imahori K (1999) Phosphorylated tau in human cerebrospinal fluid is a diagnostic marker for Alzheimer's disease. Neurosci Lett 270(3):91–94
54. Itoh N, Arai K, Urakami K, Ishiguro K, Ohno H, Kohno H, Hampel H, Bürger K, Wiltfang J, Otto M et al (2000) Clinical utility of quantification of CSF phosphorylated tau in the antemortem diagnosis of Alzheimer's disease. Ann Neurol 50:1506
55. Kapaki E, Kilidireas K, Paraskevas GP, Michalopoulou M, Patsouris E (2001) Highly increased CSF tau protein and decreased beta-amyloid (1-42) in sporadic CJD: a discrimination from Alzheimer's disease? J Neurol Neurosurg Psychiatry 71(3):401–403
56. Otto M, Wiltfang J, Cepek L, Neumann M, Mollenhauer B, Steinacker P, Ciesielczyk B, Schulz-Schaeffer W, Kretzschmar HA, Poser S (2002) Tau protein and 14-3-3 protein in the differential diagnosis of Creutzfeldt-Jakob disease. Neurology 58:192–197
57. Satoh K, Shirabe S, Tsujino A, Eguchi H, Motomura M, Honda H, Tomita I, Satoh A, Tsujihata M, Matsuo H, Nakagawa M, Eguchi K (2007) Total tau protein in cerebrospinal fluid and diffusion-weighted MRI as an early diagnostic marker for Creutzfeldt-Jakob disease. Dement Geriatr Cogn Disord 24(3):207–212
58. Buerger K, Otto M, Teipel SJ, Zinkowski R, Blennow K, DeBernardis J, Kerkman D, Schroder J, Schonknecht P, Cepek L et al (2006) Dissociation between CSF total tau and tau protein phosphorylated at threonine 231 in Creutzfedt-Jakob disease. Neurobiol Aging 27(1):10–15
59. Otto M, Wiltfang J, Tumani H, Zerr I, Lantsch M, Kornhuber J, Weber T, Kretzschmar HA, Poser S (1997) Elevated levels of tau-protein in cerebrospinal fluid of patients with Creutzfeldt-Jakob disease. Neurosci Lett 225(3):210–212
60. Galasko DCL, Mottet R, Clark CM, Kaye J, Knopman D, Thomas R, Kholodenko D, Schenk D, Lieberburg I, Miller B, Green R, Basherad R, Kertiles L, Boss MA, Seubert P (1998) High cerebrospinal fluid tau and low amyloid beta42 levels in the clinical diagnosis of Alzheimer disease and relation to apolipoprotein E genotype. Arch Neurol 55(7):937–945
61. Ladogana A, Sanchez-Juan P, Mitrová E, Green A, Cuadrado-Corrales N, Sánchez-Valle R, Koscova S, Aguzzi A, Sklaviadis T, Kulczycki J, Gawinecka J, Saiz A, Calero M, van Duijn CM, Pocchiari M, Knight R, Zerr I (2009) Cerebrospinal fluid biomarkers in human genetic transmissible spongiform encephalopathies. J Neurol 256(10):1620–1628
62. Goodall CA, Head MW, Everington D, Ironside JW, Knight RS, Green AJ (2006) Raised CSF phospho-tau concentrations in variant Creutzfeldt-Jakob disease: diagnostic and pathological implications. J Neurol Neurosurg Psychiatry 77:89–91
63. Riemenschneider M, Wagenpfeil S, Vanderstichele H, Otto M, Wiltfang J, Kretzschmar H, Vanmechelen E, Förstl H, Kurz A (2003) Phospho-tau/total tau ration in cerebrospinal fluid discriminates Creutzfeldt-Jakob disease from other dementias. Mol Psychiatry 8:343–347
64. Selkoe DJ (1994) Normal and abnormal biology of the beta amyloid precursor protein. Annu Rev Neurosci 17:489–517
65. Shoji M, Golde TE, Ghiso J, Cheung TT, Estus S, Schaffer LM, Cai XD, McKay DM, Tinter R, Frangione B (1992) Production of the Alzheimer amyloid beta protein by normal proteolytic processing. Science 258(2079):126–129
66. Seubert P, Vigo-Pelfrey C, Esch F, Lee M, Dovey H, Davic D, Sinha S, Schlossmacher M, Whaley J, Swindlehurst C et al (1992) Isolation and quantification of soluble Alzheimer's beta-peptide from biological fluids. Nature 359(6393):325–327
67. Mattson MP (1999) Impairment of membrane transport and signal transduction systems by amyloidogenic proteins. Methods Enzymol 309:733–746
68. Lauderback CM, Hackett JM, Huang FF, Keller JN, Szweda LI, Markesbery WR, Butterfield AD (2001) The glial glutamate transporter, GLT-1, is oxidatively modified by 4-hydroxy-2-nonenal in the Alzheimer's disease brain: the role of Abeta1-42. J Neurochem 78:413–416
69. Varadajaran S, Yatin S, Aksenova M, Butterfield DA (2000) Review: Alzheimer's amyloid beta-peptide-associated free radical oxidative stress and neurotoxicity. J Struct Biol 130(2–3):184–208
70. Yatin S, Varadajaran S, Link CD (1999) In vitro and in vivo oxidative stress associated with Alzheimer's amyloid beta-peptide (1-42). Neurobiol Aging 20(3):325–330
71. Dickson DW (1997) The pathogenesis of senile plaques. J Neuopathol Exp Neurol 56(4):321–339

72. Shoji M, Matsubara E, Kanai M, Watanabe M, Nakamura T, Tomidokoro Y, Shizuka M, Wakabayashi K, Igeta Y, Ikeda Y, Mizushima K, Amari M, Ishiguro K, Kawarabayashi T, Harigaya Y, Okamoto K, Hirai S (1998) Combination assay of CSF tau, a beta 1-40 and a beta 1-42(43) as a biochemical marker of Alzheimer's disease. J Neurol Sci 158(2):134–140
73. Andreasen N, Hesse C, Davidsson P, Minthon L, Wallin A, Winblad B, Vanderstichele H, Vanmechelen E, Blennow K (1999) Cerebrospinal fluid beta-amyloid(1-42) in Alzheimer disease: differences between early- and late-onset Alzheimer disease and stability during the course of disease. Arch Neurol 56(6):673–680
74. Tamaoka A, Sawamur N, Fukushima T, Shoji S, Matsubara E, Shoji M, Hirai S, Furiya Y, Endoh R, Mori H (1997) Amyloid beta protein 42(43) in cerebrospinal fluid of patients with Alzheimer's disease. J Neurol Sci 148(1):41–45
75. Otto M, Esselmann H, Schulz-Schaeffer W, Neumann M, Schröter A, Ratzka P, Cepek L, Zerr I, Steinacker P, Windl O et al (2000) Decreased beta-amyloid$_{1-42}$ in cerebrospinal fluid of patients with Creutzfeldt-Jakob disease. Neurology 54(5):1099–1102
76. Van Everbroeck B, Green A, Pals P, Martin JJ, Cras P (1999) Decreased levels of amyloid-beta 1-42 in cerebrospinal fluid of Creutzfeldt-Jakob disease patients. J Alzheimers Dis 1:419–424
77. Wiltfang J, Esselmann H, Smirnov A, Bibl M, Cepek L, Steinacker P, Mollenhauer B, Buerger K, Hampel H, Paul S et al (2003) Beta-amyloid peptides in cerebrospinal fluid of patients with Creutzfeldt-Jakob disease. Ann Neurol 54:263–267
78. Gloeckner SF, Meyne F, Wagner F, Heinemann U, Krasnianski A, Meissner B, Zerr I (2008) Quantitative analysis of transthyretin, tau and amyloid-beta in patients with dementia. J Alzheimers Dis 14: 17–25
79. Drohat AC, Baldisseri D, Rustandi RR, Weber DJ (1998) Solution structure of calcium-bound rat S100B (beta beta) as determined by nuclear magnetic resonance spectroscopy. Biochemistry 37:2729–2740
80. Kilby PM, Van Eldik LJ, Roberts GC (1996) The solution structure of the bovine S100B protein dimer in the calcium-free state. Structure 4:1041–1052
81. Donato R (1988) Calcium-independent, pH regulated effects of S-100 proteins on assembly-disassembly of brain microtubule protein in vitro. J Biol Chem 263:106–110
82. Bianchi R, Giambianco I, Donato R (1993) S-100 protein, but not calmodulin, binds to the gial fibrillary acidic protein and inhibits its polymerization in a Ca2+–dependent manner. J Biol Chem 268:12669–12674
83. Bhattacharyya A, Oppenheim R, Prevette D, Moore BW, Brackenbury R, Ratner N (1992) S100 is present in developing chicken neurons and Schwann cells and promotes motor neuron survival in vivo. J Neurobiol 23:451–466
84. Barger SW, Van Eldik LJ, Mattson MP (1995) S100β protects hippocampal neurons from damage induces by glucose deprivation. Brain Res 677:167–170
85. Huttunen HJ, Kuja-Panula J, Sorci G, Aggneletti AL, Donata R, Rauvala H (2000) Coregulation of neurite outgrowth and cell survival by amphoterin and S100 proteins through receptor for advanced glycation end products (RAGE) activation. J Biol Chem 275:40096–40105
86. Fulle S, Pietrangelo T, Mariggio MA, Lorenzon P, Racanicchi L, Mozrzymas J, Guarnieri S, D'annunzio G (2000) Calcium and fos involvement in brain-derived Ca2+–binding protein (S100)-dependent apoptosis in rat phaeochromocytoma cells. Exp Physiol 85:243–253
87. Kaiser E, Kuzmits R, Pregant P, Burghuber O, Worofka W (1989) Clinical biochemistry of neuron specific enolase. Clin Chim Acta 183(1):13–31
88. Marangos PJ, Schmechel DE (1987) Neuron specific enolase, a clinically useful marker for neurons and neuroendocrine cells. Annu Rev Neurosci 10:269–295
89. Jauch EC, Lindsell C, Broderick J, Fagan SC, Tilley BC, Levine SR, NINDS rt-PA Stroke Study Group (2006) Association of serial biochemical markers with acute ischemic stroke: the National Institute of Neurological Disorders and Stroke recombinant tissue plasminogen activator stroke study. Stroke 37(10):2508–2513
90. Zerr I, Bodemer M, Räcker S, Grosche S, Poser S, Kretzschmar HA, Weber T (1995) Cerebrospinal fluid concentration of neuron-specific enolase in diagnosis of Creutzfeldt-Jakob disease. Lancet 345:1609–1610
91. van Duijn CM, Delasnerie-Lauprêtre N, Masullo C, Zerr I, de Silva R, Wientjens DPWM, Brandel JP, Weber T, Bonavita V, Zeidler M et al (1998) Case-control study of risk factors of Creutzfeldt-Jakob disease in Europe during 1993–95. European Union (EU) collaborative study Group of Creutzfeldt-Jakob disease (CJD). Lancet 351(9109):1081–1085

92. Beaudry P, Cohen P, Brandel JP, Delasnerie-Laupretre N, Richard S, Launay JM, Laplanche JL (1999) 14-3-3 protein, neuron-specific enolase, and S-100 protein in cerebrospinal fluid of patients with Creutzfeldt-Jakob disease. Dement Geriatr Cogn Disord 10(1):40–46
93. Kohira I, Tsuji T, Ishizu H, Takao Y, Wake A, Abe K, Kuroda S (2000) Elevation of neuron-specific enolase in serum and cerebrospinal fluid of early stage Creutzfeldt-Jakob disease. Acta Neurol Scand 102:385–387
94. Otto M, Stein H, Szudra A, Zerr I, Bodemer M, Gefeller O, Poser S, Kretzschmar HA, Maeder M, Weber T (1997) S-100 protein concentration in the cerebrospinal fluid of patients with Creutzfeldt-Jakob disease. J Neurol 244(9):566–570
95. Kropp S, Zerr I, Schulz-Schaeffer WJ, Riedemann C, Bodemer M, Laske C, Kretzschmar HA, Poser S (1999) Increase of neuron-specific enolase in patients with Creutzfeldt-Jakob disease. Neurosci Lett 261:124–126
96. Jesse SSP, Cepek L, Arnim CV, Tumani H, Lehnert S, Kretzschmar HA, Baier M, Otto M (2009) Glial fibrillary acidic protein and protein S-100B: different concentration pattern of glial proteins in cerebrospinal fluid of patients with Alzheimer's disease and Creutzfeldt-Jakob disease. J Alzheimers Dis 17(3):541–551
97. Minghetti L, Cardone F, Greco A, Puopolo M, Levi G, Green AJ, Knight R, Pocchiari P (2002) Increased CSF levels of prostaglandin E(2) in variant Creutzfeldt-Jakob disease. Neurology 58:127–129
98. Sanchez JC, Guillaume E, Lescuyer P, Allard L, Carrette O, Scherl A, Burgess J, Corthals GL, Burkhard PR, Hochstrasser DF (2004) Cystatin C as a potential cerebrospinal fluid marker for the diagnosis of Creutzfeldt-Jakob disease. Proteomics 4:2229–2233
99. Schmidt H, Otto M, Niedmann P, Cepek L, Schroter A, Kretzschmar HA, Poser S (2004) CSF lactate dehydrogenase activity in patients with Creutzfeldt-Jakob disease exceeds that in other dementias. Dement Geriatr Cogn Disord 17:204–206
100. Guillaume E, Zimmermann C, Burkhard PR, Hochstrasser DF, Sanchez JC (2003) A potenial cerebrospinal fluid and plasmatic marker for the diagnosis of Creutzfeldt-Jalob disease. Proteomics 3(8):1495–1499
101. Stoeck K, Bodemer M, Ciesielczyk B, Meissner B, Bartl M, Heinemann U, Zerr I (2005) Interleukin 4 and interleukin 10 levels are elevated in the cerebrospinal fluid of patients with Creutzfeldt-Jakob disease. Arch Neurol 62(10):1591–1594
102. Silveyra MXC-CN, Marcos A, Barguero MS, Rabano A, Calero M, Saez-Valero J (2006) Altered glycosylation of acetylcholinesterase in Creutzfeldt-Jakob disease. J Neurochem 96(1):97–104
103. Schätzl HM, Da Costa M, Taylor L, Cohen FE, Prusiner SB (1995) Prion protein gene variation among primates. J Mol Biol 245(4):362–374
104. Prusiner S (1991) Molecular biology of prion diseases. Science 252(5012):1515–1522
105. Naslavsky N, Stein R, Yanai A, Friedlander G, Taraboulos A (1997) Characterization of detergent-insoluble complexes containing the cellular prion protein and its scrapie isoform. J Biol Chem 272(10):6324–6331
106. Moser M, Colello R, Pott U, Oesch B (1995) Developmental expression of the prion protein gene in glial cells. Neuron 14(3):509–517
107. Brandner S, Klein MA, Aguzzi A (1999) A crucial role for B cells in neuroinvasive scrapie. Transfus Clin Biol 6(1):17–23
108. Holada K, Mondoro TH, Muller J, Vostal JG (1998) Increased expression of phosphatidylinositol-specific phospholipase C resistant prion proteins on the surface of activated platelets. Br J Haematol 103(1):276–282
109. Pauly PC, Harris DA (1998) Copper stimulates endocytosis of the prion protein. J Biol Chem 273(50):33107–33110
110. Shyng SL, Huber MT, Harris DA (1993) A prion protein cycles between the cell surface and an endocytic compartment in cultured neuroblastoma cells. J Biol Chem 268(21):15922–15928
111. Shyng SL, Lehmann S, Moulder KL, Lesko A, Harris DA (1995) The N-terminal domain of a glycolipid-anchored prion protein is essential for its endocytosis via clathrin-coated pits. J Biol Chem 270(24):14793–14800
112. Kuwahara C, Takeuchi AM, Nishimura T, Haraguchi K, Kubosaki A, Matsumoto Y, Saeki K, Yokoyama T, Itohara S, Onodera T (1999) Prions prevent neuronal cell-line death. Nature 400(6741):225–226
113. Bounhar Y, Zhang Y, Goodyer CG, LeBlanc AC (2001) Prion protein protects human neurons against Bax-mediated apoptosis. J Biol Chem 276:39145–39149
114. Paitel E, Sunyach C, Alves da Costa C, Bourdon JC, Vincent B, Checler F (2004) Primary cultured neurons devoid of cellular prion display lower responsiveness to staurosporine through the control of p53 at both transcriptional and post-transcriptional levels. J Biol Chem 279(1):612–618
115. Wong BS, Liu T, Li R, Pan T, Petersen RB, Smith MA, Gambetti P, Perry G, Manson JC,

Brown DR et al (2001) Increased levels of oxidative stress markers detected in the brains of mice devoid of prion protein. J Neurochem 76(2):565–572
116. Brown DR, Schuzl-Schaeffer WJ, Schmidt B, Kretzschmar H (1997) Prion protein-deficient cells show altered response to oxidative stress due to decreased SOD-1 activity. Exp Neurol 146(1):104–112
117. Chiarini LB, Freitas AR, Zanata SM, Berntani RR, Martins VR, Linden R (2002) Cellular prion protein transduces neuroprotective signals. EMBO J 13:3317–3326
118. Zanata SM, Lopes MH, Mercadante AF, Hajj GN, Chiarini LB, Nomizo R, Freitas AR, Cabral AL, Lee KS, Juliano MA et al (2002) Stress-inducible protein 1 is a cell surface ligand for cellular prion that triggers neuroprotection. EMBO J 13:3307–3316
119. Kim BH, Choi JK, Kim JI, Choi EK, Carp RI, Kim YS (2004) The cellular prion protein (PrPC) prevents apoptotic neuronal cell death and mitochondrial dysfunction induced by serum deprivation. Mol Brain Res 124(1):40–50
120. Prusiner SB (1982) Novel proteinaceous infectious particles cause scrapie. Science 216:136–144
121. Prusiner SB (1998) The prion diseases. Brain Pathol 8(3):499–513
122. Wong BS, Green AJ, Li R, Xie Z, Pan T, Liu T, Chen SG, Gambetti P, Sy MS (2001) Absence of protease-resistant prion protein in the cerebrospinal fluid of Creutzfeldt-Jakob disease. J Pathol 194(1):9–14
123. Bieschke J, Giese A, Schulz-Schaeffer W, Zerr I, Poser S, Eigen M, Kretzschmar H (2000) Ultrasensitive detection of pathological prion protein aggregates by dual-color scanning for intensely fluorescent targets. PNAS 97(10):5468–5473
124. Saborio GP, Permanne B, Soto C (2001) Sensitive detection of pathological prion protein by cyclic amplification of protein misfolding. Nature 411:810–813
125. Soto C, Anderes L, Suardi S, Cardone F, Castilla J, Frossard MJ, Peano S, Saa P, Limido L, Carbonatto M, Ironside J, Torres JM, Pocchiari M, Tagliavini F (2005) Pre-symptomatic detection of prions by cyclic amplification of protein misfolding. FEBS Lett 579:638–642
126. Wroe SJ, Pal S, Siddique D, Hyare H, Macfarlane R, Joiner S, Linehan JM, Brandner S, Wadsworth JD, Hewitt P et al (2006) Clinical presentation and pre-mortem diagnosis of variant Creutzfeldt-Jakob disease associated with blood transfusion: a case report. Lancet 368(9552):2061–2067
127. Llewelyn CA, Hewitt PE, Knight RSG, Amar K, Cousens S, Mackenzie J, Will RG (2004) Possible transmission of variant Creutzfeld-Jakob disease by blood transfusion. Lancet 363:417–421
128. Hewitt PE, Llewelyn CA, Mackenzie J, Will RG (2006) Creutzfeldt-Jakob disease and blood transfusion: results of the UK transfusion medicine epidemiological review study. Vox Sang 91(3):221–230
129. Zou S, Fang CT, Schonberger LB (2008) Transfusion transmission of human prion diseases. Transfus Med Rev 22(1):58–99
130. Tagliavini F, Prelli F, Porro M, Salmona M, Bugiani O, Frangione B (1992) A soluble form of prion protein in human cerebrospinal fluid: implications for prion-related encephalopathies. Biochem Biophys Res Commun 15(18):1398–1404
131. Völkel D, Zimmermann K, Zerr I, Bodemer M, Lindner T, Turecek PL, Poser S, Schwarz HP (2001) Immunochemical determination of cellular prion protein in plasma from healthy subjects and patients with sporadic CJD or other neurologic diseases. Transfusion 41:441–448
132. Meyne FGS, Ciesielczyk B, Heinemann U, Kransianski A, Meissner B, Zerr I (2009) Total prion protein levels in the cerebrospinal fluid are reduced in patients with various neurological disorders. J Alzheimers Dis 17:863–873
133. Boesenberg-Grosse C, Schulz-Schaeffer WJ, Bodemer M, Ciesielczyk B, Meissner B, Krasnianski A, Bartl M, Heinemann U, Varges D, Eigenbrod S, Kretzschmar HA, Green A, Zerr I (2006) Brain-derived proteins in the CSF: do they correlate with brain pathology in CJD? BMC Neurol 6:35
134. Zolg JW, Langen H (2004) How industry is approaching the search for new diagnostic markers and biomarkers. Mol Cell Proteomics 3(4):345–354
135. Anderson NL, Anderson NG (1998) Proteome and proteomics: new technologies, new concepts, and new words. Electrophoresis 19(11):1853–1861
136. Unlu M, Morgan ME, Minden JS (1997) Difference gel electrophoresis: a single gel method for detecting changes in protein extracts. Electrophoresis 18(11):2071–2077
137. Wolters DA, Washburn MP, Yates JR 3rd (2001) An automated multidimensional protein identification technology for shotgun proteomics. Anal Chem 73(23):5683–5690
138. Weston AD, Hood L (2004) Systems biology, proteomics, and the future of health care: toward predictive, preventative, and personalized medicine. J Proteome Res 3(2):179–196

139. Zerr I, Bodemer M, Otto M, Poser S, Windl O, Kretzschmar HA, Gefeller O, Weber T (1996) Diagnosis of Creutzfeldt-Jakob disease by two-dimensional gel electrophoresis of cerebrospinal fluid. Lancet 348(9031):846–849
140. Piubelli C, Fiorini M, Zanusso G, Milli A, Fasoli E, Monaco S, Righetti PG (2006) Searching for markers of Creutzfeldt-Jakob disease in cerebrospinal fluid by two-dimensional mapping. Proteomics 6(Suppl 1):256–261
141. Cepek L, Brechlin P, Steinacker P, Mollenhauer B, Klingebiel E, Bibl M, Kretzschmar HA, Wiltfang J, Otto M (2007) Proteomic analysis of the cerebrospinal fluid of patients with reutzfeldt-Jakob disease. Dement Geriatr Cogn Disord 23:22–28
142. Brechlin P, Jahn O, Steinacker P, Cepek L, Kratzin H, Lehnert S, Jesse S, Mollenhauer B, Kretzschmar HA, Wiltfang J, Otto M (2008) Cerebrospinal fluid-optimized two-dimensional difference gel electrophoresis (2-D DIGE) facilitates the differential diagnosis of Creutzfeldt-Jakob disease. Proteomics 8(20):4357–4366
143. Choe LH, Green A, Knight RS, Thompson EJ, Lee KH (2002) Apopoliprotein E and other cerebrospinal fluid proteins differentiate ante mortem variant Creutzfeldt-Jakob disease from ante mortem sporadic Creutzfeldt-Jakob disease. Electrophoresis 23(14):2242–2246
144. Demeester N, Castro G, Desrumaux C, De Geitere C, Fruchart JC, Santens P, Mulleners E, Engelborghs S, De Deyn PP, Vandekerckhove J, Rosseneu M, Labeur C (2000) Characterization and functional studies of lipoproteins, lipid transfer proteins, and lecithin:cholesterol acyltransferase in CSF of normal individuals and patients with Alzheimer's disease. J Lipid Res 41(6):963–974
145. Puchades M, Hansson SF, Nilsson CL, Andreasen N, Blennow K, Davidsson P (2003) Proteomic studies of potential cerebrospinal fluid protein markers for Alzheimer's disease. Brain Res Mol Brain Res 118(1–2):140–146
146. Iqbal K, Grundke-Iqbal I (1997) Elevated levels of tau and ubiquitin in brain and cerebrospinal fluid in Alzheimer's disease. Int J Geriatr Psychiatry 9(Suppl 1):289–296
147. Green AJ, Ramljak S, Muller WE, Knight RS, Schroder HC (2002) 14-3-3 in the cerebrospinal fluid of patients with variant and sporadic Creutzfeldt-Jakob disease measured using capture assay able to detect low levels of 14-3-3 protein. Neurosci Lett 324:57–60
148. Bleich S, Otto M, Zerr I, Kropp S, Kretzschmar HA, Wiltfang J (2005) Creutzfeldt-Jakob disease and homocysteine levels in plasma and cerebrospinal fluid. Gerontology 51(2):142–144
149. Mollenhauer B, Cullen V, Kahn I, Krastins B, Outeiro TF, Pepivani I, Ng J, Schulz-Schaeffer W, Kretzschmar HA, McLean PJ, Trenkwalder C, Sarracino DA, Vonsattel JP, Locascio JJ, El-Agnaf OM, Schlossmacher MG (2008) Direct quantification of CSF alpha-synuclein by ELISA and first cross-sectional study in patients with neurodegeneration. Exp Neurol 213(2):315–325
150. Kettlun A, Collados L, Garcia L, Cartier LA, Wolf ME, Mosnaim AD, Valenzuela MA (2003) Matrix metalloproteinase profile in patients with Creutzfeldt-Jakob disease. Int J Clin Pract 57(6):475–478
151. Holsinger RM, Lee JS, Boyd A, Masters CL, Collins SJ (2006) CSF BACE1 activity is increased in CJD and Alzheimer disease versus [corrected] other dementias. Neurology 67(4):710–712
152. Cartier L, García L, Kettlun AM, Castañeda P, Collados L, Vásquez F, Giraudon P, Belin MF, Valenzuela MA (2004) Extracellular matrix protein expression in cerebrospinal fluid from patients with tropical spastic paraparesis associated with HTLV-I and Creutzfeldt-Jakob disease. Scand J Clin Lab Invest 64(2):101–107
153. Manaka H, Kato T, Kurita K, Katagiri T, Shikama Y, Kujirai K, Kawanami T, Suzuki Y, Nihei K, Sasaki H et al (1992) Marked increase in cerebrospinal fluid ubiquitin in Creutzfeldt-Jakob disease. Neurosci Lett 139(1):47–49
154. Alberti C, Gonzalez J, Maldonado H, Medina F, Barriga A, García L, Kettlun A, Collados L, Puente J, Cartier L, Valenzuela M (2009) Comparative study of CSF neurofilaments in HTLV-1-associated myelopathy/tropical spastic paraparesis and other neurological disorders. AIDS Res Hum Retrovir 25(8):803–809
155. Albrecht D, Garcia L, Cartier L, Kettlun AM, Vergara C, Collados L, Valenzuela MA (2006) Trophic factors in cerebrospinal fluid and spinal cord of patients with tropical spastic paraparesis, HIV, and Creutzfeldt-Jakob disease. AIDS Res Hum Retrovir 22(3):248–254
156. Atarashi R, Satoh K, Sano K, Fuse T, Yamaguchi N, Ishibashi D, Matsubara T, Nakagaki T, Yamanaka H, Shirabe S, Yamada M, Mizusawa H, Kitamoto T, Klug G, McGlade A, Collins SJ, Nishida N (2011) Ultrasensitive human prion detection in cerebrospinal fluid by real-time quaking-induced conversion. Nat Med 17:175–178
157. McGuire LI, Peden AH, Orrú CD, Wilham JM, Appleford NE, Mallinson G, Andrews M,

Head MW, Caughey B, Will RG, Knight RS, Green AJ (2012) Real time quaking-induced conversion analysis of cerebrospinal fluid in sporadic Creutzfeldt-Jakob disease. Ann Neurol 72:78–85

158. Sano K, Satoh K, Atarashi R, Takashima H, Iwasaki Y, Yoshida M, Sanjo N, Murai H, Mizusawa H, Schmitz M, Zerr I, Kim YS, Nishida N (2013) Early detection of abnormal prion protein in genetic human prion diseases now possible using real-time QUIC assay. PLoS One 8:e54915

159. Cramm M, Schmitz M, Karch A, Mitrova E, Kuhn F, Schroeder B, Raeber A, Varges D, Kim YS, Satoh K, Collins S, Zerr I (2016) Stability and reproducibility underscore utility of RT-QuIC for diagnosis of Creutzfeldt-Jakob disease. Mol Neurobiol 53:1896–1904

160. Schmitz M, Ebert E, Stoeck K, Karch A, Collins S, Calero M, Sklaviadis T, Laplanche JL, Golanska E, Baldeiras I, Satoh K, Sanchez-Valle R, Ladogana A, Skinningsrud A, Hammarin AL, Mitrova E, Llorens F, Kim YS, Green A, Zerr I (2016) Validation of 14-3-3 protein as a marker in sporadic Creutzfeldt-Jakob disease diagnostic. Mol Neurobiol 53(4):2189–2199

Chapter 15

The Use of Transgenic and Knockout Mice in Prion Research

Abigail B. Diack and Jean C. Manson

Abstract

Despite several decades since the identification of the prion protein (PrP), we still do not know the full extent of its normal function or its role in transmissible spongiform encephalopathies (TSEs) or prion diseases. The production of transgenic mice both devoid of PrP and expressing different forms of PrP have enabled us to study the role of PrP in health and disease. Transgenic models expressing different forms of PrP allow us to define the role of PrP disease susceptibility, model disease transmission both within and between species, and understand the impact of PrP mutations on the species barrier. Knockout or PrP null mice have been utilized in discovering the normal functions of PrP and have led to the development of large animal models devoid of PrP. This chapter will outline the role that transgenic mice play in the study of prion diseases and the insights they have provided into the normal function of PrP and its role in prion disease susceptibility.

Key words Transmissible spongiform encephalopathy (TSE), Prion disease, Transgenic mouse, Knockout mouse, *Prnp*, Disease transmission

1 Introduction

The first transgenic mice for use in prion or transmissible spongiform encephalopathy (TSE) research were developed in 1989 by Scott et al. Mice were produced which overexpressed the hamster prion protein (PrP) via random insertion of the hamster gene (*Prnp*) into the murine genome. Since then, numerous transgenic models have been produced including PrP knockout mice, mice expressing heterologous prion protein, and mice in which PrP expression can be controlled both spatially and temporally, all of which have enabled researchers to study both the physiological and the pathological roles of PrP. The ever-increasing number of transgenic models produced and the vast number of TSE transmission studies carried out are beyond the scope of a single book chapter. This chapter aims to give an overview of the use of transgenic mice

Pawel P. Liberski (ed.), *Prion Diseases*, Neuromethods, vol. 129,
DOI 10.1007/978-1-4939-7211-1_15,

in understanding the impact of PrP modifications on susceptibility to disease, modelling both inter- and intraspecies transmission and defining the function of PrP. Moreover, it will particularly focus on the use of gene-targeted models including those in which the mouse protein sequence is replaced with that of another species or with a change in a specific amino acid within the coding sequence of *Prnp*. Where possible, specific examples are given and discussed; however in many cases there are multiple examples in a range of models available. We have included additional references in the tables in order to provide a wider range of information sources and highlight the volume of data available in this field.

2 Production of Transgenic Mice

Gene-targeting is a commonly used method of disrupting a specific gene in the mouse to give a "gene knockout" model [1]. This system takes advantage of homologous recombination to replace the normal copy of an exon with a mutated one, thus disrupting the open reading frame of the gene and blocking gene expression [2]. This can be done in a number of ways including utilizing drug selection markers such as neomycin or puromycin, incorporation of the HSV thymidine kinase gene to aid in positive and/or negative gene selection [3], and use of the hypoxanthine-guanosine-phosphoribosyl transferase (hprt) system [4]. Alternatively, conditional knockout mice can be produced by Cre/loxP technology. Cre recombinase is used to excise a region of DNA placed between two loxP sites (a floxed allele); the timing of the recombination event can then be regulated by use of a tamoxifen-inducible Cre [5]. The use of conditional knockout mice can partially avoid the lethality of some knockout mice phenotypes. Another important advantage of inducible knockout models is that they allow gene deletion at specific time points during a disease process. More recent technologies to produce transgenic mice have now been developed including Zinc Finger Nucleases (ZFN) [6, 7], Transcription Activator-Like Effector Nucleases (TALEN) [8, 9], and Clustered Regularly Interspaced Short Palindromic Repeat (CRISPR) [10]. Of note is CRISPR/Cas technology which is less labor intensive and can be more time efficient. This technology has the potential to become more widely used over the next few years (reviewed in [11]).

3 PrP Expression and Disease Susceptibility

Seminal studies by Scott et al. [12] used mouse models which overexpressed hamster PrP, introduced by random integration into the mouse genome on the endogenous murine PrP background.

Inoculation of these mice with hamster scrapie resulted in clinical scrapie with significantly shorter incubation periods than control mice. Similarly, mice overexpressing murine *Prnp* also demonstrated shorter incubation periods when inoculated with murine scrapie [13]. These models were the first indication that there was a relationship between PrP expression level and incubation time of prion diseases. The production of PrP null mice produced further evidence that there is a relationship between PrP expression and incubation time. Heterozygous null mice, with one copy of *Prnp*, express lower levels of PrP (50–70% of wild-type levels) and have consistently longer incubation times when inoculated with prion agents than wild-type controls [14, 15].

Tissue-specific and temporal-specific knock down of the *Prnp* gene in neurones during the course of the disease has also demonstrated that the preclinical phase of disease can be considerably extended (Manson personal communication) or halted despite the presence of abnormal PrP [16]. These studies have suggested that PrP^C itself could be a therapeutic target and there is considerable scope for disease intervention even after the onset of pathology.

The development of PrP null mice has also allowed both the role of PrP in disease and its normal function to be investigated. A number of lines of mice in which PrP has been removed have been produced including $Prnp^{-/-}$ (Edinburgh) [14], $Prnp^{0/0}$ (Zurich I) [17], and $Prnp^{-/-}$ (Nagasaki) [18] using different targeting vectors and thus altering different regions of the *Prnp* gene. PrP null mice develop normally and have proved to be resistant to prion infection [14, 15, 17, 19].

4 Modelling Interspecies Disease Transmission

The species barrier can be described as the inefficient transmission of a prion agent to a new host species often with long incubation periods which may decrease upon subsequent passage in the new host [20, 21]. The production of the first transgenic mouse models in which hamster PrP was expressed on the endogenous murine PrP background led to the hypothesis that sequence identity between host and donor PrP is important in determining the species barrier [12]. The inability or inefficiency of prions originating from one species to infect a host of another species was attributed to lack of sequence identity between the PrP of the host and prion agent [20]. To overcome the species barrier and further understand specific diseases, a range of gene-targeted and conventional transgenic mice have been produced expressing PrP of a different species. Of particular importance are transgenic models expressing human PrP which allow the zoonotic potential of prion diseases to be assessed (Table 1). Both conventional overexpressing and gene-targeted transgenic mice expressing

Table 1
Assessment of zoonotic potential using transgenic mice

Prion agent	Transmission	Mouse line (genotype and expression level)	References
Atypical scrapie	Negative	Tg35c[a] (129Met; 2×)	[66]
	Negative	HuM (129Met; gene-targeted)	[29]
	Negative	HuV (129Met; gene-targeted)	[29]
	Negative	HuMV (129Met; gene-targeted)	[29]
	Negative	Tg152c[a] (129Val; ×6)	[66]
BASE	Negative	HuM (129Met; gene-targeted)	[29]
	Negative	HuV (129Val; gene-targeted)	[29]
	Negative	HuMV (129Met/Val; gene-targeted)	[29]
	Positive	Tg650 (129Met; 6 fold)	[67]
BSE	Positive	Tg650 (129Met; 6 fold)	[25, 68]
	Positive	Tg340 (129Met; 4 fold)	[25, 68]
	Negative	Tg361 (129Val; 4 fold)	[68]
	Negative	Tg340/Tg361 (129Met/Val; 4 fold)	[68]
	Positive	Tg35 (129Met; ×2)	[66, 69]
	Positive	Tg45 (129Met; ×4)	[31]
	Positive	Tg152 (129Val; ×6)	[31]
	Negative	HuM (129Met; gene-targeted)	[22]
	Negative	HuV (129Val; gene-targeted)	[22]
	Negative	HuMV (129Met/Val; gene-targeted)	[22]
BSE-H	Negative	HuM (129Met; gene-targeted)	[29]
	Negative	HuV (129Val; gene-targeted)	[29]
	Negative	HuMV (129Met/Val; gene-targeted)	[29]
	Negative	Tg650 (129Met; 6 fold)	[67]
CWD	Negative	Tg35 (129Met; ×2)	[31]
	Negative	HuM (129Met; gene-targeted)	[29]
	Negative	HuV (129Val; gene-targeted)	[29]
	Negative	HuMV (129Met/Val; gene-targeted)	[29]
	Negative	Tg45 (129Met; ×4)	[31]
	Negative	Tg152 (129Val; ×6)	[31]
	Negative	Tg40 (129Met; 1 fold)	[30]
	Negative	Tg1 (129Met; 2 fold)	[30]

(continued)

Table 1 (continued)

Prion agent	Transmission	Mouse line (genotype and expression level)	References
Ovine BSE	Negative	Tg35c[a] (129Met; 2×)	[66]
	Positive[b]	Tg152c[a] (129Val; ×6)	[66]
	Positive	HuM (129Met; gene-targeted)	[28]
	Negative	HuV (129Val; gene-targeted)	[28]
	Negative	HuMV (129Met/Val; gene-targeted)	[28]
	Positive	Tg650 (129Met; 6 fold)	[25]
	Positive	Tg340 (129Met; 4 fold)	[25]
Scrapie	Positive	Tg650 (129Met; 6 fold)	[68]
	Positive	Tg340 (129Met; 4 fold)	[68]
	Negative	Tg361 (129Val; 4 fold)	[68]
	Negative	Tg340/Tg361 (129Met/Val; 4 fold)	[68]
	Negative	Tg35c[a] (129Met; 2×)	[66]
	Negative	Tg152c[a] (129Val; ×6)	[66]
	Negative	HuM (129Met; gene-targeted)	[28]
	Negative	HuV (129Val; gene-targeted)	[28]
	Negative	HuMV (129Met/Val; gene-targeted)	[28]

Summary of transmission studies carried out using transgenic mice expressing human PrP
[a] Tg35 and Tg152 were used to generate FVB/NHuPrP congenic lines
[b] Positive sample found upon second mouse

human PrP^C have been produced which carry each of the *PRNP* codon 129 genotypes [22–25]. Codon 129 is of interest in humans as this codon plays a role in susceptibility to prion disease as reviewed in [26] but also shown experimentally in gene-targeted mice expressing human PrP (referred to as HuM, HuMV, and HuV). For example, when infected with vCJD, susceptibility to disease decreases in the order HuM > HuMV > HuV, similar to that found in the UK population where with the exception of one case in a 129MV individual in 2016, all clinical cases of vCJD have been identified in 129MM individuals [22].

Bovine spongiform encephalopathy (BSE) has shown an ability to transmit between species in the field with the BSE strain being identified in humans, felids, and ungulates [27]. Transmission studies have therefore focused on the host susceptibility of classical BSE utilizing gene-targeted mice expressing human PrP (HuM, HuMV, and HuV) which did not show any evidence of BSE transmission [22] whereas overexpressing mice (129VVTg-152) exhibited longer incubation periods than wild-type (FVB) mice (602 days versus 371 days, respectively) and a low transmission rate (38%) when inoculated with BSE [23]. These studies demonstrate that a significant species barrier exists between BSE and human PrP. This has been borne out by the limited numbers of identified cases of vCJD (178 cases as of June 2017 (www.cjd.ed.ac.uk) compared to the large number of cases of BSE (www.oie.int/animal-health-in-the-world/bse-specific-data/). In contrast, sheep BSE appears to transmit more efficiently than classical BSE with reduced incubation periods and increased susceptibility in overexpressing transgenic mice [25] and an increased susceptibility in gene-targeted mice with 16/23 HuM mice showing evidence of disease compared to 0/18 mice inoculated with cattle BSE [28]. These results indicate that once BSE has been transmitted through an intermediate species there could be a higher zoonotic risk than BSE derived from cattle.

The use of overexpressing versus gene-targeted mice can lead to conflicting results which can prove difficult to interpret in terms of zoonotic risk to humans. Scrapie has been endemic in the sheep population for many years with no apparent evidence of transmission to humans. However recent transmission studies utilizing mice which overexpress human PrP four or six times that found in normal human brain tissue demonstrated scrapie transmission in Tg Met/Val_{129} mice and Tg650 Met mice at primary passage and increased susceptibility at second passage. It is possible that the overexpression of the PrP gene may lead to a lowering of the species barrier and thus not be truly representative of the situation in the field; however, such studies do not allow the risk to human health to be dismissed. Studies of classical and atypical scrapie have found no transmission in gene-targeted mice [29].

In North America, chronic wasting disease (CWD), a prion disease of cervids, has become a concern due to increasing numbers of affected animals and the risk of potential infection though ingestion of contaminated meat and handling of contaminated materials by hunters and agricultural workers. Transmission studies of CWD have provided no evidence to date of positive transmissions to any transgenic mouse model expressing human PrP [29–31]. This is in contrast to in vitro model systems which suggest that CWD can have zoonotic potential [32]. These studies demonstrate the difficulties in interpreting data from both in vivo transmission and in vitro model systems and understanding what conflicting results could mean for the transmission of an agent to a new species.

5 Human Prion Transmissions

One of the advantages of producing transgenic mice that express PrP of another species is the ability to study potential for intraspecies transmission and the effects of host genetics such as codon 129 genotype in humans. This is of particular importance when assessing the risk of human-to-human transmission of prion diseases. sCJD and vCJD have been studied extensively in transgenic models with the human *PRNP* gene. Four strains of sCJD have been identified in gene-targeted models [33] with differing susceptibilities in the three genotypes. The transmissible human prion disease, vCJD, has been shown to transmit both to mice overexpressing human PrP and gene-targeted mice. These studies demonstrated that there were varying degrees of susceptibility dependent upon both agent and host genotype. Bishop et al. [22] showed that 129MV and 129VV genotypes may exhibit longer incubation periods with vCJD and indeed may remain asymptomatic during life. This study proposed that there may be significant levels of asymptomatic vCJD in the human population in all genotypes. Initial prevalence studies indeed estimated that 1 in 4000 people [34] had evidence of abnormal PrP in their lymphoreticular system. A more recent study revised this estimate to 1 in 2000 people in the UK population with evidence of abnormal PrP in their appendices [35]. Transmission studies of spleen from an asymptomatic individual in which abnormal PrP was identified only in the spleen and lymph node showed that the abnormal PrP deposited in the spleen was capable of transmitting infectivity to HuM mice [36]. Thus understanding the significance of the peripheral accumulation of PrP in the appendix study and in particular the relationship between abnormal PrP and infectivity is of great importance in assessing the risk of vCJD infectivity to the individual and the population.

6 Variance in Human Prion Disease Transmission Ability

Variably protease-sensitive prionopathy (VPSPr) is a rare and sporadic prion disease recently identified in all codon 129 genotypes. VPSPr is characterized by unusual biochemical properties including increased sensitivity to protease digestion compared with sCJD and the ladder-like appearance of protease-resistance fragments on Western blot and a prominent low molecular weight fragment of ~8 kDa. Individuals also accumulate PrP in the form of microplaques in the cerebellum and thalamus [37–39]. Transmission studies of VPSPr into gene-targeted and conventional transgenic mice expressing human PrP, respectively, showed that VPSPr was able to transmit disease but inefficiently compared to sCJD [40, 41]. Diack et al. demonstrated that unlike sCJD which transmits from all codon 129 genotypes, only VPSPr from 129VV individuals was able to transmit to mice. Although there were similarities between sCJDMM2 and VPSPr transmission with little or no clinical signs or vacuolation and limited evidence of PrP deposition (sCJDMM2; 3/17 HuV, 2/18 HuMV and VPSPr; 1/15 HuM, 2/14 HuMV, 5/14 HuV), the PrP deposition differed in form and brain areas. In particular, some of these deposits resembled the microplaques that are found in VPSPr patients and astrocytosis was identified within their vicinity [40]. Thus while some human strains are readily transmissible others with the same *PRNP* sequence have very limited ability to transmit. The key to these differences may be associated with the type of PrP deposited and the host response to the PrP deposition.

7 PrP Mutations and Disease Transmission

The PrP protein is an essential requirement for transmission of prion diseases. Alterations in either the amino acid sequence of the prion protein gene of the host or in post-translational modifications, such as glycosylation, can have the ability to affect disease transmission characteristics such as disease susceptibility, pathology, and incubation periods. As such, models expressing PrP mutations can be used to understand host susceptibility and the mechanisms of disease transmission.

The PrP gene of the host can exert a major influence over host susceptibility but the mechanisms of this are not fully understood. In order to investigate mutations associated with human prion diseases, gene-targeted mice were produced with a single amino acid alteration, proline to leucine at codon 101 of the murine PrP sequence (101LL). This is equivalent to the 102LL mutation in *PRNP* which is associated with Gerstmann–Sträussler–Scheinker (GSS) disease [42]. Inoculation with human P102L GSS produces disease in 288 days with 100% susceptibility compared with 456 days in wild-type mice suggesting that sequence homology between host and donor sequence is important in transmissibility [43]. However, when these mice were inoculated with strains of prion disease that carry a proline at the equivalent codon 101 position; a pooled natural sheep scrapie strain (SSBP/1) or hamster passaged scrapie (263K), the incubation period was similarly reduced when compared with wild-type mice; 346 days versus over 400 days and 464 days versus over 600 days, respectively [44]. In contrast, despite being of the same species, murine scrapie strains ME7, 139A, and 79A show a longer incubation period in 101LL mice compared to wild-type mice [43, 45]. Although the transmission of prion disease was originally thought to be more efficient when host and donor PrP sequences were similar, these studies have shown that sequence homology is *not* the only criteria and that other factors must be playing an important role in disease transmission.

8 PrP Glycosylation Influences Susceptibility

In vitro studies demonstrated that post-translational modifications such as glycosylation could influence disease transmission [46]. This led to the development of a range of glycosylation-deficient mice, including conventional transgenics, 3F4 tagged murine transgenics, and gene-targeted transgenics in order to address the role of glycosylation in disease transmission [47–49]. Cancellotti et al. [49] produced three lines of gene-targeted mice in which the first (G1; N180 T), second (G2; N196 T) or both (G3; N180 T and N196 T) N-glycan sites were removed. Initial transmission studies with a range of agents showed that host PrP glycosylation was not essential for disease transmission. When inoculated intracerebrally with murine passaged strains (79A, ME7, and 301C), G1 mice were only susceptible to 79A, G3 mice were only susceptible to 79A with a longer incubation period than G1 and G2 mice were susceptible to all three agents albeit with longer incubation periods than wild-type mice when inoculated with 301C [50]. Later studies investigated the role of glycosylation in cross-species transmission and demonstrated that the absence of glycosylation at the first or both N-glycan sites results in almost complete resistance to 263 K, vCJD, and sCJDMM2. In contrast, G2 mice proved susceptible to all three agents and in particular showed shorter incubation periods and a higher susceptibility rate than wild-type mice when inoculated with sCJDMM2; 404 days, 11/20 with TSE pathology versus over 650 days and 3/41, respectively [51].

In vivo studies have also shown that host PrP glycosylation could play a role in neuroinvasion. Intraperitoneal (i.p.) inoculation of ME7 resulted in a slightly lengthened incubation time in G2 mice but no disease transmission to G1 or G3 mice whereas i.p. inoculation of 79A showed increased incubation periods in both G1 and G2 mice and again, no transmission to G3 mice [52]. Thus, these series of transmission studies highlight glycosylation of host PrP as a key factor in prion disease transmission and in the outcome of disease.

9 The Function of PrP

Despite its role in prion disease and links with other neurodegenerative diseases such as Alzheimer's disease [53, 54], the physiological role of PrP^{C} has not yet been fully established. One way to study the function of a protein is to knock out the corresponding gene and analyze the subsequent phenotype of the mice under

Table 2
PrP knockout mouse models

PrP knockout mouse model	*Prnp* codon region disrupted	Notes	References
Prnp$^{-/-}$ (Edinburgh) *Prnp*$^{-/-}$ (Npu)[a]	Exon 3	No severe abnormalities detected in phenotype	[14]
Prnp$^{0/0}$ (Zurich I) (*Prnp*$^{0/0}$ (Zrch I))[a]	Exon 3	No severe abnormalities detected in phenotype Late-onset neuropathy	[17]
Prnp$^{0/0}$ (Zurich II) *Prnp*$^{0/0}$ (Zrch II)[a]	Entire coding regions and part of intron 2	Ectopic expression of Doppel occurs leading to late-onset ataxia (starting at ~20–24 months of age)	[56]
Prnp$^{-/-}$ (Nagasaki) *Prnp*$^{0/0}$ (Ngsk)[a]	Entire coding regions and part of intron 2	Ectopic expression of Doppel occurs leading to late-onset ataxia (starting at ~70 weeks of age) Late-onset neuropathy	[18]
Prnp$^{-/-}$ (Rcm0)	Entire coding regions and part of intron 2	Ectopic expression of Doppel occurs leading to late-onset ataxia	[55]
Prnp$^{-/-}$ (Rikn)	Entire coding regions and part of intron 2	Ectopic expression of Doppel occurs leading to late-onset ataxia	[57]

Summary of PrP knock out mouse models and production
[a]Indicated abbreviated or alternative names

normal and physiologically stressed conditions. As previously discussed in this chapter, six PrP knockout mouse lines have been produced and used for this purpose (Table 2). Of the lines produced four were produced in such a manner that ectopic expression of the prion-related protein Doppel (Dpl) occurred [18, 55–57] leading to development of late-onset ataxia in the PrP knockout mice. In this short summary, we will only discuss those without ectopic Dpl expression to avoid confusion.

PrP is expressed at high levels in the nervous system although there are variations between brain regions, cell types, and neuronal populations. Many other immune and peripheral tissues also express PrP at differing levels (reviewed in [58]). Despite this ubiquitous expression, PrP knockout mice develop and reproduce normally with no overt phenotype [14]. More extensive studies have now identified deficits in a number of neuronal functions. One of the first deficits identified was that loss of PrP affected both circadian rhythm and sleep. In mice devoid of PrP there was a significant increase in the length of activity compared to wild-type mice

Table 3
Phenotypes exhibited by PrP knockout mouse models

Deficit	PrP knockout mouse model	References
Altered circadian rhythm and sleep patterns	$Prnp^{-/-}$ (Edinburgh)	[59]
Reduced mitochondrial numbers	$Prnp^{-/-}$ (Edinburgh)	[62]
Increase in oxidative stress	$Prnp^{-/-}$ (Edinburgh)	[63, 70]
	$Prnp^{0/0}$ (Zurich I)	[70]
Olfactory loss	$Prnp^{0/0}$ (Zurich I)	[64]
Altered iron homeostasis	$Prnp^{0/0}$ (Zurich I)	[65]
Impairment of anti-apoptotic pathways	$Prnp^{0/0}$ (Zurich I)	[71]
Alterations in cytoskeletal organization, brain function, and age-related neuroprotection	$Prnp^{0/0}$ (Zurich I)	[72]
Behavioral deficits	$Prnp^{0/0}$ (Zurich I)	[72–74]
	$Prnp^{-/-}$ (Edinburgh)	[60]
Increased susceptibility to kainite induced seizures	$Prnp^{0/0}$ (Zurich I)	[75, 76]
Myelin degeneration	$Prnp^{0/0}$ (Zurich I)	[61]
Alterations in long-term potentiation	$Prnp^{0/0}$ (Zurich I)	[77]
	$Prnp^{-/-}$ (Edinburgh)	[78]

Common deficits exhibited by $Prnp^{-/-}$ (Edinburgh) and $Prnp^{0/0}$ (Zurich I) mice

and they showed a more prominent sleep fragmentation and larger response to sleep deprivation than wild-type mice [59]. These results proved intriguing due to the similarities to the sleep alterations observed in the inheritable prion disorder fatal familial insomnia (FFI).

Further studies have identified deficits in hippocampal spatial learning and hippocampal synaptic plasticity in vivo. These deficits could be rescued in $PrP^{-/-}$ mice expressing PrP^{C} in neurons under control of the NSE promoter indicating that they were due to neuronal loss of PrP [60]. Again, cognitive deficits are one of the main symptoms of human prion diseases. In addition to prion diseases, PrP^{C} is thought to play a role in peripheral neuropathies in humans. Studies showed that removal of PrP^{C} could trigger a chronic demyelinating polyneuropathy [61]. Moreover, Bremer et al. used mice in which neuronal PrP^{C} was selectively depleted to elegantly demonstrate that neuronal PrP^{C} is essential for peripheral myelin maintenance [61]. Additional deficits due to PrP loss include reduced mitochondrial numbers [62], roles in oxidative homeostasis [63], olfactory loss [64], and iron homeostasis [65] and are summarized with others in Table 3.

10 Summary

The use of transgenic and knockout mouse models have made a major contribution to the study of prion diseases and our understanding of PrP in health and disease. Using transgenic mice expressing different forms of PrP or knockout mice allows us to define the roles of PrP in disease susceptibility, disease transmission and understand the role PrP plays in normal homeostasis. The advent of new techniques in transgenic production can only increase our knowledge and understanding of TSEs allowing us to move towards defining strategies for the intervention and treatment of these devastating and fatal diseases.

References

1. Capecchi MR (2005) Gene targeting in mice: functional analysis of the mammalian genome for the twenty-first century. Nat Rev Genet 6(6):507–512
2. Folger KR, Wong EA, Wahl G et al (1982) Patterns of integration of DNA microinjected into cultured mammalian cells: evidence for homologous recombination between injected plasmid DNA molecules. Mol Cell Biol 2(11):1372–1387
3. Mansour SL, Thomas KR, Capecchi MR (1988) Disruption of the proto-oncogene int-2 in mouse embryo-derived stem cells: a general strategy for targeting mutations to non-selectable genes. Nature 336(6197):348–352
4. Melton D (2002) Gene-targeting strategies. In: Clarke A (ed) Transgenesis techniques, Methods in molecular biology, vol 180 Springer New York, pp 151–173. doi:10.1385/1-59259-178-7
5. Sauer B (1998) Inducible gene targeting in mice using the Cre/lox system. Methods 14(4):381–392
6. Meyer M, de Angelis MH, Wurst W et al (2010) Gene targeting by homologous recombination in mouse zygotes mediated by zinc-finger nucleases. Proc Natl Acad Sci U S A 107(34):15022–15026
7. Carbery ID, Ji D, Harrington A et al (2010) Targeted genome modification in mice using zinc-finger nucleases. Genetics 186(2):451–459
8. Wang H, Y-C H, Markoulaki S et al (2013) TALEN-mediated editing of the mouse Y chromosome. Nat Biotechnol 31(6):530–532
9. Sommer D, Peters A, Baumgart A-K et al (2015) TALEN-mediated genome engineering to generate targeted mice. Chromosom Res 23(1):43–55
10. Yang H, Wang H, Jaenisch R (2014) Generating genetically modified mice using CRISPR/Cas-mediated genome engineering. Nat Protoc 9(8):1956–1968
11. Singh P, Schimenti JC, Bolcun-Filas E (2015) A mouse geneticist's practical guide to CRISPR applications. Genetics 199(1):1–15
12. Scott M, Foster D, Mirenda C et al (1989) Transgenic mice expressing hamster prion protein produce species-specific scrapie infectivity and amyloid plaques. Cell 59(5):847–857
13. Westaway D, Mirenda CA, Foster D et al (1991) Paradoxical shortening of scrapie incubation times by expression of prion protein transgenes derived from long incubation period mice. Neuron 7(1):59–68
14. Manson JC, Clarke AR, Hooper ML et al (1994) 129/Ola mice carrying a null mutation in PrP that abolishes mRNA production are developmentally normal. Mol Neurobiol 8(2–3):121–127
15. Bueler H, Aguzzi A, Sailer A et al (1993) Mice devoid of PrP are resistant to scrapie. Cell 73(7):1339–1347
16. Mallucci G, Dickinson A, Linehan J et al (2003) Depleting neuronal PrP in prion infection prevents disease and reverses spongiosis. Science (New York, NY) 302(5646):871–874
17. Bueler H, Fischer M, Lang Y et al (1992) Normal development and behaviour of mice lacking the neuronal cell-surface PrP protein. Nature 356(6370):577–582
18. Sakaguchi S, Katamine S, Nishida N et al (1996) Loss of cerebellar Purkinje cells in aged mice homozygous for a disrupted PrP gene. Nature 380(6574):528–531

19. Manson JC, Clarke AR, McBride PA et al (1994) PrP gene dosage determines the timing but not the final intensity or distribution of lesions in scrapie pathology. Neurodegeneration 3(4):331–340
20. Kimberlin RH, Cole S, Walker CA (1987) Temporary and permanent modifications to a single strain of mouse scrapie on transmission to rats and hamsters. J Gen Virol 68(7):1875–1881
21. Kimberlin RH, Walker CA (1979) Pathogenesis of scrapie: agent multiplication in brain at the first and second passage of hamster scrapie in mice. J Gen Virol 42(1):107–117
22. Bishop MT, Hart P, Aitchison L et al (2006) Predicting susceptibility and incubation time of human-to-human transmission of vCJD. Lancet Neurol 5(5):393–398
23. Hill AF, Desbruslais M, Joiner S et al (1997) The same prion strain causes vCJD and BSE. Nature 389(6650):448–450, 526
24. Beringue V, Le Dur A, Tixador P et al (2008) Prominent and persistent extraneural infection in human PrP transgenic mice infected with variant CJD. PLoS One 3(1):e1419
25. Padilla D, Beringue V, Espinosa JC et al (2011) Sheep and goat BSE propagate more efficiently than cattle BSE in human PrP transgenic mice. PLoS Pathog 7(3):e1001319
26. Kobayashi A, Teruya K, Matsuura Y et al (2015) The influence of PRNP polymorphisms on human prion disease susceptibility: an update. Acta Neuropathol 130(2):159–170
27. Bruce M, Chree A, McConnell I et al (1994) Transmission of bovine spongiform encephalopathy and scrapie to mice: strain variation and the species barrier. Philos Trans R Soc Lond Ser B Biol Sci 343(1306):405–411
28. Plinston C, Hart P, Chong A et al (2011) Increased susceptibility of human-PrP transgenic mice to bovine spongiform encephalopathy infection following passage in sheep. J Virol 85(3):1174–1181
29. Wilson R, Plinston C, Hunter N et al (2012) Chronic wasting disease and atypical forms of bovine spongiform encephalopathy and scrapie are not transmissible to mice expressing wild-type levels of human prion protein. J Gen Virol 93(Pt 7):1624–1629
30. Kong Q, Huang S, Zou W et al (2005) Chronic wasting disease of elk: transmissibility to humans examined by transgenic mouse models. J Neurosci 25(35):7944–7949
31. Sandberg MK, Al-Doujaily H, Sigurdson CJ et al (2010) Chronic wasting disease prions are not transmissible to transgenic mice overexpressing human prion protein. J Gen Virol 91(Pt 10):2651–2657
32. Barria MA, Balachandran A, Morita M et al (2014) Molecular barriers to zoonotic transmission of prions. Emerg Infect Dis 20(1):88–97
33. Bishop MT, Will RG, Manson JC (2010) Defining sporadic Creutzfeldt-Jakob disease strains and their transmission properties. Proc Natl Acad Sci U S A 107(26):12005–12010
34. Hilton DA, Ghani AC, Conyers L et al (2004) Prevalence of lymphoreticular prion protein accumulation in UK tissue samples. J Pathol 203(3):733–739
35. Gill ON, Spencer Y, Richard-Loendt A et al (2013) Prevalent abnormal prion protein in human appendixes after bovine spongiform encephalopathy epizootic: large scale survey. BMJ 347:f5675
36. Bishop MT, Diack AB, Ritchie DL et al (2013) Prion infectivity in the spleen of a PRNP heterozygous individual with subclinical variant Creutzfeldt-Jakob disease. Brain 136(Pt 4):1139–1145
37. Head MW, Knight R, Zeidler M et al (2009) A case of protease sensitive prionopathy in a patient in the UK. Neuropathol Appl Neurobiol 35(6):628–632
38. Jansen C, Head MW, van Gool WA et al (2010) The first case of protease-sensitive prionopathy (PSPr) in The Netherlands: a patient with an unusual GSS-like clinical phenotype. J Neurol Neurosurg Psychiatry 81(9):1052–1055
39. Zou WQ, Puoti G, Xiao X et al (2010) Variably protease-sensitive prionopathy: a new sporadic disease of the prion protein. Ann Neurol 68(2):162–172
40. Diack AB, Ritchie DL, Peden AH et al (2014) Variably protease-sensitive prionopathy, a unique prion variant with inefficient transmission properties. Emerg Infect Dis 20(12):1969–1979
41. Notari S, Xiao X, Espinosa JC et al (2014) Transmission characteristics of variably protease-sensitive prionopathy. Emerg Infect Dis 20(12):2006–2014
42. Manson JC, Jamieson E, Baybutt H et al (1999) A single amino acid alteration (101L) introduced into murine PrP dramatically alters incubation time of transmissible spongiform encephalopathy. EMBO J 18(23):6855–6864
43. Manson JC, Barron R, Jamieson E et al (2000) A single amino acid alteration in murine PrP dramatically alters TSE incubation time. Arch Virol (16):95–102
44. Barron RM, Thomson V, Jamieson E et al (2001) Changing a single amino acid in the N-terminus of murine PrP alters TSE incubation time across three species barriers. EMBO J 20(18):5070–5078

45. Barron RM, Thomson V, King D et al (2003) Transmission of murine scrapie to P101L transgenic mice. J Gen Virol 84(Pt 11):3165–3172
46. Priola SA, Lawson VA (2001) Glycosylation influences cross-species formation of protease-resistant prion protein. EMBO J 20(23):6692–6699
47. Neuendorf E, Weber A, Saalmueller A et al (2004) Glycosylation deficiency at either one of the two glycan attachment sites of cellular prion protein preserves susceptibility to bovine spongiform encephalopathy and scrapie infections. J Biol Chem 279(51):53306–53316
48. DeArmond SJ, Sanchez H, Yehiely F et al (1997) Selective neuronal targeting in prion disease. Neuron 19(6):1337–1348
49. Cancellotti E, Wiseman F, Tuzi NL et al (2005) Altered glycosylated PrP proteins can have different neuronal trafficking in brain but do not acquire scrapie-like properties. J Biol Chem 280(52):42909–42918
50. Tuzi NL, Cancellotti E, Baybutt H et al (2008) Host PrP glycosylation: a major factor determining the outcome of prion infection. PLoS Biol 6(4):e100
51. Wiseman FK, Cancellotti E, Piccardo P et al (2015) The glycosylation status of PrPC is a key factor in determining transmissible spongiform encephalopathy transmission between species. J Virol 89(9):4738–4747
52. Cancellotti E, Bradford BM, Tuzi NL et al (2010) Glycosylation of PrPC determines timing of neuroinvasion and targeting in the brain following transmissible spongiform encephalopathy infection by a peripheral route. J Virol 84(7):3464–3475
53. Parkin ET, Watt NT, Hussain I et al (2007) Cellular prion protein regulates beta-secretase cleavage of the Alzheimer's amyloid precursor protein. Proc Natl Acad Sci U S A 104(26):11062–11067
54. Lauren J, Gimbel DA, Nygaard HB et al (2009) Cellular prion protein mediates impairment of synaptic plasticity by amyloid-beta oligomers. Nature 457(7233):1128–1132
55. Moore RC, Lee IY, Silverman GL et al (1999) Ataxia in prion protein (PrP)-deficient mice is associated with upregulation of the novel PrP-like protein doppel1. J Mol Biol 292(4):797–817
56. Rossi D, Cozzio A, Flechsig E et al (2001) Onset of ataxia and Purkinje cell loss in PrP null mice inversely correlated with Dpl level in brain. EMBO J 20(4):694–702
57. Yokoyama T, Kimura KM, Ushiki Y et al (2001) In vivo conversion of cellular prion protein to pathogenic isoforms, as monitored by conformation-specific antibodies. J Biol Chem 276(14):11265–11271
58. Linden R, Martins VR, Prado MAM et al (2008) Physiology of the prion protein. Physiol Rev 88(2):673–728
59. Tobler I, Gaus SE, Deboer T et al (1996) Altered circadian activity rhythms and sleep in mice devoid of prion protein. Nature 380(6575):639–642
60. Criado JR, Sanchez-Alavez M, Conti B et al (2005) Mice devoid of prion protein have cognitive deficits that are rescued by reconstitution of PrP in neurons. Neurobiol Dis 19(1–2):255–265
61. Bremer J, Baumann F, Tiberi C et al (2010) Axonal prion protein is required for peripheral myelin maintenance. Nat Neurosci 13(3):310–318
62. Miele G, Jeffrey M, Turnbull D et al (2002) Ablation of cellular prion protein expression affects mitochondrial numbers and morphology. Biochem Biophys Res Commun 291(2):372–377
63. Wong BS, Liu T, Li R et al (2001) Increased levels of oxidative stress markers detected in the brains of mice devoid of prion protein. J Neurochem 76(2):565–572
64. Le Pichon CE, Valley MT, Polymenidou M et al (2009) Olfactory behavior and physiology are disrupted in prion protein knockout mice. Nat Neurosci 12(1):60–69
65. Singh A, Kong Q, Luo X et al (2009) Prion protein (PrP) knock-out mice show altered iron metabolism: a functional role for PrP in iron uptake and transport. PLoS One 4(7):e6115
66. Wadsworth JD, Joiner S, Linehan JM et al (2013) Atypical scrapie prions from sheep and lack of disease in transgenic mice overexpressing human prion protein. Emerg Infect Dis 19(11):1731–1739
67. Beringue V, Herzog L, Reine F et al (2008) Transmission of atypical bovine prions to mice transgenic for human prion protein. Emerg Infect Dis 14(12):1898–1901
68. Cassard H, Torres J-M, Lacroux C et al (2014) Evidence for zoonotic potential of ovine scrapie prions. Nat Commun 5:5821
69. Asante EA, Linehan JM, Desbruslais M et al (2002) BSE prions propagate as either variant CJD-like or sporadic CJD-like prion strains in transgenic mice expressing human prion protein. EMBO J 21(23):6358–6366
70. Brown DR, Nicholas RSJ, Canevari L (2002) Lack of prion protein expression results in a neuronal phenotype sensitive to stress. J Neurosci Res 67(2):211–224
71. Weise J, Sandau R, Schwarting S et al (2006) Deletion of cellular prion protein results in reduced Akt activation, enhanced postischemic caspase-3 activation, and exacerbation of

ischemic brain injury. Stroke 37(5): 1296–1300
72. Schmitz M, Greis C, Ottis P et al (2014) Loss of prion protein leads to age-dependent behavioral abnormalities and changes in cytoskeletal protein expression. Mol Neurobiol 50(3):923–936
73. Gadotti VM, Bonfield SP, Zamponi GW (2012) Depressive-like behaviour of mice lacking cellular prion protein. Behav Brain Res 227(2):319–323
74. Büdefeld T, Majer A, Jerin A et al (2014) Deletion of the prion gene Prnp affects offensive aggression in mice. Behav Brain Res 266:216–221
75. Rangel A, Burgaya F, Gavin R et al (2007) Enhanced susceptibility of Prnp-deficient mice to kainate-induced seizures, neuronal apoptosis, and death: role of AMPA/kainate receptors. J Neurosci Res 85(12):2741–2755
76. Walz R, Amaral OB, Rockenbach IC et al (1999) Increased sensitivity to seizures in mice lacking cellular prion protein. Epilepsia 40(12):1679–1682
77. Collinge J, Whittington MA, Sidle KCL et al (1994) Prion protein is necessary for normal synaptic function. Nature 370(6487):295–297
78. Manson JC, Hope J, Clarke AR et al (1995) PrP gene dosage and long term potentiation. Neurodegeneration 4(1):113–114

Chapter 16

Transgenic Mouse Models of Prion Diseases

Julie Moreno and Glenn C. Telling

Abstract

Seminal studies showed that optimal disease progression requires matching of the primary structures of PrP^{Sc} and PrP^{C} paved the way for the development of Tg models in which to study human prions and subsequently other naturally occurring mammalian prions. Here we review how transgenic mouse models are being used to elucidate basic molecular mechanisms of prion propagation and strain variation, and to address the likelihood of interspecies prion transmissions, in particular the zoonotic potential of various animal prion strains.

Key words Quasispecies, Species barriers, Strains, Transgenic models

1 Introduction

Prions challenge fundamental concepts of inheritance and infection. Prion-mediated phenotypes, in organisms such as yeast and Aplasia, as well as a variety of mammalian neurodegenerative diseases result from the ability of the prion conformation to interact with and induce further structural conversion of counterpart endogenous proteins in their normal states. The prototype prion diseases are the transmissible spongiform encephalopathies (TSEs) of animals and humans, which include scrapie in sheep, bovine spongiform encephalopathy (BSE), chronic wasting disease (CWD) of deer and elk, transmissible mink encephalopathy (TME), and various human disorders, the most common being Creutzfeldt Jakob disease (CJD). They share a number of common features, the most consistent being that disease is generally naturally and/or experimentally transmissible; the extraordinary physical properties of the infectious agent that account for its extreme resistance to conventional disinfection procedures; and the neuropathologic changes that accompany disease in the central nervous system (CNS). These features typically consist of neuronal vacuolation and degeneration, a reactive proliferation

Pawel P. Liberski (ed.), *Prion Diseases*, Neuromethods, vol. 129,
DOI 10.1007/978-1-4939-7211-1_16, © Springer Science+Business Media LLC 2017

of astroglia, and, in some conditions such as the human prion diseases kuru, Gerstmann Straussler Scheinker (GSS) syndrome, and variant CJD (vCJD), the deposition of amyloid plaques.

Central to all prion states are the protean conformational properties of particular host-encoded proteins. In the case of TSEs, this is the prion protein (PrP). The normal form, referred to as PrP^C, is a sialoglycoprotein of molecular weight ~ 33 kDa that is attached to the surface of neurons and other cell types by means of a glycophosphatidyl inositol (GPI) anchor. Experimentally, PrP^C is sensitive to protease treatment, can be released from cell surfaces by treatment with phosphoinositide phospholipase C (PIPLC), and is soluble in detergents as a monomer. During disease, a conformational variant of PrP^C, referred to as PrP^{Sc}, accumulates in infected brains and, in most cases, tissues of the lymphoreticular system. PrP^{Sc} is partially resistant to protease treatment, is resistant to PIPLC treatment, insoluble in detergents, and is prone to aggregate. In most examples of prion disease, experimental protease cleavage of the amino-terminal 66, or so, amino acids of PrP^{Sc} gives rise to a protease-resistant core, originally referred to as PrP27–30. Considerable evidence now supports the once unorthodox hypothesis that prions lack nucleic acid, are composed largely, if not entirely, of PrP^{Sc}, and that replication involves corruption of the benign cellular form of the prion protein by its abnormally conformed infective counterpart, which results in the exponential accumulation of prions, and inevitable demise of the infected host as a result of profound CNS neurodegeneration [1].

2 Prion Transmission Barriers

A crucial aspect of prion disorders is their transmissibility. Inoculation of prion diseased brain material into individuals of the same species will typically reproduce disease with a remarkably synchronous time to disease onset. Although transmission efficacy between species is, by comparison, generally significantly less than when prions propagate within their natural hosts, cross-species infection is nevertheless known to have featured in the etiologies of several epidemics, in particular BSE, vCJD, and most likely prions disease in mink, referred to as transmissible mink encephalopathy (TME).

2.1 PrP Primary Structure, Prion Strains, and the Species Barrier

Although the elements controlling interspecies prion transmission are not completely understood, seminal studies in Tg mice [2, 3] and cell-free systems [4] suggested that minimal amino acid divergences may have a major impact on the transmission efficiency, and that barriers to transmission between species resulted from primary structural incompatibilities between PrP^{Sc} constituting the prion, and substrate PrP^C expressed in the newly infected host.

Experimentally, the barrier to prion transmission between species may be absolute, in which case no transmission is recorded, or, more commonly, primary interspecies transmissions are characterized by variable rates of infection, and protracted, inconstant intervals to disease onset. Mechanistically, this is thought to reflect a two-step process in which initial stochastic conversion of host PrP^C by structurally mismatched PrP^{Sc} is followed by effective PrP^C conversion by the resulting structurally compatible PrP^{Sc}, leading to efficient replication of adapted prions, neurodegeneration, and inevitable death of the host. Consistent with this notion, serial passage of such adapted prions to additional animals expressing the same PrP^C results in a relatively short, synchronous time to disease in all inoculated recipients.

Transgenic approaches have also been used to address the notion that PrP primary structures expressed in particular species are inherently resistant to infectious conversion. Vidal and coworkers generated Tg mice overexpressing rabbit PrP (TgRab) under the control of the mouse prion protein gene promoter [5]. Their rationale for this approach was that rabbits have generally been considered to be resistant to prion infection. This group showed, however, that PMCA-derived rabbit PrP^{Sc} caused a prion disease in rabbits [6]. In subsequent studies, the susceptibility of TgRab to different prions was assessed. Susceptible prions included BSE, sheep BSE, L-type BSE, in vitro generated rabbit prions, and the same passaged through rabbits, ME7, and RML. SSBP-1, CWD, and atypical scrapie did not cause disease.

The influence of intraspecies PrP polymorphisms on prion disease susceptibility in mice [7], sheep [8], and humans [9–11] supported the concept that the primary structure of PrP was a crucially important determinant of prion species barriers. Nonetheless, while the foregoing mechanism is in accordance with the outcomes of the majority of prion interspecies transmission barriers, other circumstances indicate that our knowledge of the parameters controlling interspecies prion transmission remains incomplete. For example, the criteria governing PrP primary structural control over transmission appear to be moot in the case of the bank voles, which are susceptible to prions from a number of mammals with divergent PrP primary structures [12].

Mammalian prion strains are an equally important component affecting prion transmission. Seminal studies indicated that distinct heritable prion strain phenotypes, including the incubation time to disease, and the profile of CNS lesions, are enciphered within the conformation of PrP^{Sc} [13, 14]. This provided investigators with the ability to use the biochemical properties of PrP^{Sc} as a means of tracking, and to some extent identifying strains by assessing of PrP^{Sc} conformation, and/or the extent of PrP^{Sc} glycosylation [15–18]. Since a change in PrP^{Sc} conformation accompanies the emergence of newly adapted prions following passage across species barriers [16], the concept of a strain barrier was introduced to account for the influence of distinct conformations of PrP^{Sc} molecules [19]. The unexpectedly wide host range of BSE prions, exemplified by transmission to humans as variant CJD (vCJD), as well as the restricted transmission of vCJD prions in human PrP Tg mice [20, 21] compared to bovine PrP Tg mice [22] exemplify the capacity of particular prion strains to overcome the influence of PrP primary structure.

While mutational events in agent-associated nucleic acid were originally cited as the cause of strain instability [23], more recently, changes in the conformation of PrP^{Sc} have been shown to be associated with the acquisition of new strain properties [16]. To account for the phenomena of prion transmission barriers, strain instability, heterogeneity, and adaptation in the context of PrP^{Sc} conformation, the conformational selection model [24, 25] was proposed to reconcile how PrP primary structure and prion strain conformations interact to control transmission barriers to prion propagation. This model postulates that only a subset of PrP^{Sc} conformations is compatible with each individual PrP primary structure [24, 26]. By extension, this leads to the notion that prions exist as an array of quasispecies conformations, a model first applied to populations of a virus within its host [27]. There is a mounting body of experimental evidence for this concept [28–30]. Quasispecies acquire fitness when populations of prion conformers are subjected to selective pressure, for example, during propagation in a host expressing a different PrP primary structure following interspecies transmission. From this perspective, PrP^C primary structure influences the portfolio of thermodynamically preferred PrP^{Sc} conformations that are kinetically selected during propagation, such that only a subset of such PrP^{Sc} conformers are optimized for fitness in a particular host [31, 32].

The issue of strain effects on the efficacy of cross-species transmission was elegantly addressed in transgenic experiments by Béringue and colleagues who compared the ability of brain and lymphoid tissues from ovine and human PrP Tg mice to replicate foreign, inefficiently transmitted prions [33]. They observed that lymphoid tissue was consistently more permissive than the brain for CWD and BSE prions. Furthermore, when the transmission barrier was overcome through strain shifting in the brain, a distinct agent propagated in the spleen, which retained the ability to infect the original host.

An additional factor affecting the efficiency of interspecies transmissions was revealed by recent experiments using Tg mice expressing PrP^C lacking the GPI moiety that tethers the protein to the cell surface. Previous in vitro studies showed that an artificially mutant version of PrP that lacks the terminal sequence for addition of the GPI lipid anchor could acquire resistance to protease digestion and therefore resembled PrP^{Sc} [4, 34]. To determine whether this mutant PrP can support prion propagation and PrP^{Sc} production in vivo, Chesebro and colleagues produced Tg mice expressing "anchorless" PrP [35]. As expected, anchorless PrP was not expressed on the surface of cells derived from Tg mice; instead it was secreted. When inoculated with prions, the mice accumulated mutant PrP in the form of abundant amyloid plaques throughout the brain as early as 70 days after inoculation. But despite their accumulation of amyloid-forming protease-resistant PrP (in many cases at levels higher than PrP^{Sc} found in the brains of clinically sick wild-type mice), these Tg mice failed to develop neurologic signs of prion disease up to 600 days following prion inoculation—long after inoculated control mice succumbed to disease. To test the effect of lack of GPI anchoring on a species barrier model anchorless 22L mouse prions derived from $TgGPI^-$ mice were more infectious than wild-type 22L prions in tg66 mice expressing HuPrP at levels 8- to 16-fold above normal, but not in tgRM transgenic mice, which expressed human PrP at 2- to 4-fold above normal.

2.2 Variable Effects of Endogenous Mouse PrP Expression in Transgenic Mouse Models

The availability of *Prnp*$^{0/0}$ knockout mice and the characterization of increasing numbers of Tg mice expressing different PrP alleles revealed interesting protective functions of wild-type PrP on various PrP-related pathologies. In some cases disease can be partially or fully suppressed by co-expression of wild-type PrP. This effect was first observed during the characterization of Tg mice expressing human (Hu) PrP, referred to as Tg(HuPrP) mice, which only became susceptible to CJD prions when endogenous wild-type mouse PrP was eliminated by crossing the HuPrP transgene array to *Prnp*$^{0/0}$ knockout mice [36, 37]. Because of this dominant negative effect, subsequent Tg mouse models expressing foreign PrP coding sequences have generally been either produced in or ultimately crossed with *Prnp*$^{0/0}$ knockout mice. Coincidentally, these observations also formed the partial basis for the model of prion replication involving an auxiliary factor, referred to as protein X, which was proposed to bind to PrP^C and facilitate conversion to PrP^{Sc} [21, 37, 38]. While subsequent experimental evidence questioned the requirement for protein X [39], this model remains a guiding principle that has informed structural analyses of PrP^C [40, 41].

Interactions between wild-type and mutant prion proteins have also been shown to modulate neurodegeneration in Tg mouse models expressing mutated PrP, the seminal observations being made in Tg mouse models of the inherited human prion disease called Gerstmann Strauusler Sheinker (GSS) syndrome, caused by mutation of codon 102 which results in substitution of proline for leucine [42]. In contrast, neurodegeneration induced by expression of a nine octapeptide insertion associated with familial Creutzfeldt Jakob disease (CJD), designated Tg(PG14), is unaffected by wild-type mouse PrP expression [43]. These studies are also consistent with a model in which mutated and wild-type PrP compete for a hypothetical binding partner that, in this case, serves to transduce neurotoxic signals. Interestingly, while Tg mice expressing PrP tagged at its amino terminus with green fluorescent protein (GFP) supported compromised prion replication, prion propagation was facilitated by co-expression of wild-type PrP, suggesting that wild-type PrP rescued an altered amino terminal function in the tagged PrP [44]. In contrast, the effect of tagging PrP at the C-terminus with GFP was that Tg mice expressing this construct were incapable of sustaining prion infection and the PrP-GFP chimera acted as a dominant-negative inhibitor of wild-type PrP conversion to PrP^{Sc} [45]. In similar fashion, co-expression of wild-type mouse PrP was shown to rescue the neurodegenerative phenotype of mice expressing amino-terminal deletions [46–48].

2.3 Transgenic Mouse Models of Mammalian Prion Diseases

Seminal experiments by Scott and coworkers abrogated the resistance of mice to hamster prions by transgenic expression of hamster PrP^{C} in mice [2, 49] and paved the way for the development of a variety of facile Tg mouse models that authentically recapitulate known transmission barriers [50]. The lack of species barrier during homotypic transmission, i.e., when the host expresses a PrP gene identical to that of the infecting species, led to the development of many different facile mouse models in which to study the biology of mammalian prions, including sheep, bovine, human, cervid, and mink, by transgenic expression in mice of PrP coding sequences from these various species. These Tg resources largely circumvented the imprecise approach of studying the biology of various mammalian prion diseases by transmission to generally non-susceptible experimental animals, or, where possible, the natural host, but has also greatly contributed to our understanding of various mechanistic aspects of these extraordinary pathogens [51]. Generally, an inverse correlation exists between the length of prion incubation time in these Tg mouse models and transgene expression level. Transgenes are usually expressed on a *Prnp* knockout background (*Prnp*$^{0/0}$) [52] in order to avoid partial or full suppression of disease caused by co-expression of wild-type PrP. A variety of different transgenic mice expressing chimeric versions of PrP in which specific regions of mouse PrP primary structure were replaced by the corresponding elements from human, sheep, and bovine PrP have also been created [36, 37, 53–55]. Inclusion of certain mouse PrP primary structural elements in such constructs, in particular MHu2MPrP, also countermanded the inhibitory effect of mouse PrP co-expression [36, 37].

In an alternative approach, expression of foreign PrP genes in mice has been accomplished by gene replacement methods. This approach ensures that the PrP coding sequence is controlled by the same regulatory elements as wild-type mouse PrP, in which case gene expression is expected to recapitulate authentic PrP^C expression. In both cases, PrP sequence identity between the transgenic host and donor usually lead to a higher transmission rate as compared to wild-type mice. When considering such transmission models, it is important to consider that outcomes might be limited by the life span of the recipient host, which might be exceeded by the incubation period, although pathological examination, PrP^{Sc} detection, and secondary passage may identify infected animals. These problems can be partially overcome by transgenic mice overexpressing PrP^C which might be desirable to fully assess the extent of a species barrier, since it results in highly reduced incubation times. For example, while human PrP knock-in mice did not register disease when challenged with prions from cattle affected with BSE [56], infection did occur in Tg mice expressing higher levels of human PrP [20].

3 Human Prion Diseases

The initial transmission of human prions to experimental primates has a rich history beginning with William Hadlow's recognition of the similarity between kuru and scrapie, and his prediction that patient brain extracts would cause disease in inoculated non-human primates after a prolonged incubation period [57]. Seven years later, Gajdusek, Gibbs, and Alpers demonstrated the transmissibility of kuru to chimpanzees after incubation periods ranging from 18 to 21 months [58], and subsequently showed that CJD was similarly infectious [59]. For many years the transmission of human prion diseases was investigated using this approach [60], but the expense of housing these animals for long time periods, as well as ethical concerns surrounding their use, severely limited these studies. Inoculations of laboratory rodents produced variable results [61–65]. Generally, only ~10% of intracerebrally inoculated mice developed CNS dysfunction with incubation times of >500 days (d) [36], limiting the ability of this approach to characterize specific agent strains.

While seminal transgenic investigations by Prusiner and colleagues study the transmission properties of scrapie prions experimentally adapted to hamsters or mice [3, 49], Tg mice expressing human PrP^C, referred to as Tg(HuPrP) mice, were the first to abrogate species barriers to naturally occurring prions, in this case prions causing human diseases such as human disorders such as sporadic and iatrogenic CJD (sCJD and iCJD) [36, 37]. As previously described for the animal prion diseases, human prion disease susceptibility is strongly influenced by polymorphic variation of *PRNP*. In particular, homozygosity at *PRNP* codon 129, which encodes methionine (M) or valine (V), predisposes to the development of sporadic and acquired CJD [10, 11]. Surprisingly, two lines of Tg(HuPrP) mice expressing HuPrP with V at residue 129 (HuPrP-V129), referred to as Tg(HuPrP)152 and Tg(HuPrP)110, inoculated with CJD prions failed to develop CNS dysfunction more frequently than non-transgenic controls [36]. Subsequently, mice expressing a chimeric human/mouse PrP transgene, designated MHu2M, were constructed, because earlier studies had shown that a chimeric hamster/mouse PrP gene supported transmission of either mouse or hamster prions [3, 66]. These Tg(MHu2M)5378 mice were found to be highly susceptible to human prions suggesting that Tg(HuPrP) mice have considerable difficulty converting $HuPrP^{C}$ into PrP^{Sc} [36]. However, Tg(HuPrP)152 mice, and another line designated Tg(HuPrP)440, which expresses HuPrP with M at 129, when crossed with $Prnp^{0/0}$ [52], were rendered susceptible to human prions [37]. These observations demonstrated that Tg(HuPrP) mice were resistant to human prions because mouse PrP^{C} inhibited the conversion of $HuPrP^{C}$ into PrP^{Sc}. In contrast, Tg(MHu2M)5378 mice crossed onto the null background were only slightly more susceptible to human prions compared to Tg(MHu2M)5378 mice that expressed both chimeric and $MoPrP^{C}$. Furthermore, Tg(MHu2M) mice inoculated with either Hu or chimeric MHu2M prions exhibited similar incubation times.

The availability of such susceptible Tg mice made possible the rapid and relatively inexpensive transmission of human prion diseases for the first time. Based on these findings, several additional similar Tg mouse models expressing HuPrP have been produced, with identical results [20, 67, 68]. Gene-targeting approaches have also been employed to produce mice expressing human PrP [69]. This approach has the obvious advantage of expressing "normal" levels of transgene-encoded PrP under the control of the *Prnp* transcriptional elements, and, for example, to conveniently model the effects of heterozygosity at codon 129. Transmission of a limited number of sporadic CJD cases in these mice has provided evidence for four distinct prion strains. Evidence of strain variation in sCJD has also come from laboratory studies in bank voles [12].

Early transmission studies to Tg(MHu2M)5378 mice provided evidence that different human prion strains, namely in patients with fatal familial insomnia (FFI), caused by mutation at codon 178 (D178N), and familial CJD, caused by mutation at codon 200 (E200K), are enciphered by different conformational states of PrP^{Sc} [13], a concept first elaborated following transmission of the hamster-adapted strains of TME called hyper (HY) and drowsy (DY) [14]. While the conformational enciphering hypothesis is supported by considerable additional experimental evidence, defining human prion strain prevalence has been hampered by difficulties in arriving at an internationally accepted classification system for human prion strains [70, 71], and by the observation that multiple PrP^{Sc} subtypes coexist in the same brain [29, 72]. Coincidentally, there transmissions also supported the concept that genetically programmed prion diseases are also transmissible. Other examples of inherited human prion diseases that have been transmitted under similar circumstances include transmission of an inherited form of CJD caused by mutation of codon 210 which changes valine to isoleucine at this residue [73].

Given the importance of infectious transmission in prion diseases, the availability of Tg mice with susceptibility to human prions has increased the number of known sporadic prion diseases. Transmission to Tg(MHu2M) mice of an unusual case of human prion diseases that presented with insomnia but no *PRNP* abnormalities led to the discovery of a novel prion disease referred to as sporadic fatal insomnia [74]. More recently transmission of a new prion disorder, referred to as variably protease-sensitive prionopathy (VPSPr), a seemingly sporadic disease that is distinct from CJD but shares features of GSS, confirmed that VPSPr might be the long-sought sporadic form of GSS [75]. While inefficient transmission of VPSPr was recorded on first passage to Tg mice overexpressing HuPrP, infectivity was not serially transmissible. Thus, while VPSPr is an authentic prion disease, Tg(HuPrP) mice do not appear able to sustain replication beyond the first passage. Similar findings using gene-targeted mice expressing HuPrP confirmed the interpretation that VPSPr has limited potential for human-to-human transmission [76].

A significant goal of Tg development was to produce mice in which the incubation time of prions was as rapid as possible using additional chimeric mouse/human PrP transgenes [21]. Korth and coworkers refined the MHu2M PrP approach, and optimized human prion transmission by replacing key human PrP residues with the equivalent residues from mouse. The resulting chimera, referred to as Tg(MHu2M,M165 V,E167Q) mice resulted in shortening the incubation time to approximately 110 days for prions from sCJD patients and divergence into two strain types [21]. Even shorter incubation times and CJD strain evolution were also observed in another line, termed Tg1014 in which a single additional residue (M111 V) was reverted to mouse [77].

Development of Tg mice with susceptibility to human prions was timely, as it occurred in the context of significant human exposure to BSE prions, at least in the UK, and the consequential occurrence of vCJD in young adults and teens [78]. Tg(HuPrP) created by Prusiner and colleagues, and similar mice subsequently generated by other groups [20] were used to characterize vCJD prions, and to model human susceptibility to BSE [17, 18, 22, 79].

As previously described for the animal prion diseases, human prion disease susceptibility is strongly influenced by polymorphic variation of *PRNP*. In particular, homozygosity at *PRNP* codon 129, which encodes M or V, predisposes to the development of sporadic and acquired CJD [10, 11]. Transmission studies of human CJD cases to transgenic mice confirm the influence of this polymorphism. While mice expressing V129 are susceptible to all PrP^{Sc} types and PrP 129 genotypes [17, 18, 21, 37, 79], mice expressing the HuPrP-129 M allele are susceptible to prions from M129 homozygous patients, transmissions from patients mismatched at this codon, or heterozygotes are generally more inefficient [20, 21, 37].

Strikingly, all neuropathologically confirmed vCJD cases studied so far have been homozygous for M at codon 129 [80]. Transmission studies of human CJD cases to Tg mice confirm the influence of this polymorphism. While mice expressing V129 are susceptible to all PrP^{Sc} types and PrP 129 genotypes [17, 18, 21, 37, 79], mice expressing the HuPrP-129 M allele are susceptible to prions from M129 homozygous patients, transmissions from patients mismatched at this codon, or heterozygotes are generally more inefficient [20, 21, 37]. Although initial BSE transmissions to Tg(HuPrP)152 mice were uniformly negative, suggesting a substantial species barrier in humans to BSE prions [79], subsequent BSE transmission to Tg mice expressing M at human PrP codon 129 revealed inefficient transmission, characterized by low attack rates and long incubation times. Moreover, a strain shift was occasionally observed in these transmissions, producing a sCJD-like phenotype in a proportion of inoculated Tg mice [20, 56]. In contrast, gene-targeted transgenic mice expressing human PrP were not susceptible to BSE prions [56]. However, these mice were susceptible to sheep-adapted BSE prions suggesting increased susceptibility of humans to BSE prions following passage through sheep [81], an effect that is unrelated to increased titer of BSE prions in sheep brain [82]. The effect of codon 129 heterozygosity on human prion susceptibility has been further examined in Tg mice co-expressing transgene arrays expressing HuPrP-M129 and HuPrP-V129 [83].

The emergence of vCJD in the late twentieth century renewed interest in kuru, another acquired human prion disease among the Fore peoples of the inner highlands of Papua New Guinea. Kuru exemplifies the epidemic nature of prion diseases. By mid-twentieth century it was the leading cause of death in this region, killing over 3000 people in the exposed population of 30,000. While its etiology was initially puzzling, studies showing that kuru patient brain extracts produced a progressive neurodegenerative condition in inoculated chimpanzees after a prolonged incubation period [58], supporting the notion that kuru was an infectious disorder. Despite neuropathological similarities, scrapie was clearly not a candidate for the origin of this infectious disorder. Moreover, kuru predated the BSE/vCJD epidemic by several decades. In fact kuru was a devastating consequence of ritualistic endocannibalism, where brains and other body parts of tribal elders were consumed as an act of mourning.

Remarkably, a novel *PRNP* variant was found uniquely among unaffected individuals in the kuru-exposed population, in which V at residue 127 was replaced by glycine (G) [84]. This polymorphism was hypothesized to be an acquired prion disease resistance factor, selected in response to the kuru epidemic. Asante and colleagues modeled this kuru resistance polymorphism, referred to as HuPrP V^{127}, in Tg mice [85]. Tg mice having the genotype associated with disease resistance, specifically heterozygosity (G/V) at position 127, and M/M homozygosity at 129, referred to as $G^{127}M^{129}/V^{127}M^{129}$, were completely protected against kuru prions. In contrast all kuru inoculated mice expressing G/G at 127 and either M/M or V/V at 129 ($G^{127}M^{129}/G^{127}M^{129}$ and $G^{127}V^{129}/G^{127}V^{129}$) developed disease after ~200 days. Complete recapitulation of genetic susceptibility to kuru therefore validated this transgenic modeling approach. Given the previously recognized similarities between kuru and sCJD, it came as no surprise that $G^{127}M^{129}/V^{127}M^{129}$ were also protected from CJD prions. Interestingly however, these same mice were incompletely protected against vCJD prions. Remarkably, mice homozygous for G^{127} M^{129} were completely protected against all forms of human prion diseases, including vCJD, and failed to manifest either clinical or subclinical signs of prion disease. As such, they behaved like mice which fail to express PrP as a result of disruption of the PrP gene, and yet still express HuPrP. The availability of Tg mice expressing different levels of HuPrP M^{129} and V^{129} showed that the mechanism underlying the profound inhibitory effects of HuPrP V^{127} occur independently from M129 V, which is thought to influence PrP interactions during the replicative process. Moreover, HuPrP V^{127} acts as a dominant negative inhibitor of prion conversion.

4 Transgenic Models of Inherited Human Prion Diseases

Approximately 10–20% of human prion diseases exhibit an autosomal dominant mode of inheritance resulting from missense or insertion mutations in the coding sequence of the human PRNP. Several mutations are genetically linked to loci controlling familial CJD, GSS syndrome, and FFI. GSS syndrome, which is characterized clinically by ataxia and dementia and neuropathologically by the deposition of PrP amyloid, most commonly results from mutation at codon 102 of PRNP resulting in the substitution of leucine (L) for proline

(P) [86]. GSS linked to this mutation is transmissible to non-human primates [60], wild-type mice [63], Tg mice expressing a chimeric mouse-human PrP gene expressing the GSS mutation [37], and *Prnp* gene-targeted mice referred to as 101LL [87].

Initial studies in Tg(GSS) mice which attempted to understand how an inherited disease could also be infectious suggested that prions in the brains of spontaneously sick Tg(GSS) mice could be transmitted to Tg mice expressing lower levels of mutant protein, referred to as Tg196 mice [42, 88]. Disease was also induced in Tg196 mice by a mutant synthetic peptide comprising MoPrP residues 89–103 refolded into a beta-sheet conformation [89] and this disease was subsequently propagated to additional Tg196 mice [90]. While these studies lent support to the prion hypothesis, since they suggested that pathogenic PrP gene mutations resulted in the spontaneous formation of PrP^{Sc} and de novo production of prions [91], this explanation was controversial for several reasons. Although protease-sensitive forms of PrP^{Sc} have been identified using biochemical and immunological methods [15, 92], the lack of protease-resistant PrP^{Sc} in the brains of spontaneously sick or recipient mice eliminated a property that, to some, was synonymous with prion infectivity. Moreover, *Prnp* gene-targeted 101LL mice expressing MoPrP-P101L failed to spontaneously develop neurodegenerative disease [87]. Finally, disease transmission from spontaneously sick mice to wild-type mice did not occur and spontaneous disease was eventually registered in aged Tg196 mice [89, 90] complicating the interpretation of the original transmission experiments.

To address the apparent dissociation of prion infectivity and PrP^{Sc} in this well-established Tg model, subsequent studies attempted to use means other than differential resistance to proteinase K treatment to detect disease-associated forms of PrP in spontaneously sick Tg mice expressing MoPrP-P101L. Using the prototype PrP^{Sc}-specific monoclonal antibody (Mab) reagent referred to as 15B3 [93], Nazor and coworkers showed that disease in Tg mice overexpressing MoPrP-P101L results from the spontaneous conversion of mutant PrP^{C} to protease-sensitive $MoPrP^{Sc}$-P101L, defined by its reactivity with 15B3, that accumulates as aggregates in the brains of sick Tg mice [94]. To understand the influence of mutant PrP expression levels on the transmissibility of spontaneously generated pathogenic MoPrP-P101L, they produced mice in which transgene copy numbers and levels of MoPrP-P101L expression were carefully defined. While inoculation of disease-associated MoPrP-P101L accelerated disease in Tg mice expressing MoPrP-P101L from multiple transgenes, disease transmission neither occurred to wild-type nor Tg mice expressing MoPrP-P101L from two transgene copies that did not develop disease spontaneously in their natural life span.

Since disease transmission from spontaneously sick Tg(GSS) mice depended on recipient mice expressing MoPrP-P101L at levels greater than that produced by two transgene copies, and since such levels of overexpression ultimately resulted in spontaneous disease in older uninoculated recipients, these results suggest that the phenomenon of disease transmission from spontaneously sick Tg(GSS) mice might be more appropriately viewed as disease acceleration whereby inoculation of disease-associated MoPrP-P101L promotes the aggregation of precursors of pathological MoPrP-P101L that result from transgene overexpression. Such a scheme is consistent with a nucleated polymerization mechanism of prion replication originally postulated from cell-free conversion systems [4] and subsequently demonstrated to be the basis of prion propagation in lower eukaryotes [95]. According to this model, PrP^C is in equilibrium with PrP^{Sc}, or its precursor, and the equilibrium normally favors PrPC. Also, PrP^{Sc} is stable only in its aggregated form that can "seed" polymerization of additional PrP^C, thus converting it into additional PrP^{Sc}. Further, more definitive work combining studies in Tg(GSS) and *Prnp* gene-targeted 101LL mice indicated that clinical or profound neuropathological changes were absent in gene-targeted mice inoculated with brain extracts of spontaneously sick Tg(GSS) mice, indicating that de novo formation of abnormally aggregated PrP in the host does not always result in a transmissible prion disease [96]. In a related issue, the role of PrP overexpression in the production and transmission of synthetic mammalian prions [97] originating from *E. coli*-derived recombinant MoPrP remains to be determined. While the transmission properties and protease-resistance of MoPrP (89–230) SMPs are clearly different from disease-associated MoPrP-P101L, it may be significant that the Tg mice in which these SMPs were initially derived expressed MoPrP(89–231) at levels 16 times higher than normal.

Interestingly, unlike mice expressing the GSS P102L mutation in the context of the mouse PrP primary structure, Tg mice expressing human PrP with the P102L mutation failed to develop disease spontaneously with increasing age. However, these mice were susceptible to infection from patients with the homotypic pathogenic mutation, as well as CJD, producing distinct prion strains with transmission properties distinct from sporadic and acquired human prion disease [98]. In contrast to reports in the gene-targeted mouse model, GSS-102 L prions produced in this study were incapable of transmitting disease to wild-type mice [87].

Transgenic expression of other disease-associated mutations in the context of mouse or human PrP has been met with varying success. While Tg mice expressing mouse PrP containing the most common E200K fCJD mutation (E199K in mouse PrP) did not develop disease spontaneously [42], a Tg mouse expressing chimeric

MHu2M PrP [37] containing the E199K mutation PrP developed neurological signs at 5–6 months of age and deteriorated to death several months thereafter [99]. Inoculation of brain extracts from diseased Tg(MHu2M-E199K) mice induced a distinct fatal prion disease in wt mice. Mice expressing a mouse PrP version of a nine octapeptide insertion associated with familial CJD, designated Tg(PG14), exhibited a slowly progressive neurological disorder characterized by apoptotic loss of cerebellar granule cells, gliosis but no spongiosis [100]. Whether the brains of sick Tg(PG14) mice, like sick Tg(GSS) mice, contain 15B3-immunoprecipitable PrP has not yet been reported; however, in both models, mutated PrP adopts different pathologic conformations either spontaneously or following inoculation with authentic prions [94, 101]. Like Tg(GSS) mice, brain homogenates from spontaneously sick Tg(PG14) mice failed to transmit disease to Tg mice that express low levels of mutated PrP that do not become sick spontaneously. Whether differences in the state of aggregation of PG14spon compared to MoPrP-P101L will affect its ability to accelerate disease progression in overexpressor Tg(PG14) mice remains to be determined.

Chiesa's group also described a Tg mouse model of inherited CJD expressing the mouse homolog of the D178N in combination with the V129 polymorphism. These Tg(CJD) mice had EEG and sleep abnormalities, memory impairment, motor dysfunction, and striking morphological alterations of the neuronal endoplasmic reticulum (ER) associated with ER retention of mutant PrP [102]. This study was followed by reports of the properties of Tg mice expressing the mouse PrP homolog of the same D178N mutation in cis with M129, which is associated with FFI [103]. Spontaneous disease in these so-called Tg(FFI) mice was different from Tg(CJD) mice. Tg(FFI) synthesize misfolded mutant PrP in their brains and, like FFI, illness is associated with sleep disruption. However, unlike this form of fCJD and FFI, bioassay and protein misfolding cyclic amplification (PMCA) showed that prions were not produced in Tg(FFI) and Tg(CJD) brains, suggesting to the authors that the disease-encoding properties of mutant PrP do not depend on its ability to propagate its misfolded conformation. Tg(FFI) mice complement a gene-targeted model of this disease that provided a different outcome. Knock-in mice carrying the FFI mutation [104] developed biochemical, physiological, behavioral, and neuropathological abnormalities similar to FFI, and this spontaneous disease could be transmitted, and serially passaged, to mice expressing physiological amounts of PrP without the mutation. Likewise, the same investigators produced knock-in mouse models of CJD caused by the N178/V129 variants. These mice differed phenotypically from FFI knock-in mice, producing a spontaneous, transmissible CJD-like disease [105]. The reasons for the discrepant properties of Tg and knock-in models with respect to spontaneous prion formation are unclear.

Transgenic mice expressing a mutation at codon 117 associated with a telencephalic form of GSS [106] also spontaneously developed neurodegenerative disease and accumulated an aberrant, neurotoxic form of PrP termed CtmPrP, which appears to be distinct from conventional protease-resistant PrP^{Sc} [107]. Mastrianni and colleagues also constructed Tg mice that express PrP carrying the mouse homolog of this GSS mutation. These Tg(A116V) mice express approximately six times the endogenous levels of PrP, and recapitulate many clinicopathologic features of GSS(A117V) that are distinct from CJD [108]. More recently, Asante and coworkers produced Tg mice expressing the A117V mutation in the context of HuPrP. Unlike previous attempts at transmission, prions from human patients with GSS A117V transmitted to these Tg mice, producing appropriate neuropathology, and accumulation of PrP^{Sc} [109].

5 Bovine Prion Disease

Several lines of Tg mice expressing bovine PrP, referred to as Tg(BoPrP) mice, have been independently produced [55, 110, 111]. All Tg(BoPrP) mice are susceptible to BSE prions, and their availability during the peak years of the BSE and vCJD epidemics were invaluable for ascertaining the pathogenesis of BSE and the properties of these and other bovine prion diseases.

While BSE initially appeared to be a homogeneous disease, the large-scale testing of livestock nervous tissues for the presence of PrP^{Sc} led to the recognition, in Europe, Japan, and the USA, of two additional bovine PrP^{Sc} variants termed H- and L-types [112]. The molecular signature of bovine PrP^{Sc} from animals with the bovine amyloidotic spongiform encephalopathy (BASE) variant corresponds to L-type and appears similar to a distinct subtype of sporadic CJD [113]. L-type has a tendency to form amyloid plaques in cattle brain and has a distribution of brain pathology distinct from BSE [113]. These "atypical BSE" cases have been detected in aged asymptomatic cattle during systematic testing at slaughterhouse. The etiology of these atypical forms remains unexplained but could involve either (1) a change in the biological properties of the BSE agent, (2) infection of cattle with prions from another source, such as scrapie or CWD, or (3) previously unrecognized sporadic forms of prion disease in cattle. A case of BSE in the USA with an H-type PrP^{Sc} signature in an approximately 10-year-old cow from Alabama was also associated with mutation of glutamate (E) to lysine (K) at 211, referred to as E211K [114]. Of particular significance, the identical substitution at the equivalent codon 200 in human *PRNP* is linked to the most frequent form of familial CJD with clusters described in Chileans, Oravian Slovaks, Libyan Jews, Britons, and Japanese.

The development of various Tg mouse models expressing bovine PrP were invaluable for characterizing and titrating BSE infectivity in a variety of tissues [55, 111, 115], and provided compelling evidence for a relationship between vCJD and BSE [22]. While these Tg mouse models were characterized by rapid incubation times and 100% attack rates, mice expressing bovine PrP generated by gene replacement of the mouse PrP coding sequence had long incubation times (>500 days) and incomplete attack rates [56]. The experimental transmission of H- and L-type cases to bovine PrP Tg mice unambiguously demonstrated their infectious nature and revealed strain properties distinct from BSE [116–119]. While BSE and BASE transmitted readily to Tgbov XV mice, they produced different clinical, neuropathological, and molecular disease phenotypes [119]. Interestingly, the same study indicated that BASE prions were able to convert into BSE prions upon serial transmission in inbred mice. The relationship of this finding to the apparently protean nature of BSE prions in aforementioned transmission studies [20, 21, 120] remains to be determined.

Strikingly, serial passage of the L-type strain to wild-type mice, and mice expressing the VRQ allele of ovine PrP induced a disease phenotype indistinguishable from that of BSE [117, 119], suggesting a possible etiological relationship between atypical and classical BSE. The relevance of these findings to studies in Tg mice, which consistently reveal the existence of more than one molecular type of PrP^{Sc} [20, 21, 120] and suggest that more than one BSE prion strain might infect humans [20], remains to be determined.

Challenge of two lines of Tg mice expressing human PrP with M at codon 129 with L-type isolates produced a molecular phenotype distinct from classical BSE [67, 68]. In one case, L-type transmitted with no transmission barrier [68], and in both cases the L-type PrP^{Sc} biochemical signature was conserved upon transmission. In contrast, the transmission efficiency of classical BSE and H-type isolates to transgenic mice expressing human PrP is relatively low [20, 68, 121]. Increased pathogenicity of sheep-passaged BSE occurred in Tg mice expressing porcine PrP [122] or human PrP [82] raising the possibility that BSE may gain virulence by passage in another species.

6 Ovine Prion Disease

Transgenic mice expressing ovine PrP are susceptible to prions from scrapie-affected sheep [123–128]. A clear link to codon 136 genotype and susceptibility/resistance to different sheep scrapie isolates has been described in multiple previous studies. Generally, increased susceptibility to scrapie is associated with expression of sheep PrP with valine (V) at residue 136, referred to as OvPrP-V136 compared to alanine (A) at residue 136, referred to as OvPrP-A136, with A/A136 being the most resistant, and V/V136 the most susceptible genotypes. In the case of SSBP/1 incubation periods are ~170 days in V/V136 sheep, while transmission to A/A136 sheep is relatively inefficient, with no disease recorded after >1000 days [129]. The most widely characterized models include Tg mice expressing OvPrP-V136, including tg338 [127], and Tgov59 mice [130], or Tgov4 [125] mice expressing OvPrP-A136. In the case of tg338 mice, the transgene comprised a bacterial artificial chromosome insert of 125 kb of sheep DNA, while in the case of Tgov59 and Tgov4 mice the neuron-specific enolase promoter was used to drive OvPrP expression. These lines are maintained on different heterogeneous genetic backgrounds, and CNS expression levels in tg338 mice are ~8- to 10-fold higher than wild type, while Tgov59 and Tgov4 lines each overexpress OvPrP-ARQ at levels ~2- to 4-fold higher than those found in sheep brain. Spontaneous neurological dysfunction has been reported in Tg lines overexpressing OvPrP [123, 127].

Subsequently, two additional Tg models, referred to as Tg(OvPrP-A136)3533$^{+/-}$ and Tg(OvPrP-V136)4166$^{+/-}$ mice were produced and characterized that express either OvPrP-V136 or OvPrP-A136 approximately equivalent to PrP levels normally expressed in the CNS of wild-type mice [131]. Importantly, the influence of residue 136 on the transmission of SSBP/1 and CH1641 prions in Tg(OvPrP-A136)3533$^{+/-}$ and Tg(OvPrP-V136)4166$^{+/-}$ mice is in accordance with the properties of these isolates in sheep of various genotypes [132]. While SSBP/1 eventually transmits to Tg(OvPrP-A136)3533$^{+/-}$ mice with incubation times exceeding 400 days, the general effects of the A/V136 dimorphism on SSBP/1 transmission observed in sheep are recapitulated in Tg(OvPrP-A136)3533$^{+/-}$ and Tg(OvPrP-V136)4166$^{+/-}$ mice. Similarly, CH1641, which propagates efficiently in A/A136 sheep [129], preferentially propagates in Tg(OvPrP-A136)3533$^{+/-}$ mice. In other studies, CH1641 transmitted to TgOvPrP4 mice with an ~250 d mean incubation time [133].

Tg(OvPrP-A136)3533$^{+/-}$ and Tg(OvPrP-V136)4166$^{+/-}$ lines were produced using the cosSHa.Tet cosmid vector which drives expression from the PrP gene promoter [66], and therefore it was expected that expression of OvPrP^{C}-A136 and OvPrP^{C}-V136 occurs in identical neuronal populations. Accordingly, homozygous Tg mice were mated to produce mice, referred to as Tg(OvPrP-A/V) mice, that co-express both alleles [131]. Previous studies reported on Tg mice expressing OvPrP with V at 136, referred to as Tg(OvPrP)14,882$^{+/-}$ mice, that were also produced in a *Prnp*$^{0/0}$/FVB background using the cosSHa.Tet cosmid vector [128]. However, in that study, comparable Tg mice expressing OvPrP-A136 were not reported. While SSBP/1 incubation times are prolonged in A/V136 compared to V/V 136 sheep [129], incubation times were shorter in Tg(OvPrP-A/V) than in Tg(OvPrP-V136)4166$^{+/-}$ mice.

A novel mAb PRC5, the epitope of which comprises residue 136 [134], was used to monitor conversion of OvPrP^{C}-A136 in compound heterozygous. Surprisingly, in contrast to its relatively slow conversion when OvPrP^{C}-A136 is expressed in isolation, co-expression with OvPrP^{C}-V136 in Tg(OvPrP-A/V136) mice facilitated rapid conversion of OvPrP^{C}-A136 to OvPrP^{Sc}-A136. The conformation and diffuse CNS distribution of the resulting OvPrP^{Sc}-A136(U) were equivalent to that of OvPrP^{Sc}-V136(U) and not OvPrP^{Sc}-A136(S). These results demonstrate that under conditions of allele co-expression a dominant conformer may alter the conversion potential of an otherwise resistant PrP polymorphic variant to an unfavorable prion strain [131].

Ovine Tg mice have been shown to be a useful tool for discriminating scrapie strains [135] and in particular for differentiating BSE in sheep from natural scrapie isolates [126, 130, 136]. Tg mouse models are also at the forefront of characterizing a relatively newly emerging "atypical" scrapie strain [137] related to so-called Nor98 cases first identified in Norwegian sheep. Atypical scrapie was confirmed to be a prion disease following transmission to Tg338 mice expressing OvPrP-V136, and revealed the uniformity of features between atypical scrapie cases [137], confirming limited studies in the natural host [138]. In one study, Tg mice that overexpress human prion protein were found to be susceptible to BSE prions, but not classical or atypical scrapie prions [139]. In contrast, Andrioletti and coworkers show that a panel of classical sheep scrapie prions transmits to several Tg mice expressing human PrP mouse models with an efficiency comparable to that of cattle BSE indicating that classical scrapie prions have zoonotic potential [140].

7 Cervid Prion Disease

Some 15 years ago, CWD was perhaps the least understood of all the prion diseases of animals and humans. Known to be highly contagious, its origins and mode of transmission were unclear, and it was not known whether multiple CWD strains exist or whether CWD prions pose a risk to other animals or humans. The main objectives at that time were to develop rapid and sensitive bioassays for CWD prions, and to experimentally address the risks that CWD prions pose to humans and other species. The development of Tg mice that were the first reliable bioassay for rapid and sensitive detection of CWD prions was a significant advance. With these resources in hand, investigators have obtained important information about CWD pathogenesis, and the molecular mechanisms of prion propagation, species barriers, and strains. Using CWD-susceptible Tg mice it became possible to bioassay CWD prions in tissues, body fluids, and secretions of deer and elk, which provided insights into the mode of transmission of this highly contagious disease. The study of inter-mammalian species barriers in Tg mice allows investigators to model the risks posed to humans and livestock from exposure to CWD prions, and this information helps facilitate management decisions designed to minimize interspecies prion transmission. During this time, the development of more facile, sensitive approaches to amplify prions in vitro, such as PMCA and RT-QuIC, have revolutionized our ability to detect prions, even at extremely low titers. In concert, cell culture models have provided an alternate means of CWD titration that has largely superseded bioassay in Tg mice and provided insights into strategies for developing compounds that inhibit CWD propagation. The development of compounds such as IND24 provides significant optimism for treating this currently incurable disease. Finally, studies of CWD using these newly developed tools has provided unexpected mechanistic insights into PrP^{C}-to-PrP^{Sc} conversion, particularly the role of the $\beta2$–$\alpha2$ loop/α-helix 3 epitope, and the proposal that prion strains exist as a continuum of conformational quasispecies.

The expense of housing cervids under prion-free conditions for long periods and the highly communicable nature of CWD present significant challenges for using deer as experimental hosts [141]. As noted above, transmission to other species has yielded mixed results. The resistance of mice [142] and the inefficient transmission of CWD to ferrets [143] are examples of species barriers to CWD prions, albeit of varying extent. The prototype Tg(CerPrP) mice developed signs of neurological dysfunction ~230 days following intracerebral inoculation with four CWD isolates [142]. The brains of sick Tg mice recapitulated the cardinal neuropathological features of CWD. As part of a larger study of CWD pathogenesis, Tg(CerPrP) mice were used as a sensitive means to show that skeletal muscles of CWD-infected deer harbor infectious prions, demonstrating that humans consuming or handling meat from CWD-infected animals are at risk to prion exposure [144]. Similar analyses of skeletal muscle BSE-affected cattle in a larger study of BSE pathogenesis using Tgbov XV mice did not reveal high levels of BSE infectivity [115]. Since the seminal reports of accelerated CWD transmission from deer and elk to Tg(CerPrP) mice [142], several other groups have reported similar results using comparable Tg mouse models [145–148]. CWD has also been transmitted, albeit with less efficiency, to Tg mice expressing mouse [149] or Syrian hamster PrP [150].

The generation of CWD-susceptible Tg mice, in concert with the development of PMCA-based approaches for amplifying CWD infectivity using PrP^{C} expressed in the CNS of those mice [151, 152], has also provided crucial information about the biology of CWD and cervid prions. Not only was amplification in vitro shown to maintain CWD prion strain properties, it also provided a means of generating novel cervid prion strains [151–154]. Transgenic and in vitro amplification approaches have also facilitated our understanding of the mechanism of CWD transmission among deer and elk [141, 155–157]. Transmission studies in Tg(CerPrP)1536$^{+/-}$ and similar Tg mice demonstrated that CWD prions were present in urine and feces and saliva [147, 155], and these findings are substantiated by in vitro amplification techniques [155, 158, 159]. Tg approaches have been essential for assessing the potential risk of human exposure to CWD prions [144, 160, 161]. The availability of CWD-susceptible transgenic mouse models, for the first time, also provided a means of quantifying CWD infectivity by end-point titration [160].

As demonstrated in other species in which prion diseases occur naturally, susceptibility to CWD is highly dependent on polymorphic variation in deer and elk *PRNP*. Polymorphisms at codons 95 [glutamine (Q) or histidine (H)] [162], 96 [glycine (G) or serine (S)] [162, 163], and 116 [alanine (A) or glycine (G)] [164] in white-tailed deer have been reported. While all major genotypes were found in deer with CWD, the Q96, G96, A116 allele (QGA) was more frequently found in CWD-affected deer than the QSA allele [162, 165], suggesting a protective effect of the counterpart polymorphisms. The elk *PRNP* coding sequence is polymorphic at codon 132 encoding either methionine (M) or leucine (L) [166, 167]. This position is equivalent to human *PRNP* codon 129. Studies of free-ranging and captive elk with CWD [168], as well as oral transmission experiments [169, 170], indicate that the L132 allele protects against CWD.

Transgenic mouse modeling provided a means of assessing the role of these cervid PrP gene polymorphisms on CWD pathogenesis. In recent work combining studies in Tg mice, the natural host, cell-free prion amplification, and molecular modeling approaches, we analyzed the effects of deer polymorphic amino acid variations on CWD propagation and susceptibility to prions from different species [50]. Reflecting the general authenticity of the Tg modeling approach, the properties of CWD prions were faithfully maintained in deer following their passage through Tg mice expressing cognate PrP. Moreover, the protective influences of naturally occurring PrP polymorphisms on CWD susceptibility were accurately reproduced in Tg mice or during cell-free amplification. The resistance of Tg mice expressing deer PrP S96 to CWD, referred to as Tg(DeerPrP-S96)7511 mice, is consistent with previously generated tg60 mice expressing serine at residue 96 [145]. In the studies of Angers and colleagues, whereas substitutions at residues 95 and 96 affected CWD propagation, their protective effects were overridden during replication of sheep prions in Tg mice and, in the case of residue 96, deer.

To more fully address the influence of the elk 132 polymorphism, transmissibility of CWD prions was assessed in Tg mice expressing cervid PrP^{C} with L or M at residue 132 [171]. While Tg mice expressing CerPrP-L132 afforded partial resistance to CWD, SSBP/1 sheep scrapie prions transmitted efficiently to Tg mice expressing CerPrP-L132, suggesting that the elk 132 polymorphism also controls prion susceptibility at the level of prion strain selection. The contrasting ability of CWD and SSBP/1 prions to overcome the inhibitory effects of the CerPrPL132 allele is reminiscent of studies describing the effects of the human codon 129 methionine (M)/valine (V) polymorphism on vCJD/BSE prion propagation in Tg mice expressing human PrP, which concluded that human PrP V129 severely restricts propagation of the BSE prion strain [120]. It therefore appears that amino acid substitutions in the unstructured region of PrP affect PrP^{C}-to-PrP^{Sc} conversion in a strain-specific manner.

The susceptibility of Tg(DeerPrP-S96)7511 mice, albeit with incomplete attack rates and long incubation times, are at odds with previous work showing complete resistance of tg60 mice [145, 172]. This apparent discrepancy is most likely related to the low transgene expression in tg60 mice, reported to be 70% the levels found in deer. CWD occurs naturally in deer homozygous for the PrP-S96 allele [173], which is clearly inconsistent with a completely protective effect of this substitution, suggesting that Tg(DeerPrP-S96)7511 mice represent an accurate Tg model in which to assess the effects of the S96 substitution.

In accordance with a role for this region in strain selection, in subsequent studies, Tg mice expressing wild-type deer PrP (tg33) or tg60 were challenged with CWD prions from experimentally infected deer with varying polymorphisms at residues 95 and 96 [174]. Passage of deer CWD prions into tg33 mice expressing wild-type deer PrP resulted in 100% attack rates, with CWD prions from deer expressing H95 or S96 having significantly longer incubation periods. Remarkably, otherwise resistant tg60 mice [145, 172] developed disease only when inoculated with prions from deer expressing H95/Q95 and H95/S96 PrP genotypes. Serial passage in tg60 mice resulted in propagation of a novel CWD strain, referred to as H95(+), while transmission to tg33 mice produced two disease phenotypes consistent with propagation of two strains.

High resolution structural studies showed the loop region linking the second beta-sheet (β2) with the alpha2-helix (α2) of cervid PrP to be extremely well defined compared to most other species, raising the possibility that this structural characteristic correlates with the unusually facile contagious transmission of CWD [175]. Tg mice expressing mouse PrP in which the β2-α2 loop was replaced by the corresponding region from cervid species, spontaneously developed prion disease [176]. Additional studies consistently point to the importance of the β2-α2 loop in regulating transmission barriers, including that of humans to CWD [154, 177–179]. Subsequent work suggested a more complex mechanism in which the β2–α2 loop participates with the distal region of α-helix 3 to form a solvent-accessible contiguous epitope [41]. These and later studies [40] ascribed greater importance to the plasticity of this discontinuous epitope. For example, substitution of D found in horse, which contains a similarly structured loop region, by S at residue 170 of mouse (elk PrP numbering) loop, increased not only the structural order of the loop, but also the long-range interaction with Y228 in α-helix 3. Underscoring the importance of long-range β2–α2 loop/α-helix 3 interactions, similar structural connections occur between this residue, which is alanine (A) in Tammar wallaby PrP, and residue 169, which is isoleucine (I) in this species (19). Stabilizing long-range interactions between the β2–α2 loop and α-helix 3 also occur in rabbit PrP, a species generally regarded as resistant to prion infection. X-ray crystallographic analyses showed the rabbit β2–α2 loop to be clearly ordered and indicated that hydrophobic interactions between the side chains of V169 and, in this case Y221 (elk PrP numbering) of α-helix 3, contributed to the stability of the β2–α2 loop/α helix 3 epitope [180].

In mule deer, polymorphism at codon 225 encoding serine (S) or phenylalanine (F) influences CWD susceptibility, the 225F allele being relatively protective. The occurrence of CWD was found to be 30-fold higher in deer homozygous for serine at position 225 (225SS) than in heterozygous (225SF) animals; the frequency of 225SF and 225FF genotypes in CWD-negative deer was 9.3%, but only 0.3% in CWD-positive deer [181]. Recent studies comparing CWD susceptibility in mule deer of the two residue 225 genotypes (225SS, 225FF) showed that 225FF mule deer had differences in clinical disease presentation, as well as more-subtle, atypical traits [182]. Immediately adjacent to the protective mule deer PrP polymorphism at 225, residue 226 encodes the singular primary structural difference between Rocky Mountain elk and deer PrP. Elk PrP contains glutamate (E), and deer PrP Q at this position.

Recent findings show that residues 225 and 226 play a critical role in PrP^{C}-to-PrP^{Sc} conversion and strain propagation, but that their effects are distinct from those produced by the H95Q, G95S, and M132 L polymorphisms [50]. Structural analyses confirm that residues 225 and 226 are located in the distal region of α-helix 3 that participates with the β2–α2 loop to form a solvent-accessible contiguous epitope [41]. Consistent with a role for this epitope in PrP conversion, these polymorphisms severely impact replication of both SSBP/1 and, to variable degrees, CWD. In the case of Tg mice expressing deerPrP-F225, referred to as Tg(DeerPrP-F225),

SSBP/1 incubation times were prolonged threefold, whereas inoculation with CWD produced incomplete attack rates or prolonged and variable incubation times in small numbers of mice. In those Tg(DeerPrP-F225) mice that did succumb to CWD, PrP^{Sc} distribution patterns were altered compared with Tg(DeerPrP) mice.

To address the effects of substitution of E for Q at residue 226, we assessed whether Tg mice expressing wild-type elk or deer PrP differed in their responses to CWD. These studies showed that differences at residue 226 also affected CWD replication, but to a lesser degree than the residue 225 polymorphism, with disease onset prolonged by 20–46% in CWD-inoculated Tg(DeerPrP) compared with Tg(ElkPrP) mice, and PrP^{Sc} distribution and neuropathology varying in each case [160]. In contrast to Tg(DeerPrP) mice which are susceptible to SSBP/1 [171], Tg(ElkPrP) were completely resistant [160], although the resistance of elk PrP^{C} to propagation of SSBP/1 was overcome following adaptation in deer or Tg(DeerPrP) mice. Passage in Tg mice expressing E226 or Q226 profoundly affected the ability of SSBP/1 to reinfect Tg mice expressing sheep PrP^{C}. These studies paralleled aspects of our previously published studies indicating that amino acid differences at residue 226 controlled the manifestation of CWD quasispecies or closely related strains [29]. These findings therefore collectively point to an important role for residues 225 and 226 in PrP^{C}-PrP^{Sc} conversion and the manifestation of prion strain properties, and substantiate the view that long-range interactions between the β2–α2 loop and α-helix 3 provide protection against prion infection and suggest a likely mechanism to account for the protective effects of the F225 polymorphism. Molecular dynamics analyses [50] showed that the S225F and E226Q substitutions in deer alter the orientations of D170 in the β2–α2 loop and Y228 in α-helix 3. This structural change allows hydrogen bonding between the side chains of these residues, which results in reduced plasticity of the β2–α2 loop/α-helix 3 epitope compared with deer or elk PrP structures. This suggests that the increased stability of this tertiary structural epitope precludes PrP^{C}-to-PrP^{Sc} conversion of deerPrP-F225.

Although our seminal studies in Tg mice [142] and subsequent work [146] raised the possibility of CWD strain variation, the limited number of isolates and the lack of detailed strain analyses in those studies meant that this hypothesis remained speculative. Subsequent studies supported the feasibility of using Tg(CerPrP)$1536^{+/-}$ mice for characterizing naturally occurring CWD strains, and novel cervid prions generated by PMCA [152]. To address whether different CWD strains occur in various geographic locations or in different cervid species, bioassays in transgenic mice were used to analyze CWD in a large collection of captive and wild mule deer, white-tailed deer, and elk from various geographic locations in North America [29]. These findings provided substantial evidence for two prevalent CWD prion strains, referred to as CWD1 and CWD2, with different clinical and neuropathological properties. Remarkably, primary transmissions of CWD prions from elk produced either CWD1 or CWD2 profiles, while transmission of deer inocula favored the production of mixed intra-study incubation times and CWD1 and CWD2 neuropathologies. These findings indicate that elk may be infected with *either* CWD1 *or* CWD2, while deer brains tend to harbor CWD1/CWD2 strain mixtures.

The different primary structures of deer and elk PrP at residue 226 provides a framework for understanding these differences in strain profiles of deer and elk. Because of the role played by residue 226, the description of a lysine polymorphism at this position in deer [183] and its possible role on strain stability may be significant. It is unknown whether CWD1 and CWD2 interfere or act synergistically, or whether their co-existence contributes to the unparalleled efficiency of CWD transmission. Interestingly, transmission results reported in previous studies suggested that cervid brain inocula might be composed of strain mixtures [147].

Additional studies support the existence of multiple CWD strains. CWD has also been transmitted, albeit with varying efficiency, to transgenic mice expressing mouse PrP [147, 149]. In the former study, a single mule deer isolate produced disease in all inoculated Tga20 mice, which express mouse PrP at high levels. On successive passages, incubation times dropped to ~160 d. In the second study, one elk isolate from a total of eight deer and elk CWD isolates induced disease in 75% of inoculated Tg4053 mice, which also overexpress mouse PrP. The distribution of lesions in both studies appeared to resemble the CWD1 pattern. Low efficiency CWD prion transmission was also recorded in hamsters and transgenic mice expressing Syrian hamster PrP [150]. In that study, during serial passage of mule deer CWD, fast and slow incubation time strains with different patterns of brain pathology and PrP^{Sc} deposition were also isolated. In yet other studies, serial passages of CWD from white-tailed deer into transgenic mice expressing hamster PrP, and then Syrian golden hamsters, produced a strain, referred to as "wasting" (WST), characterized by a prominent preclinical wasting disease, similar to cachexia, which the authors propose is due to a prion-induced endocrinopathy [184]. These same investigators identified a second strain, defined as "cheeky" (CKY), derived from infection of Tg mice that express hamster PrP [185]. The CKY strain had a shorter incubation period than WST, but after transmission to hamsters, the incubation period of CKY became ~150 days longer than WST. In this case, PK digestion revealed strain-specific PrP^{Sc} signatures that were maintained in both hosts, but the solubility and conformational stability of PrP^{Sc} differed for the CWD strains in a host-dependent manner. In addition to supporting the view that there are multiple CWD strains, these findings suggest the importance of host-specific pathways, independent of PrP, that participate in the selection and propagation of distinct strains.

Studies in cell culture also support the existence of CWD strains. Quinacrine altered the transmission properties of CWD prions, as well as the biochemical characteristics of the constitutive PrP^{Sc} [30]. Despite accumulating significantly higher prion titers, quinacrine treated (Q-CWD) prions from $Elk21^{+}$ cells produced prolonged incubation times in Tg(DeerPrP) and Tg(ElkPrP) mice compared to CWD from untreated $Elk21^{+}$ cells. Since kinetics of disease onset in prion-infected animals is inversely related to the titer of a given prion strain, this unusual outcome is consistent with quinacrine affecting the intrinsic properties of the CWD prion. In accordance with this notion, while the deposition patterns of PrP^{Sc} in the brains of diseased Tg(DeerPrP) and Tg(ElkPrP) mice receiving prions from $Elk21^{+}$ are concordant with our previously published descriptions following transmissions of naturally occurring or PMCA-generated CWD prions [152], these distinctive patterns were not recapitulated in either line of Tg mice receiving Q-CWD prions. Prion incubation times and neuronal targeting are the biological criteria by which prion strains are defined. Previous studies showed that strain properties are enciphered within the conformation of PrP^{Sc} [13, 152], and that cervid prion incubation times positively correlate with PrP^{Sc} conformational stability [152]. It is therefore significant that the longer incubation times of Q-CWD prions and altered patterns of PrP^{Sc} deposition in both Tg models were associated with an increase in the relative stability of PrP^{Sc} conformers constituting Q-CWD prions.

The properties of Q-CWD prions provided convincing evidence for conformational mutation that were suggested, but could not be observed, in previous studies of mouse-adapted scrapie prions in cell cultures treated with swainsonine, an inhibitor of Golgi α-mannosidase II [28]. Since the swainsonine-induced properties of cell-derived mouse prions reverted to their original characteristics when propagated in vivo, and there were no detectable differences in the conformations of PrP^{Sc} generated under conditions of swainsonine treatment compared to untreated cells, claims of strain mutation were indirectly based on the differing properties of prions from swainsonine treated and untreated cells using a cell panel assay. In contrast, CWD and Q-CWD prions comprise distinct PrP^{Sc} conformers that produce discernable phenotypes in two lines of Tg mice.

In accordance with Western blotting-based analyses showing higher conformational stabilities of PrP^{Sc} produced in the brains of CWD-infected Tg(DeerPrP) mice compared with Tg(ElkPrP) mice [29], C-CSA confirmed that the conformation of deer PrP^{Sc} in RK13 cells expressing deer PrP (RKD^{+}) was more stable than elk PrP^{Sc} produced during infection of $Elk21^{+}$. Because $Elk21^{+}$ and RKD^{+} propagated CWD prions from the same source differences in conformational stability result from the effects of residue 226, the sole primary structural difference between elk and deer PrP.

The identification and characterization of distinct CWD strains, and the influence of PrP primary structure on their stabilities, is of importance when considering the potential for interspecies transmission. The appearance of vCJD following human exposure to BSE [18, 186], place the human species barrier to other animal prion diseases, particularly CWD, at the forefront of public health concerns. Since North American hunters harvest thousands of deer and elk each year, and it is not currently mandatory to have these animals tested, the demonstration of CWD prions in skeletal muscle and fat of deer [144, 161] makes it is likely that humans consume CWD prions. The substantial market for elk antler velvet in traditional Asian medicine also warrants concern [160].

Estimates of the zoonotic potential of CWD are currently mixed. Surveillance currently shows no evidence of transmission to humans [187, 188]. While initial cell-free conversion studies suggested that the ability of CWD prions to transform human PrP^C into protease-resistant PrP was low [163], subsequent results showed that cervid PrP^{Sc} induced the conversion of human PrP^C by protein misfolding cyclic amplification (PMCA), following CWD prion strain stabilization by successive passages in vitro or in vivo [189]. These results have implications for the human species barrier to CWD, and underscore the role of strain adaptation on interspecies transmission barriers. Additional studies using transgenic mice expressing human PrP^C showed that CWD failed to induce disease following intracerebral CWD infection [147, 148, 190]. However, CWD transmission was reported to nonhuman primates through the intracerebral inoculation of squirrel monkeys (*Saimiri sciureus*) [191, 192]. Moreover, using RT-QuIC to model the transmission barrier of CWD to human rPrP, the work of Davenport and coworkers results suggest that, at the level of protein-protein interactions, CWD adapts to a new species more readily than does BSE, and that the barrier preventing transmission of CWD to humans may be less robust than estimated [193].

References

1. Prusiner SB (1998) Prions (Les Prix Nobel Lecture). In: Frängsmyr T (ed) Les prix Nobel. Almqvist & Wiksell International, Stockholm, pp 268–323
2. Prusiner SB et al (1990) Transgenetic studies implicate interactions between homologous PrP isoforms in scrapie prion replication. Cell 63(4):673–686
3. Scott M et al (1993) Propagation of prions with artificial properties in transgenic mice expressing chimeric PrP genes. Cell 73(5):979–988
4. Kocisko DA et al (1994) Cell-free formation of protease-resistant prion protein. Nature 370(6489):471–474
5. Vidal E et al (2015) Transgenic mouse bioassay: evidence that rabbits are susceptible to a variety of prion isolates. PLoS Pathog 11(8):e1004977
6. Chianini F et al (2012) Rabbits are not resistant to prion infection. Proc Natl Acad Sci U S A 109(13):5080–5085
7. Westaway D et al (1987) Distinct prion proteins in short and long scrapie incubation period mice. Cell 51(4):651–662

8. Hunter N et al (2000) Sheep and goats: natural and experimental TSEs and factors influencing incidence of disease. Arch Virol Suppl 16:181–188
9. Baker HF et al (1991) Amino acid polymorphism in human prion protein and age at death in inherited prion disease. Lancet 337(8752):1286
10. Collinge J, Palmer MS, Dryden AJ (1991) Genetic predisposition to iatrogenic Creutzfeldt-Jakob disease. Lancet 337(8755):1441–1442
11. Palmer MS et al (1991) Homozygous prion protein genotype predisposes to sporadic Creutzfeldt-Jakob disease. Nature 352(6333):340–342
12. Nonno R et al (2006) Efficient transmission and characterization of Creutzfeldt-Jakob disease strains in bank voles. PLoS Pathog 2(2):e12
13. Telling GC et al (1996) Evidence for the conformation of the pathologic isoform of the prion protein enciphering and propagating prion diversity. Science 274(5295):2079–2082
14. Bessen RA, Marsh RF (1994) Distinct PrP properties suggest the molecular basis of strain variation in transmissible mink encephalopathy. J Virol 68(12):7859–7868
15. Safar J et al (1998) Eight prion strains have PrP^{Sc} molecules with different conformations. Nat Med 4(10):1157–1165
16. Peretz D et al (2002) A change in the conformation of prions accompanies the emergence of a new prion strain. Neuron 34(6):921–932
17. Collinge J et al (1996) Molecular analysis of prion strain variation and the aetiology of "new variant" CJD. Nature 383(6602): 685–690
18. Hill AF et al (1997) The same prion strain causes vCJD and BSE. Nature 389(6650): 448–450
19. Scott MR et al (2005) Transmission barriers for bovine, ovine, and human prions in transgenic mice. J Virol 79(9):5259–5271
20. Asante EA et al (2002) BSE prions propagate as either variant CJD-like or sporadic CJD-like prion strains in transgenic mice expressing human prion protein. EMBO j 21(23):6358–6366
21. Korth C et al (2003) Abbreviated incubation times for human prions in mice expressing a chimeric mouse-human prion protein transgene. Proc Natl Acad Sci U S A 100(8): 4784–4789
22. Scott MR et al (1999) Compelling transgenetic evidence for transmission of bovine spongiform encephalopathy prions to humans. Proc Natl Acad Sci U S A 96(26): 15137–15142
23. Bruce ME, Dickinson AG (1987) Biological evidence that the scrapie agent has an independent genome. J Gen Virol 68(1):79–89
24. Collinge J, Clarke AR (2007) A general model of prion strains and their pathogenicity. Science 318(5852):930–936
25. Haldiman T et al (2013) Co-existence of distinct prion types enables conformational evolution of human PrPSc by competitive selection. J Biol Chem 288(41):29846–29861
26. Collinge J (1999) Variant Creutzfeldt-Jakob disease. Lancet 354(9175):317–323
27. Eigen M (1996) On the nature of virus quasispecies. Trends Microbiol 4(6):216–218
28. Li J et al (2010) Darwinian evolution of prions in cell culture. Science 327(5967):869–872
29. Angers RC et al (2010) Prion strain mutation determined by prion protein conformational compatibility and primary structure. Science 328(5982):1154–1158
30. Bian J, Kang HE, Telling GC (2014) Quinacrine promotes replication and conformational mutation of chronic wasting disease prions. Proc Natl Acad Sci U S A 111(16): 6028–6033
31. Makarava N, Baskakov IV (2008) The same primary structure of the prion protein yields two distinct self-propagating states. J Biol Chem 283(23):15988–15996
32. Rezaei H et al (2002) Amyloidogenic unfolding intermediates differentiate sheep prion protein variants. J Mol Biol 322(4):799–814
33. Beringue V et al (2012) Facilitated cross-species transmission of prions in extraneural tissue. Science 335(6067):472–475
34. Rogers M et al (1993) Conversion of truncated and elongated prion proteins into the scrapie isoform in cultured cells. Proc Natl Acad Sci U S A 90(8):3182–3186
35. Chesebro B et al (2005) Anchorless prion protein results in infectious amyloid disease without clinical scrapie. Science 308(5727):1435–1439
36. Telling GC et al (1994) Transmission of Creutzfeldt-Jakob disease from humans to transgenic mice expressing chimeric human-mouse prion protein. Proc Natl Acad Sci U S A 91(21):9936–9940
37. Telling GC et al (1995) Prion propagation in mice expressing human and chimeric PrP transgenes implicates the interaction of cellular PrP with another protein. Cell 83(1):79–90
38. Perrier V et al (2002) Dominant-negative inhibition of prion replication in transgenic mice. Proc Natl Acad Sci U S A 99(20): 13079–13084

39. Geoghegan JC et al (2009) Trans-dominant inhibition of prion propagation in vitro is not mediated by an accessory cofactor. PLoS Pathog 5(7):e1000535
40. Christen B et al (2009) Prion protein NMR structure from tammar wallaby (Macropus eugenii) shows that the beta2-alpha2 loop is modulated by long-range sequence effects. J Mol Biol 389(5):833–845
41. Perez DR, Damberger FF, Wuthrich K (2010) Horse prion protein NMR structure and comparisons with related variants of the mouse prion protein. J Mol Biol 400(2):121–128
42. Telling GC et al (1996) Interactions between wild-type and mutant prion proteins modulate neurodegeneration in transgenic mice. Genes Dev 10(14):1736–1750
43. Chiesa R et al (2000) Accumulation of protease-resistant prion protein (PrP) and apoptosis of cerebellar granule cells in transgenic mice expressing a PrP insertional mutation. Proc Natl Acad Sci U S A 97(10):5574–5579
44. Bian J et al (2006) GFP-tagged PrP supports compromised prion replication in transgenic mice. Biochem Biophys Res Commun 340(3):894–900
45. Barmada SJ, Harris DA (2005) Visualization of prion infection in transgenic mice expressing green fluorescent protein-tagged prion protein. J Neurosci 25(24):5824–5832
46. Shmerling D et al (1998) Expression of amino-terminally truncated PrP in the mouse leading to ataxia and specific cerebellar lesions. Cell 93(2):203–214
47. Baumann F et al (2007) Lethal recessive myelin toxicity of prion protein lacking its central domain. EMBO j 26(2):538–547
48. Li A et al (2007) Neonatal lethality in transgenic mice expressing prion protein with a deletion of residues 105-125. EMBO j 26(2): 548–558
49. Scott M et al (1989) Transgenic mice expressing hamster prion protein produce species-specific scrapie infectivity and amyloid plaques. Cell 59(5):847–857
50. Angers R et al (2014) Structural effects of PrP polymorphisms on intra- and interspecies prion transmission. Proc Natl Acad Sci U S A 111(30):11169–11174
51. Telling GC (2010) Nucleic acid-free mutation of prion strains. Prion 4(4):252–255
52. Büeler H et al (1992) Normal development and behaviour of mice lacking the neuronal cell-surface PrP protein. Nature 356(6370):577–582
53. Gombojav A et al (2003) Susceptibility of transgenic mice expressing chimeric sheep, bovine and human PrP genes to sheep scrapie. J Vet Med Sci 65(3):341–347
54. Kupfer L et al (2007) Amino acid sequence and prion strain specific effects on the in vitro and in vivo convertibility of ovine/murine and bovine/murine prion protein chimeras. Biochim Biophys Acta 1772(6):704–713
55. Scott MR et al (1997) Identification of a prion protein epitope modulating transmission of bovine spongiform encephalopathy prions to transgenic mice. Proc Natl Acad Sci U S A 94(26):14279–14284
56. Bishop MT et al (2006) Predicting susceptibility and incubation time of human-to-human transmission of vCJD. Lancet Neurol 5(5):393–398
57. Hadlow WJ (1959) Scrapie and kuru. Lancet 2:289–290
58. Gajdusek DC, Gibbs CJ Jr, Alpers M (1966) Experimental transmission of a kuru-like syndrome to chimpanzees. Nature 209(5025): 794–796
59. Gibbs CJ Jr et al (1968) Creutzfeldt-Jakob disease (spongiform encephalopathy): transmission to the chimpanzee. Science 161(3839):388–389
60. Brown P et al (1994) Human spongiform encephalopathy: the National Institutes of Health series of 300 cases of experimentally transmitted disease. Ann Neurol 35(5): 513–529
61. Gibbs CJ Jr, Gajdusek DC, Amyx H (1979) Strain variation in the viruses of Creutzfeldt-Jakob disease and kuru. In: Prusiner SB, Hadlow WJ (eds) Slow transmissible diseases of the nervous system, vol 2. Academic, New York, pp 87–110
62. Tateishi J, Sato Y, Ohta M (1983) Creutzfeldt-Jakob disease in humans and laboratory animals. In: Zimmerman HM (ed) Progress in neuropathology, vol 5. Raven Press, New York, pp 195–221
63. Tateishi J, Kitamoto T (1995) Inherited prion diseases and transmission to rodents. Brain Pathol 5(1):53–59
64. Manuelidis E et al (1976) Serial propagation of Creutzfeldt-Jakob disease in guinea pigs. Proc Natl Acad Sci U S A 73(1):223–227
65. Manuelidis E, Gorgacz EJ, Manuelidis L (1978) Interspecies transmission of Creutzfeldt-Jakob disease to Syrian hamsters with reference to clinical syndromes and strains of agent. Proc Natl Acad Sci U S A 75: 3422–3436
66. Scott MR et al (1992) Chimeric prion protein expression in cultured cells and transgenic mice. Protein Sci 1(8):986–997
67. Kong Q et al (2008) Evaluation of the human transmission risk of an atypical bovine spongi-

form encephalopathy prion strain. J Virol 82(7):3697–3701
68. Beringue V et al (2008) Transmission of atypical bovine prions to mice transgenic for human prion protein. Emerg Infect Dis 14(12):1898–1901
69. Bishop MT, Will RG, Manson JC (2010) Defining sporadic Creutzfeldt-Jakob disease strains and their transmission properties. Proc Natl Acad Sci U S A 107(26):12005–12010
70. Hill AF et al (2003) Molecular classification of sporadic Creutzfeldt-Jakob disease. Brain 126(6):1333–1346
71. Parchi P et al (1996) Molecular basis of phenotypic variability in sporadic Creutzfeldt-Jakob disease. Ann Neurol 39(6):767–778
72. Polymenidou M et al (2005) Coexistence of multiple PrPSc types in individuals with Creutzfeldt-Jakob disease. Lancet Neurol 4(12):805–814
73. Mastrianni JA et al (2001) Inherited prion disease caused by the V210I mutation: transmission to transgenic mice. Neurology 57(12):2198–2205
74. Mastrianni J et al (1997) Fatal sporadic insomnia: fatal familial insomnia phenotype without a mutation of the prion protein gene. Neurology 48(Suppl):A296
75. Notari S et al (2014) Transmission characteristics of variably protease-sensitive prionopathy. Emerg Infect dis 20(12):2006–2014
76. Diack AB et al (2014) Variably protease-sensitive prionopathy, a unique prion variant with inefficient transmission properties. Emerg Infect Dis 20(12):1969–1979
77. Giles K et al (2010) Human prion strain selection in transgenic mice. Ann Neurol 68(2):151–161
78. Will RG et al (1996) A new variant of Creutzfeldt-Jakob disease in the UK. Lancet 347(9006):921–925
79. Collinge J et al (1995) Unaltered susceptibility to BSE in transgenic mice expressing human prion protein. Nature 378(6559): 779–783
80. Mead S et al (2009) Genetic risk factors for variant Creutzfeldt-Jakob disease: a genome-wide association study. Lancet Neurol 8(1):57–66
81. Plinston C et al (2011) Increased susceptibility of human-PrP transgenic mice to bovine spongiform encephalopathy infection following passage in sheep. J Virol 85(3):1174–1181
82. Plinston C et al (2014) Increased susceptibility of transgenic mice expressing human PrP to experimental sheep bovine spongiform encephalopathy is not due to increased agent titre in sheep brain tissue. J Gen Virol 95(Pt_8):1855–1859
83. Asante EA et al (2006) Dissociation of pathological and molecular phenotype of variant Creutzfeldt-Jakob disease in transgenic human prion protein 129 heterozygous mice. Proc Natl Acad Sci U S A 103(28): 10759–10764
84. Mead S et al (2009) A novel protective prion protein variant that colocalizes with kuru exposure. N Engl J Med 361(21): 2056–2065
85. Asante EA et al (2015) A naturally occurring variant of the human prion protein completely prevents prion disease. Nature 522(7557):478–481
86. Hsiao K et al (1989) Linkage of a prion protein missense variant to Gerstmann-Sträussler syndrome. Nature 338(6213):342–345
87. Manson JC et al (1999) A single amino acid alteration (101L) introduced into murine PrP dramatically alters incubation time of transmissible spongiform encephalopathy. EMBO 18(23):6855–6864
88. Hsiao KK et al (1994) Serial transmission in rodents of neurodegeneration from transgenic mice expressing mutant prion protein. Proc Natl Acad Sci U S A 91(19): 9126–9130
89. Kaneko K, Ball HL, Wille H, Zhang H, Groth D, Torchia M, Tremblay P, Safar J, Prusiner SB, SJ DA, Baldwin MA, Cohen FE (2000) A synthetic peptide initiates Gerstmann-Straussler-Scheinker (GSS) disease in transgenic mice. J Mol Biol 295(4):997–1007
90. Tremblay P et al (2004) Mutant PrPSc conformers induced by a synthetic peptide and several prion strains. J Virol 78(4):2088–2099
91. Cohen FE et al (1994) Structural clues to prion replication. Science 264(5158): 530–531
92. Tzaban S et al (2002) Protease-sensitive scrapie prion protein in aggregates of heterogeneous sizes. Biochemistry 41(42): 12868–12875
93. Korth C et al (1997) Prion (PrP^{Sc})-specific epitope defined by a monoclonal antibody. Nature 389:74–77
94. Nazor KE et al (2005) Immunodetection of disease-associated mutant PrP, which accelerates disease in GSS transgenic mice. EMBO j 24(13):2472–2480
95. Uptain SM, Lindquist S (2002) Prions as protein-based genetic elements. Annu Rev Microbiol 56(1):703–741
96. Piccardo P et al (2013) Dissociation of prion protein amyloid seeding from transmission of a spongiform encephalopathy. J Virol 87(22):12349–12356
97. Legname G et al (2004) Synthetic mammalian prions. Science 305(5684):673–676

98. Asante EA et al (2015) Transmission properties of human PrP 102L prions challenge the relevance of mouse models of GSS. PLoS Pathog 11(7):e1004953
99. Friedman-Levi Y et al (2011) Fatal prion disease in a mouse model of genetic E200K Creutzfeldt-Jakob disease. PLoS Pathog 7(11):e1002350
100. Chiesa R et al (1998) Neurological illness in transgenic mice expressing a prion protein with an insertional mutation. Neuron 21(6):1339–1351
101. Chiesa R et al (2003) Molecular distinction between pathogenic and infectious properties of the prion protein. J Virol 77(13):7611–7622
102. Dossena S et al (2008) Mutant prion protein expression causes motor and memory deficits and abnormal sleep patterns in a transgenic mouse model. Neuron 60(4):598–609
103. Bouybayoune I et al (2015) Transgenic fatal familial insomnia mice indicate prion infectivity-independent mechanisms of pathogenesis and phenotypic expression of disease. PLoS Pathog 11(4):e1004796
104. Jackson WS et al (2009) Spontaneous generation of prion infectivity in fatal familial insomnia knockin mice. Neuron 63(4):438–450
105. Jackson WS et al (2013) Profoundly different prion diseases in knock-in mice carrying single PrP codon substitutions associated with human diseases. Proc Natl Acad Sci U S A 110(36):14759–14764
106. Hsiao KK et al (1991) A prion protein variant in a family with the telencephalic form of Gerstmann-Sträussler-Scheinker syndrome. Neurology 41(5):681–684
107. Hegde RS et al (1998) A transmembrane form of the prion protein in neurodegenerative disease. Science 279(5352):827–834
108. Yang W et al (2009) A new transgenic mouse model of Gerstmann-Straussler-Scheinker syndrome caused by the A117V mutation of PRNP. J Neurosci 29(32):10072–10080
109. Asante EA et al (2013) Inherited prion disease A117V is not simply a proteinopathy but produces prions transmissible to transgenic mice expressing homologous prion protein. PLoS Pathog 9(9):e1003643
110. Buschmann A et al (2000) Detection of cattle-derived BSE prions using transgenic mice overexpressing bovine PrP(C). Arch Virol Suppl 16:75–86
111. Castilla J et al (2003) Early detection of PrPres in BSE-infected bovine PrP transgenic mice. Arch Virol 148(4):677–691
112. Jacobs JG et al (2007) Molecular discrimination of atypical bovine spongiform encephalopathy strains from a geographical region spanning a wide area in Europe. J Clin Microbiol 45(6):1821–1829
113. Casalone C et al (2004) Identification of a second bovine amyloidotic spongiform encephalopathy: molecular similarities with sporadic Creutzfeldt-Jakob disease. Proc Natl Acad Sci U S A 101(9):3065–3070
114. Richt JA, Hall SM (2008) BSE case associated with prion protein gene mutation. PLoS Pathog 4(9):e1000156
115. Buschmann A, Groschup MH (2005) Highly bovine spongiform encephalopathy-sensitive transgenic mice confirm the essential restriction of infectivity to the nervous system in clinically diseased cattle. J Infect Dis 192(5):934–942
116. Beringue V et al (2006) Isolation from cattle of a prion strain distinct from that causing bovine spongiform encephalopathy. PLoS Pathog 2(10):e112
117. Beringue V et al (2007) A bovine prion acquires an epidemic bovine spongiform encephalopathy strain-like phenotype on interspecies transmission. J Neurosci 27(26): 6965–6971
118. Buschmann A et al (2006) Atypical BSE in Germany—proof of transmissibility and biochemical characterization. Vet Microbiol 117(2-4):103–116
119. Capobianco R et al (2007) Conversion of the BASE prion strain into the BSE strain: the origin of BSE? PLoS Pathog 3(3):e31
120. Wadsworth JD et al (2004) Human prion protein with valine 129 prevents expression of variant CJD phenotype. Science 306(5702):1793–1796
121. Beringue V et al (2008) Prominent and persistent extraneural infection in human PrP transgenic mice infected with variant CJD. PLoS One 3(1):e1419
122. Espinosa JC et al (2009) Transgenic mice expressing porcine prion protein resistant to classical scrapie but susceptible to sheep bovine spongiform encephalopathy and atypical scrapie. Emerg Infect Dis 15(8): 1214–1221
123. Westaway D et al (1994) Degeneration of skeletal muscle, peripheral nerves, and the central nervous system in transgenic mice overexpressing wild-type prion proteins. Cell 76(1):117–129
124. Rubenstein R et al (2012) PrP(Sc) detection and infectivity in semen from scrapie-infected sheep. J Gen Virol 93(Pt_6):1375–1383
125. Crozet C et al (2001) Efficient transmission of two different sheep scrapie isolates in transgenic mice expressing the ovine PrP gene. J Virol 75(11):5328–5334
126. Crozet C et al (2001) Florid plaques in ovine PrP transgenic mice infected with an experimental ovine BSE. EMBO Rep 2(10):952–956

127. Vilotte JL et al (2001) Markedly increased susceptibility to natural sheep scrapie of transgenic mice expressing ovine prp. J Virol 75(13):5977–5984
128. Tamguney G et al (2009) Transmission of scrapie and sheep-passaged bovine spongiform encephalopathy prions to transgenic mice expressing elk prion protein. J Gen Virol 90(4):1035–1047
129. Goldmann W et al (1994) PrP genotype and agent effects in scrapie: change in allelic interaction with different isolates of agent in sheep, a natural host of scrapie. J Gen Virol 75(5):989–995
130. Cordier C et al (2006) Transmission and characterization of bovine spongiform encephalopathy sources in two ovine transgenic mouse lines (TgOvPrP4 and TgOvPrP59). J Gen Virol 87(12): 3763–3771
131. Saijo E et al (2013) Epigenetic dominance of prion conformers. PLoS Pathog 9(10):e1003692
132. Foster JD, Dickinson AG (1988) The unusual properties of CH 1641, a sheep-passaged isolate of scrapie. Vet Rec 123(1):5–8
133. Baron T, Biacabe AG (2007) Molecular behaviors of "CH1641-like" sheep scrapie isolates in ovine transgenic mice (TgOvPrP4). J Virol 81(13):7230–7237
134. Kang HE et al (2012) Characterization of conformation-dependent prion protein epitopes. J Biol Chem 287(44):37219–37232
135. Bencsik A et al (2007) Scrapie strain transmission studies in ovine PrP transgenic mice reveal dissimilar susceptibility. Histochem Cell Biol 127(5):531–539
136. Baron T et al (2004) Molecular analysis of the protease-resistant prion protein in scrapie and bovine spongiform encephalopathy transmitted to ovine transgenic and wild-type mice. J Virol 78(12):6243–6251
137. Le Dur A et al (2005) A newly identified type of scrapie agent can naturally infect sheep with resistant PrP genotypes. Proc Natl Acad Sci U S A 102(44):16031–16036
138. Simmons MM et al (2007) Experimental transmission of atypical scrapie to sheep. BMC Vet Res 3(1):20
139. Wadsworth JD et al (2013) Atypical scrapie prions from sheep and lack of disease in transgenic mice overexpressing human prion protein. Emerg Infect Dis 19(11):1731–1739
140. Cassard H et al (2014) Evidence for zoonotic potential of ovine scrapie prions. Nat Commun 5:5821
141. Mathiason CK et al (2006) Infectious prions in the saliva and blood of deer with chronic wasting disease. Science 314(5796):133–136
142. Browning SR et al (2004) Transmission of prions from mule deer and elk with chronic wasting disease to transgenic mice expressing cervid PrP. J Virol 78(23):13345–13350
143. Bartz JC et al (1998) The host range of chronic wasting disease is altered on passage in ferrets. Virology 251(2):297–301
144. Angers RC et al (2006) Prions in skeletal muscles of deer with chronic wasting disease. Science 311(5764):1117–1117
145. Meade-White K et al (2007) Resistance to chronic wasting disease in transgenic mice expressing a naturally occurring allelic variant of deer prion protein. J Virol 81(9):4533–4539
146. LaFauci G et al (2006) Passage of chronic wasting disease prion into transgenic mice expressing Rocky Mountain elk (Cervus elaphus Nelsoni) PrPC. J Gen Virol 87(12): 3773–3780
147. Tamguney G et al (2006) Transmission of elk and deer prions to transgenic mice. J Virol 80(18):9104–9114
148. Kong Q et al (2005) Chronic wasting disease of elk: transmissibility to humans examined by transgenic mouse models. J Neurosci 25(35):7944–7949
149. Sigurdson CJ et al (2006) Strain fidelity of chronic wasting disease upon murine adaptation. J Virol 80(24):12303–12311
150. Raymond GJ et al (2007) Transmission and adaptation of chronic wasting disease to hamsters and transgenic mice: evidence for strains. J Virol 81(8):4305–4314
151. Meyerett C et al (2008) In vitro strain adaptation of CWD prions by serial protein misfolding cyclic amplification. Virology 382(2):267–276
152. Green KM et al (2008) Accelerated high fidelity prion amplification within and across prion species barriers. PLoS Pathog 4(8):e1000139
153. Kurt TD et al (2007) Efficient in vitro amplification of chronic wasting disease PrPRES. J Virol 81(17):9605–9608
154. Kurt TD et al (2009) Trans-species amplification of PrP(CWD) and correlation with rigid loop 170N. Virology 387(1):235–243
155. Haley NJ et al (2009) Detection of CWD prions in urine and saliva of deer by transgenic mouse bioassay. PLoS One 4(3):e4848
156. Tamguney G et al (2009) Asymptomatic deer excrete infectious prions in faeces. Nature 461(7263):529–532
157. Haley NJ et al (2009) Detection of sub-clinical CWD infection in conventional test-negative deer long after oral exposure to urine and feces from CWD+ deer. PLoS One 4(11):e7990
158. Pulford B et al (2012) Detection of PrPCWD in feces from naturally exposed Rocky

Mountain elk (Cervus elaphus Nelsoni) using protein misfolding cyclic amplification. J Wildl Dis 48(2):425–434
159. Henderson DM et al (2013) Rapid Antemortem detection of CWD prions in deer saliva. PLoS One 8(9):e74377
160. Angers RC et al (2009) Chronic wasting disease prions in elk antler velvet. Emerg Infect Dis 15(5):696–703
161. Race B et al (2009) Prion infectivity in fat of deer with chronic wasting disease. J Virol 83(18):9608–9610
162. Johnson C et al (2003) Prion protein gene heterogeneity in free-ranging white-tailed deer within the chronic wasting disease affected region of Wisconsin. J Wildl Dis 39(3):576–581
163. Raymond GJ et al (2000) Evidence of a molecular barrier limiting susceptibility of humans, cattle and sheep to chronic wasting disease. EMBO J 19(17):4425–4430
164. Heaton MP et al (2003) Prion gene sequence variation within diverse groups of U.S. sheep, beef cattle, and deer. Mamm Genome 14(11):765–777
165. O'Rourke KI et al (2004) Polymorphisms in the prion precursor functional gene but not the pseudogene are associated with susceptibility to chronic wasting disease in white-tailed deer. J Gen Virol 85(5):1339–1346
166. Schatzl HM et al (1997) Is codon 129 of prion protein polymorphic in human beings but not in animals? Lancet 349(9065):1603–1604
167. O'Rourke KI et al (1998) Monoclonal antibody F89/160.1.5 defines a conserved epitope on the ruminant prion protein. J Clin Microbiol 36(6):1750–1755
168. O'Rourke KI et al (1999) PrP genotypes of captive and free-ranging Rocky Mountain elk (Cervus elaphus Nelsoni) with chronic wasting disease. J Gen Virol 80(10):2765–2679
169. Hamir AN et al (2006) Preliminary observations of genetic susceptibility of elk (Cervus elaphus Nelsoni) to chronic wasting disease by experimental oral inoculation. J Vet Diagn Investig 18(1):110–114
170. O'Rourke KI et al (2007) Elk with a long incubation prion disease phenotype have a unique PrPd profile. Neuroreport 18(18):1935–1938
171. Green KM et al (2008) The elk PRNP codon 132 polymorphism controls cervid and scrapie prion propagation. J Gen Virol 89(2):598–608
172. Race B et al (2011) In vivo comparison of chronic wasting disease infectivity from deer with variation at prion protein residue 96. J Virol 85(17):9235–9238
173. Keane DP et al (2008) Chronic wasting disease in a Wisconsin white-tailed deer farm. J Vet Diagn Investig 20(5):698–703
174. Duque Velasquez C et al (2015) Deer prion proteins modulate the emergence and adaptation of chronic wasting disease strains. J Virol 89(24):12362–12373
175. Gossert AD et al (2005) Prion protein NMR structures of elk and of mouse/elk hybrids. Proc Natl Acad Sci U S A 102(3):646–650
176. Sigurdson CJ et al (2009) De novo generation of a transmissible spongiform encephalopathy by mouse transgenesis. Proc Natl Acad Sci U S A 106(1):304–309
177. Kurt TD et al (2014) Prion transmission prevented by modifying the beta2-alpha2 loop structure of host PrPC. J Neurosci 34(3):1022–1027
178. Sigurdson CJ et al (2010) A molecular switch controls interspecies prion disease transmission in mice. J Clin Invest 120(7):2590–2599
179. Kurt TD et al (2015) Human prion protein sequence elements impede cross-species chronic wasting disease transmission. J Clin Invest 125(4):1485–1496
180. Khan MQ et al (2010) Prion disease susceptibility is affected by beta-structure folding propensity and local side-chain interactions in PrP. Proc Natl Acad Sci U S A 107(46): 19808–19813
181. Jewell JE et al (2005) Low frequency of PrP genotype 225SF among free-ranging mule deer (Odocoileus hemionus) with chronic wasting disease. J Gen Virol 86(8):2127–2134
182. Wolfe LL, Fox KA, Miller MW (2014) "atypical" chronic wasting disease in PRNP genotype 225FF mule deer. J Wildl Dis 50(3):660–665
183. Johnson C et al (2006) Prion protein polymorphisms in white-tailed deer influence susceptibility to chronic wasting disease. J Gen Virol 87(7):2109–2114
184. Bessen RA et al (2011) Transmission of chronic wasting disease identifies a prion strain causing cachexia and heart infection in hamsters. PLoS One 6(12):e28026
185. Crowell J et al (2015) Host determinants of prion strain diversity independent of prion protein genotype. J Virol 89(20): 10427–10441
186. Bruce ME et al (1997) Transmissions to mice indicate that 'new variant' CJD is caused by the BSE agent. Nature 389(6650): 498–501
187. Belay ED et al (2004) Chronic wasting disease and potential transmission to humans. Emerg Infect Dis 10(6):977–984

188. Mawhinney S et al (2006) Human prion disease and relative risk associated with chronic wasting disease. Emerg Infect Dis 12(10):1527–1535
189. Barria MA et al (2011) Generation of a new form of human PrP(Sc) in vitro by interspecies transmission from cervid prions. J Biol Chem 286(9):7490–7495
190. Sandberg M et al (2010) Chronic wasting disease prions are not transmissible to transgenic mice over-expressing human prion protein. J Gen Virol 91(10):2651–2657
191. Marsh RF et al (2005) Interspecies transmission of chronic wasting disease prions to squirrel monkeys (Saimiri sciureus). J Virol 79(21):13794–13796
192. Race B et al (2009) Susceptibilities of nonhuman primates to chronic wasting disease. Emerg Infect Dis 15(9):1366–1376
193. Davenport KA et al (2015) Insights into chronic wasting disease and bovine spongiform encephalopathy species barriers by use of real-time conversion. J Virol 89(18):9524–9531

Index

A

Adrenal ... 115
Anthropology ... 33–35, 40, 42, 43, 47
Apoptosis ... 112, 139, 141, 145, 146, 148, 149, 152, 154, 237
Atypical scrapie ... 110–112, 116, 258, 271, 286
Autophagy ... 112, 117, 133, 141, 145, 147–149, 151–154, 168

B

Biological reference materials ... 57–60
Biomarker ... 229–243
Bovine spongiform encephalopathy (BSE) ... 43, 51, 54–56, 58, 66, 67, 80, 110, 141, 148, 176–178, 209, 230, 256–258, 269–272, 275, 277–279, 283, 284, 286, 287, 289, 294
Brain ... 4–6, 8, 17, 21, 50–59, 67, 70–72, 75, 79, 81–84, 86–89, 91–95, 100, 109, 110, 115, 123–125, 128, 129, 138, 139, 148, 149, 160–162, 166, 169, 174, 176, 177, 186, 187, 189, 210–213, 215, 225, 226, 230, 231, 234, 236, 239, 241, 258, 259, 262, 263, 270, 272, 273, 275, 277, 278, 281–283, 285, 292

C

Cell membrane ... 106, 108, 109, 113–115, 117
Cerebral amyloid angiopathy (CAA) ... 108–110, 117
Cerebrospinal fluid (CSF) proteins ... 74, 230
Chimpanzee ... 1, 8, 10–12, 21, 50–53, 275, 278
Chronic wasting disease (CWD) ... 53, 54, 108, 113, 138, 160, 162, 169, 177, 209, 256, 258, 269, 271, 272, 283, 286–294
Clinical diagnosis ... 67, 230, 231
Clinical features ... 43, 65–76, 230
Creutzfeldt-Jakob disease (CJD) ... 1, 10, 12, 17, 20, 22–24, 43, 45, 50–53, 57, 58, 65–67, 71, 74, 76, 80, 81, 83, 85, 86, 95, 106, 107, 123–125, 130, 135, 137, 140, 141, 148, 149, 152, 153, 160, 161, 174–176, 209, 230–239, 242, 243, 269, 273–279, 281–283

D

Diagnostic test ... 60, 67, 72–76, 93, 160, 175, 238, 239
Disease transmission ... 115, 239, 255–261, 264, 280, 281

E

EEG ... 66, 72, 74–76, 124–126, 230, 282
Electron microscopy ... 2, 21, 24, 100–116, 123, 133, 138, 139, 147, 152, 170, 202
Exosomes ... 198–205

F

Follicular dendritic cells ... 24, 113
14-3-3 ... 74–76, 175, 230–234, 236, 242–244

G

Gajdusek, D.C. ... 1, 3, 13, 15, 16, 18, 19, 40, 50, 210
Genetic tests ... 72, 76
Gerstmann–Sträussler–Scheinker disease ... 24
Glycosylation ... 90, 184, 214, 237, 238, 260, 261, 271

H

History ... 34, 35, 44, 47, 69, 70, 76, 275
Human prion disease ... 1, 65–76, 80–82, 84, 86, 88, 90–93, 95, 174, 259, 260, 273, 275, 278, 281
Humanism ... 33

I

Immunogold electron microscopy ... 100, 106
Immunohistochemistry ... 20, 56, 84–88, 93, 94, 111, 133, 161

K

Knockout mouse ... 262–264
Kuru ... 1–24, 35, 39–47, 49–51, 53, 57, 107, 108, 133, 135, 137, 138, 209, 210, 270, 275, 278, 279

Pawel P. Liberski (ed.), *Prion Diseases*, Neuromethods, vol. 129,
DOI 10.1007/978-1-4939-7211-1, © Springer Science+Business Media LLC 2017

L

Lymphoid tissues....115, 272

M

Macaque....11, 53–56, 59
Medical anthropology....40, 42, 43, 47
Mesaxon....111, 112
Monkey....12, 52–54, 59
MRI....72–76, 230

N

Natural science....33, 47
Neurodegeneration....24, 55–57, 160, 270, 271, 273
Neuron....7, 12, 17, 21, 76, 100, 112, 115, 141, 145–148, 152–154, 197, 198, 205, 230, 237, 263, 270, 285
Neuropathology....12, 17, 65, 160, 283, 291
Neurotoxicity....205

P

Papua New Guinea....2, 5, 24, 34, 41–43, 46, 67, 210, 278
Paraffin-embedded tissue (PET) blot....85, 87, 88, 94
Pathogenesis....8, 43, 50, 53–57, 84, 88, 108–110, 117, 140, 152, 160, 168, 231, 236, 283, 286–288
Prion....43, 44, 46, 65–76, 79, 80, 84, 99–117, 123, 160, 162, 197, 209, 237–239, 253–264, 269–275, 293
Prion diseases....1, 43, 65, 67, 68, 70–73, 79–85, 88, 90–94, 100, 106, 107, 109, 110, 116, 129, 139, 145–149, 152–154, 168, 173, 174, 176, 177, 197, 212, 238, 255, 259, 260, 263, 264, 269, 270, 274–283, 286, 288, 294
Prion protein (PrP)....1, 19, 20, 22–24, 39, 44, 45, 51, 54–57, 59, 65, 67, 70, 71, 75, 76, 79, 81, 84–94, 99–117, 125, 126, 133, 136, 141, 147, 148, 152, 153, 174, 176–178, 183–185, 195, 198, 211–214, 216, 219, 223–225, 230, 233, 237–240, 253–255, 257–264, 270–294
PRNP....9, 23, 67, 69–72, 75, 76, 106, 107, 109, 111, 126, 211, 230, 235, 257, 259, 260, 276–279, 283, 288
Protein aggregation....56, 147
Proteomics....230, 239–243
Protocol....59, 82–86, 88–90, 92–95, 127, 168, 169, 174, 184, 199, 200, 202–204, 219, 224

R

Real-time quaking-induced conversion (RT-QuIC)....72, 75, 76, 92, 174–179, 239, 240, 287, 294

S

Scrapie....7, 17, 49, 53, 54, 58, 80, 105, 106, 108–115, 123, 135, 136, 139–141, 148, 149, 153, 160–164, 166–170, 176–179, 184, 186, 187, 189, 190, 209–211, 233, 237, 255, 257, 258, 260, 269, 275, 279, 283, 284, 286, 289, 293
Sialic acid....184, 185
Sialylation....183–186, 193
Spiroplasma....160–164, 167–170
Squirrel monkey....51–56, 294

T

Tau....55, 56, 175, 230, 234–236
Transgenic mouse....106, 108, 109, 211, 255, 258, 273, 274, 288
Transmissible mink encephalopathy (TME)....269, 270, 276
Transmissible spongiform encephalopathy (TSE)....1, 23, 50–60, 100, 106, 107, 112, 113, 140, 148, 160–164, 169, 170, 211, 230, 232, 237, 253, 261, 270
Tubulovesicular bodies....112, 117
Two-dimensional electrophoresis....184

U

Ultrastructure....21

V

Variant CJD (vCJD) prion protein....75

W

Western blot....55, 82, 83, 87–92, 95, 177, 184, 186, 188, 189, 259

Printed in the United States
By Bookmasters